Industrial Bioseparations

Principles and Practice

Industrial Bioseparations
Principles and Practice

Daniel Forciniti

Blackwell
Publishing

Daniel Forciniti, Ph.D., is a Professor at the Department of Chemical and Biological Engineering at the University of Missouri, Rolla. Dr. Forciniti joined the University of Missouri, Rolla, after receiving his Ph.D. from North Carolina State University in 1991. Dr. Forciniti has published and presented extensively on chromatography, biomacromolecules, and biotechnology, including peer-reviewed articles, contributed chapters, and invited lectures. Dr. Forciniti has also worked in the industry as a chemist for the Argentinian Glass Company.

Blackwell Publishing Professional
2121 State Avenue, Ames, Iowa 50014, USA

Orders: 1-800-862-6657
Office: 1-515-292-0140
Fax: 1-515-292-3348
Web site: www.blackwellprofessional.com

Blackwell Publishing Ltd
9600 Garsington Road, Oxford OX4 2DQ, UK
Tel.: +44 (0)1865 776868

Blackwell Publishing Asia
550 Swanston Street, Carlton, Victoria 3053, Australia
Tel.: +61 (0)3 8359 1011

First edition, 2008

Library of Congress Cataloging-in-Publication Data

Forciniti, Daniel.
Industrial bioseparations: principles and practice Daniel Forciniti.
p. cm.
Includes bibliographical references.
ISBN 978-0-8138-2085-9 (alk. paper)
1. Biotechnology-Technique. 2. Separation (Technology) I. Title.

TP248.25.S47F67 2008
660.6028-dc22

2007013212

Dedication

To Gladys, Eliana, and Leandro,
and to my motherland, Argentina

Contents

Preface

A few years ago, universities throughout the United States and the world started education programs that would create a new kind of engineer to support biotechnology advances—one who could work for multinational pharmaceutical companies as well as small biotech ventures dedicated to developing and bringing a new product into the market. Bio-related topics are being included in core engineering courses, and numerous universities have included general biology requirements at the same level and intensity as general mathematics and English requirements. Another consequence of this new trend has been the change in the name of numerous engineering departments in the United States to reflect the new emphasis in biology. At the center of this bio-revolution is the production of compounds of biological origin for their use in the treatment of disease, in agriculture, and in household products.

The first step in the production of new bioproducts is the genetic modification of an organism (normally *Escherichia coli,* but also yeast, mammalian, insect, and plant cells) to produce, for example, a protein-based therapeutic drug. Then, the organism is grown under conditions that favor the expression of the desired product. Finally, the product is isolated, purified, and made ready for distribution and delivery.

This book is aimed at the scientist or engineering practitioner who has an interest in the purification of biologically manufactured molecules as well as graduate students and university professors who are involved in the research or the teaching of bioprocesses. Both established practitioners and newcomer engineers and scientists who work in the field of bioprocessing should benefit from the material covered in this book, as it is current and thus a valuable reference tool. The numerous references at the end of each chapter should allow the reader to dig deeper into a topic of interest. Newcomers with little background in engineering or in biology should find the material easy to read. They should find enough background information to accelerate their learning curve. In addition, senior undergraduate and graduate students who are participating in bioengineering curricula or are involved in research in bioseparations can use the book as a reference text. Finally, it should be relatively easy for a bioseparations professor to generate notes from this book.

Although currently there are comprehensive monographs on each unit operation commonly used for the purification of biomolecules, this is the first comprehensive treatise covering both the scientific and practical aspects of bioseparations. There are several textbooks dedicated to this engineering discipline, but most emphasize the science and underestimate practical considerations. Moreover, most available books fail to provide the necessary background information about biological products and engineering principles.

In spite of the undisputable importance of filtration and chromatography in bioseparations, I have tried to produce a well-balanced book. This was challenging, because the large number of different unit operations used in bioseparations makes it difficult to cover all topics in the appropriate depth without creating an encyclopedic work. We resolved that paradigm by focusing exclusively on biomacromolecules, and thus, are able to include more details in the same number of pages. That decision was made because the separation of small molecules can be effected by using methods that are more traditional.

The book consists of a healthy mix of in-depth mathematical descriptions of the various units operations used in bioseparations, a clear description of the different units in both laboratory and industrial scale, and a set of experiments that can be included in traditional units operations courses, in dedicated bioseparations classes, or as general guidelines for the protocol developer. The mathematical descriptions of the units are limited to those approaches that either facilitate the scaling up of the unit or allow the reader to better understand its functioning. The various units are described in depth with the help of drawings and graphics. The experiments are presented in enough detail as to allow the novice to be able to perform them without consulting additional sources.

The material in the book is presented in the order in which the various units operations are found in bioprocessing beginning with Chapter 5. Chapter 1 includes a general description of the problems encountered in bioseparations, as well as general approaches followed in the isolation of commercially important biomacromolecules. Chapter 2 focuses on process thermodynamics and its applications to bioseparations. Chapter 3 provides the necessary basic mass transfer principles that are needed to properly describe each of the units. The examples are focused on real case studies. Chapter 4 is devoted to a review of the main physical chemistry properties of biomacromolecules. Because those properties are not always available, experimental approaches that can be used to measure them are described and a set of experiments are presented at the end of the chapter. Chapter 5 focuses on cell and tissue disruption. Appropriate experiments are described to accentuate the concepts presented in the bulk of the chapter. Chapter 6 covers the separation of solids from liquid suspensions. The main topic of the chapter is ultrafiltration, whereas the other units (conventional filtration, centrifugation and liquid/liquid extraction) are briefly reviewed, emphasizing operating variables and scaling up. Chapter 7 covers chromatography, the main unit operation that is found in all bioseparation cascades. The chapter is comprehensive and the main approaches used to describe or analyze the operation are formally introduced. This chapter presents experiments that cover the various types of chromatography. In these experiments, emphasis is put on obtaining the scaling up parameters as well as on commonly used operating conditions. Chapters 8 and 9 cover protein precipitation, crystallization, and freeze drying. In Chapter 10, recent advances in micro-devices used for protein separation are presented. Finally, in Chapter 11, general rules of thumb used for the design of separations trains are presented. Basic concepts of experimental design are included in the Appendix for completeness.

Chapters 2, 3, 10, the Appendix, and the experiments included at the end of several chapters are unique for a bioseparations book. Scientists venturing in the bioseparations area often lack a formal education in transport phenomena and process thermodynamics. Therefore, Chapters 2 and 3 are dedicated to providing the necessary knowledge that the scientist will need to fully profit from the material presented in this book. Chapter 2 introduces the principles of thermodynamics and its application to bioprocesses. Chapter 3 introduces the reader to the principles of mass transfer. Emphasis is placed on the theoretical calculation and experimental measurement of diffusion coefficients of biomacromolecules. Chapter 10 presents the principles for the fabrication of microdevices and their use in bioseparations. Emphasis is placed on chromatography and capillary electrophoresis in a chip. Finally, principles of statistics are presented in the Appendix and the rudiments of experimental design are discussed. In addition to these four chapters, experiments are presented in detail at the end of each chapter. The experiments presented in each chapter build on each other to complete three different separations cascades using three different raw materials: tissue, yeast, and plant cells.

Many years ago, Professor D. Ollis, who was my graduate advisor at North Carolina State University, proposed that I co-author a book on bioseparations to serve as a companion to his and Professor J. Bailey's biochemical engineering seminal textbook. For various reasons, we never put together a manuscript, but the idea stayed in my mind for many years and finally crystallized with the encouragement of Blackwell Publishing. This book contains material that I have taught to more than ten generations of students that have passed through my bioseparations classes. I am in debt to all of them. I would like to thank my son (Leandro Forciniti, who is currently a graduate student at University of Texas-Austin) for reading my manuscript and making valuable suggestions and modifications. Finally, I am in debt to my wife (Gladys), and to my daughter (Eliana) for all the time that I took from our family to be able to write this book in a reasonable time.

Industrial Bioseparations
Principles and Practice

Chapter 1

Introduction to Bioseparations

Introduction

Humans have been collecting, growing, hunting, or raising living organisms for millennia. Often the products are consumed without processing; that is, without isolating a particular component. Exceptions to that practice familiar to all of us are extracting oil from seeds or whales, milking farm animals, and extracting juice from fruits. Later in humankind history, simple fermentation processes were developed to produce alcoholic beverages, such as the malting of barley and fermentation of beer in Egypt in 2500 BC; to make leavened bread about 8,000 years ago; or to treat illnesses, such as the use of moldy soybean curd to treat skin infections in China some 5,000 years ago (Prouty 1991). The production of natural or non-natural products by using chemical reactions started much more recently, approximately a century ago. However, we have only recently realized that either single cells or tissue (a conglomeration of organized cells) can be used as a chemical reactor to produce a variety of products, something that has been used (sometimes unknowingly) for centuries. This recognition led to the discipline that today is known as biotechnology.

Biotechnology is a relatively new word for our vocabulary. We now know that the magnificent art of winemaking, which was spread by Romans throughout Europe, was a fermentation process, a unit operation of biotechnology. Of course, the Romans did not know that they were employing biotechnology techniques, nor did they have protein chemists, enzymologists, geneticists, or microbiologists. They did have engineers, who were more dedicated to building roads, bridges, aqueducts, or monumental structures than to designing an optimum fermentation and separations process. Today, the ancient Romans are gone and the world is very different. Although biotechnology is still more of an art than a quantifiable science, engineers, fatigued by hydrocarbons and chemicals, have decided to venture into the life sciences.

Biotechnology has industrial, agricultural, and pharmaceutical impact. Industrial enzymes are used in most households in detergents to wash clothes and dishes. For example, thermostable proteases (enzymes that degrade proteins) and lipases (enzymes that degrade lipids or fats) are included in laundry detergents. They also are used in the production and processing of agricultural products. For example, enzymes are used to facilitate extracting juice from fruits, to improve the fermentation process in the elaboration of beer, or to tenderize meat. Biotechnology has been used in the production of agricultural products to improve the resistance to insects and other pests of the plant. For example, most of the soy cultivated worldwide has been genetically engineered to resist insects. The transgenic modification of crops is making considerable progress. For example, *Bacillus thuringiensis* expresses an anti-worm protein that can be engineered into corn or cotton, thus reducing the need for pesticides. Industrial enzymes are frequently sold as crude extracts, that is, without much purification, whereas food-grade enzymes are partially purified. Perhaps the area with more potential for growth is the use of biotechnology-related products in medicine. They include small molecules, such as antibiotics, or large complex molecules, such as hormones (insulin), antibodies, or ribonucleic acid (RNA). Antibodies and RNA are considered by many as the future of the industry because of their impact in our fight against devastating illnesses such as cancer or acquired immune deficiency syndrome (AIDS).

The genomic revolution combined with the study and manipulation of metabolic pathways open limitless opportunities to produce very complex products, from short peptides to complex monoclonal antibodies. Often, the production process starts with the cultivation of a particular organism whose biological machinery naturally or artificially, via genetic manipulation, produces the substance of interest. Once the substance of interest has been produced, it must be separated from the organism. On a few occasions, the cells secrete the product and, therefore, its purification is relatively easy. More often, the product either is in the cytoplasm of the cells or, even worse, "entangled" or embedded within the cell or organelles' membrane, making its purification quite difficult. Both the cell culture step and the purification of the product fall into the realm of biotechnology.

It has been accepted that biotechnology has revolutionized the 160-year-old pharmaceutical industry. Biopharmaceuticals are defined by the Food and Drug Administration (FDA) as recombinant proteins and monoclonal antibodies or nucleic acid-based drugs. There are over 60 FDA-approved biotech medicines and more than 360 in clinical trials. These medicines target more than two hundred diseases, such as heart disease, stroke, multiple sclerosis, breast cancer, diabetes, cystic fibrosis, and many more affecting 250 million patients worldwide. There has been a change in the structure of the market in the last six years. In 2000, big pharmaceutical companies represented 76% of total drug research expenditures and introduced into the market 69% of new drugs available on the market. The situation changed drastically by the end of 2004, as big pharmaceutical companies accounted for 75% of the research expenditures, but only 35% of the new drugs available to consumers. This change in the market suggests that even small companies have an opportunity to become important players in this new technology.

Pharmaceutical products require high purity (> 95%); and therefore, are the most challenging from a separations viewpoint. Table 1.1 includes a list of commercially important bioproducts (protein-based) used in the medical field. We discuss a few representative ones below.

Table 1.1. Partial List of Commercial Biopharmaceuticals (Ainsworth 2005).

Class	Generic	Brand	Company	Sales ($Millions)
Erythropoietin	Epoetin α	Epogen/Procrit	Amgen/J&J/Sankyo	6,191
Interferon α	Interferon α-2a	Intron A	Schreing-Plough	1,168
Insulin	Insulin	Novolin	Novo Nordisk	2,405
Colony-stimulating factor	Filgrastim	Neupogen	Amgen	1,175
Interferon α	Interferon α-1a	Avonex	Biogen	1,418
Growth hormone	Human growth hormone	Genotropin	Pfizer	1,005
Insulin	Insulin	Humulin	Eli Lilly	998
Other	Imiglucerase	Cerezyme	Genzyme	835
Growth hormone	Somatropin (rDNA origin)	Norditropin/ Norditropin SimpleXx	Novo Nordisk	387
Growth hormone	Human growth hormone	Humatrope	Eli Lilly	385
Growth hormone	Somatrem/somatropin	Protropin/Nutropin	Genentech	354

Insulin is the world's first biotechnological product. It is the most important hormone in the body to regulate the levels of glucose in blood. Figure 1.1 illustrates the timeline for the developing of insulin as a biotechnological product from the isolation of animal sources (mainly pigs) to the currently used recombinant insulin product. After its first isolation and purification, it took nearly 60 years for

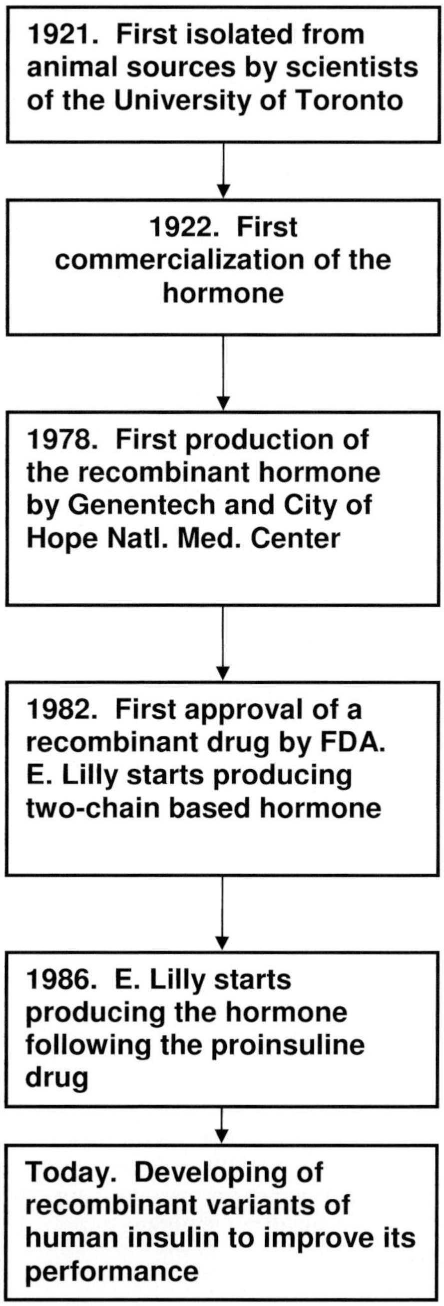

Figure 1.1. Chronology of the development of insulin—the first biotech product to reach the market.

the first recombinant product to appear. The growth from that point on has been exponential. The evolution in the insulin market is a wonderful proof of the advantages of producing pharmaceutical products in micro-organisms. Currently, there are more than 20 insulin and insulin derivatives in the market. They include insulin analogs that differ in only a few amino acids from the native variety, which have been developed to control the onset and duration of the activity of the hormone. Patent protection laws are the main reasons for the commercialization of a few animal-based products worldwide. Insulin's market is also a good example of the economical impact of biotechnology. Today, insulin products (including only those by Novo, Lilly, and Sanofi-Aventis) generate approximately 8 billion dollars in sales. The process for the production of human insulin is also a good example of the intrinsic complexity of biotechnological processes. Its production requires 31 major processing steps, 27 of which are associated with product recovery and purification (Prouty 1991).

Erythropoietin (EPO) is a glycoprotein that regulates the production of erythrocytes. EPO contains 165 amino acids and four carbohydrate chains. The human variety was first isolated and then purified from urine in the 1970s. Recombinant EPO is currently used in kidney failure treatment, HIV infections in patients treated with azidothymidine (zidovudine) (AZT), and in cancer chemotherapy. Unfortunately, EPO is better known by the general public because it is used by athletes to improve performance with adverse health affects. The sales for various EPO forms reached several billion dollars in the past few years, including $2.6 billion in sales of Amgen's Epogen and $3.6 million by Johnson & Johnson's Eprex.

An antibody is a type of protein produced by white blood cells in response to a foreign molecule, also known as an antigen disease. Antibodies are called immunoglobulins because of their relation to the immune system. Antibodies destroy antigens directly or they facilitate phagocytosis by white blood cells. The basic building blocks of antibodies are four subunits: two light chains (~23 kD each) and two heavy chains (53 to 75 kD each), which form a Y-shaped symmetric dimer. Both disulfide bonds and noncovalent interactions connect the subunits. Each light chain and heavy chain has a variable and constant region. The variable region (110 to 130 amino acids) of the antibody binds to the antigen. Antibodies are glycoproteins; each heavy chain has an *N*-linked oligosaccharide. Monoclonal antibodies are an important group of biotech-based products that are coming of age. Kohler and Milstein (1976) produced the first monoclonal antibodies in mice, but they were of little use because they were immunogenic. The solution to that problem was to use a human constant region and a variable mouse region (chimeric protein). Now, scientists are developing antibodies completely from humans. There are currently about twenty monoclonal antibodies in the market and about one hundred antibodies in various stages of clinical trials. The best-known success story is Rituximab (anti-human CD20 antigen) immunoglobulin G1 (human-mouse monoclonal IDEC-C2BB γ1-chain) disulfide with human-mouse monoclonal IDEC C2b8 κ-chain dimer. Rituximab is a chimeric protein (part mouse part human antibody) that kills normal and malignant B cells by binding to the cells' CD20 receptor and inducing either complement-dependent cytotoxicity, antibody-dependent cell mediated cytotoxicity, apoptosis, or all of the above.

Several other monoclonal antibodies that are currently used are murine OKT3, which is used in acute transplant rejection (Ortho Biotech), chimeric 7E3 Fab, which is used in postcoronary angioplasty (Centocor), and murine 17-1A, which is used in the treatment of colorectal cancer (Centocor). Antibody-based therapy covers a wide range of devastating human illnesses such as breast cancer (rhuMAb HER2 [Genentech]), rheumatoid arthritis (RA) (CAMPATH-1H [Wellcome]) or graft versus host disease (GVHD)(humanized anti-Tac [Hoffman La Roche]). Beyond their use in therapy, monoclonal antibodies are used also for diagnostic purposes. Well-known examples are the detection of prostate-specific membrane antigen by 7E11-C5.3 (prostate cancer) or the use of antibody 17-1A for the detection of epithelial membrane antigen (colorectal cancer).

Synthetic oligonucleotides, vectors, and plasmids have been increasingly used in gene therapy. The production and purification of these compounds bring new challenges. One such challenge is that industrial standards have yet to be defined well. This is a result of inherent problems in the production and purification of these compounds. For example, gene therapy is largely based on retroviruses whose purification to homogeneity may not be reachable. This needs to be contrasted with the purity requirements of protein-based pharmaceuticals. Governmental guidelines for plasmid deoxyribonucleic acid (DNA) production are more specific. For example, stringent standards have been set for the main contaminants of concern: nucleic acids, endotoxins, and host proteins. Because of the incipient development of nucleic acid-based therapies, scaling up of the processes needed is still in its infancy. Because the boundaries of variation have not been established, there are questions about the products' consistency. Moreover, because "gene therapy" is mostly in clinical trial stages, the production level that a new plant should have has not been established. This is also a reflection of the priorities of the industry. The race is to have a product accepted by the FDA and subsequently patented. Industrial-scale production follows, either by the same company or by one of the large biotech or pharmaceutical companies.

The push toward miniaturization, that is, the interest in micron and nanoscale particles, which has swept the scientific community, has also reached bioprocesses. A few years ago, scientists discovered that particles below the micron range have unique properties. From that moment on, it has been an explosion in research on the behavior of particles a few nanometers in size, and nanotechnology was born. The same scientists seem to have ignored that biochemists, biologists, biophysicists, and bioengineers have been dealing with submicron particles for many years. Soon after, nanotechnology discovered biology (or vice versa) and bionanotechnology was born, which is nothing but biotechnology on the nanoscale. Bionanotechnology is just a manifestation of the generalized trend of investigating the macroscopic world on a submicroscopic scale and hoping to improve processes on that scale.

The Role of Engineering in Biotechnology

The above discussion makes clear that the field of biotechnology has a very broad scope, and that new areas are being added daily. Stem cells, gene therapy, functional foods, edible vaccines, pharmacogenomics, biosensors, and microfluidic devices are on the horizon for biotechnology. Bioprocess engineers will have to design robust manufacturing processes to produce large quantities of some of these bioproducts, because some of them are required in large amounts for therapeutical purposes. A typical example is the use of human antibodies. Although some engineers have the necessary basic knowledge, others will have to acquaint themselves with microfluidics and self-assembly processes. The challenge to engineers will be to make the transition between the macroscopic world to the micro-scale devices built from submicron and nanoscale components. The challenge to scientists will be to research alternatives with processes in mind.

The market drives the main questions that engineers and scientists must answer, "How much we need to produce?", "How fast it needs to be done?", and "How pure does the product need to be?" Unfortunately, level of production, speed, and purity are frequently mutually exclusive. Moreover, sometimes the developing team does not have the necessary expertise, which must consist of a healthy mix of scientists with a strong biological background and engineers with a "taste" for biological products. Good knowledge about the material to be purified and its impurities is a necessity to approach the problem rationally. Although it may sound trivial, people involved in developing the process often ignore the basic makeup of the material they are handling. On the other hand, the process is sometimes developed by scientists in the laboratory who do not have a good grasp on large-scale processes, or by engineers who lack an adequate understanding of biochemistry or biology. Therefore, this book strives to provide a general engineering and biochemistry foundation to satisfy both scientists and engineers in order to facilitate the communication between people with different backgrounds.

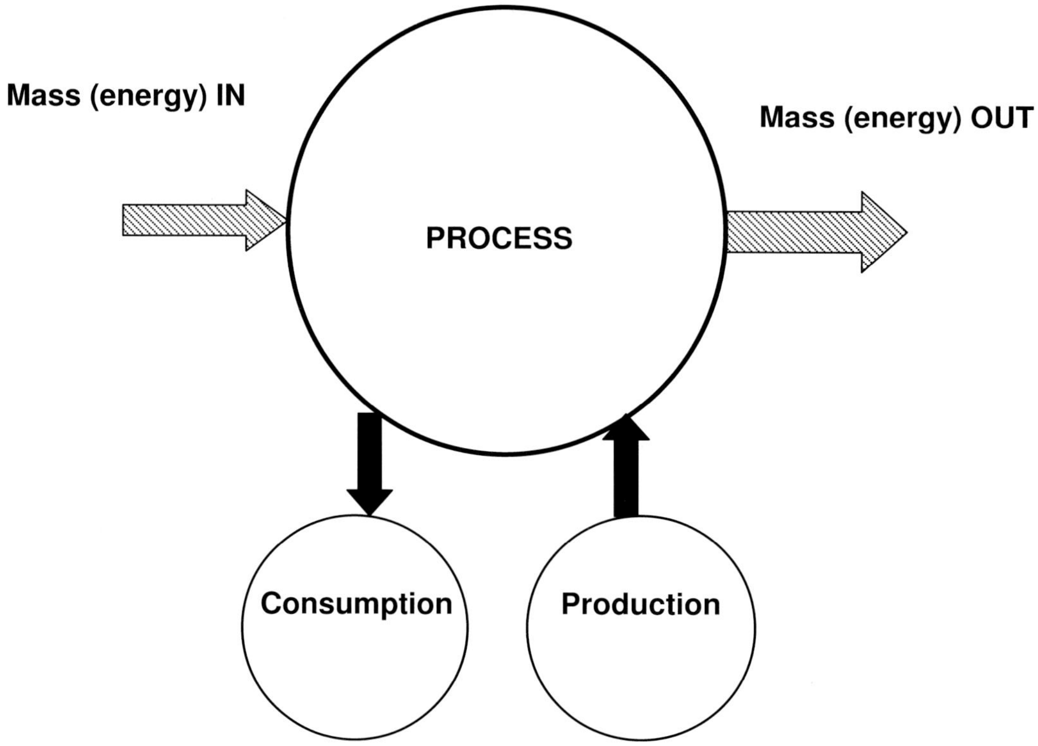

Figure 1.2. Generic mass or energy balances for an arbitrary process. Note that there would be several ports for mass (or energy) to enter or to leave the process and that there would be several means for mass (or energy) to be produced/destroyed.

Most of the engineering knowledge needed in biotechnology comes from disciplines that are core subjects in chemical engineering, that is, mass and energy balances, transport phenomena, thermodynamics, reactors design, and separations. In engineering terms, we have to "account" for the mass and energy flows associated with each process unit and around the entire process. The diagram shown in Figure 1.2 conceptualizes a mass or an energy balance.

Diagrams and drawings are useful for the understanding of the concept, but they are of little use to quantify the observations. The easier way to make the transition between conceptual drawings and equations is to first "translate" them into words,

$$\{Accumulation\} = \{In\ Flow\} - \{Out\ Flow\} + \{Amount\ produced\} - \{Amount\ consumed\} \quad 1.1$$

Mass and energy balances can be written for the overall mass, or for a specific component. Chapters 2 and 3 discuss this in greater detail.

Separation process is a discipline unique to chemical engineers and includes a healthy mix of transport phenomena and thermodynamics. The simplest way of understanding the philosophy of separation process is the recognition that our objective is to distribute solutes between two phases in such a way that the target product is enriched in one phase while the impurities are enriched in the other phase. This distribution process may be an equilibrium process or a non-equilibrium process (diffusion). Table 1.2 shows the main characteristics for the most common bioseparation units.

Table 1.2. Characteristics of the Contacting Phases in the Most Common Bioseparations Unit Operations.

Main Operation	Phases	Interface	Limited By
Ultrafiltration	L/L	Membrane	Mass transfer
Chromatography	L/S/L	Solid/liquid	Mass transfer
Precipitation	L/S	Solid/liquid	Equilibrium
Liquid/liquid extraction	L/L	Liquid/liquid	Equilibrium
Freeze Drying	S/V	Solid/vapor	Equilibrium

Equilibrium distribution between the phases is an important concept, because any separation process consists of the distribution of solutes between two totally or partially miscible phases. Equilibrium is defined here as the state of a system that is invariant as time goes by. When solutes distribute between two phases, equilibrium is rarely reached in practice; however, the state of equilibrium determines the optimum (or upper limit) that can be reached. A basic science that allows us to understand and use the concept of equilibrium is thermodynamics, which is explored in the next chapter. Of course, as the process attempts to reach equilibrium or even after equilibrium has been reached, mass (or energy) will be transferred from one phase to another through an interface (or a membrane). Therefore, we must understand, for example, how mass moves from one phase to another, and how we can affect the flux across that interface to our benefit. Chapter 3 discusses the topic of mass transfer. The time needed for a given separation is not only important because of the old equation: "time equals money," but because proteins denature at interfaces it is also risky to leave a protein at that interface (as will be the case in all separation units) for a long time. Proteins do not understand human needs; they are demanding and fragile, therefore, we need to take care of them properly.

Classification of Bioproducts

The approaches followed to isolate a particular product depend on the characteristics of the raw materials (type of cells or tissue) and on the type of product. Bioproducts may be classified either by their final use or by size. The final use may be in the food industry (aspartame), in medicine (EPO), or in agriculture (*Bacillus thuringiensis*). The examples given above also give the reader a good idea of the different sizes that we may encounter (from a few angstroms in the case of aspartame to microns in the case of *B. thuringiensis*). Therefore, another way of classifying bioproducts is according to size. Examples of small molecules are very abundant and were found well before people even talked about biochemical engineering. Some examples are ethanol, citric acid, amino acids, sugars, antibiotics, some hormones, and vitamins.

Large molecule production is relatively new (in an earth age framework!). In general, large molecules are polymers of some kind. The two most important polymers in biotechnology are nucleic acids and proteins. The examples mentioned above make clear that proteins are currently the focal point for a relatively broad sector of the industry, but that there is also a growing interest for DNA fragments, for RNA, viruses, and even bacteria (they are not polymers, of course). There is a market also for other "biopolymers"—xantan gum and dextran are just two examples. Some industries, like Dupont, are moving into the enzymatic synthesis of polymers (like polyesters). The use of enzymes instead of inorganic catalysts is attractive because of the specificity of enzymes (e.g., possibility to produce only one type of stereoisomer eliminating the need for the separation of chiral compounds) and because the process takes place at moderate temperatures and in the absence of organic solvents hence making the process innately green. It is anticipated that as enzymatic processes replace inorganic catalytic ones, there will be a need for the production of relatively pure enzymes in large quantities.

Because the approaches to separate small molecules and large molecules are quite different, we concentrate our efforts on the separation of proteins, a subclass of biomacromolecules.

Overview of Bioseparations

Although there is not a well-established separations protocol that can be applied to biological molecules of interest, the following steps should be considered as a first trial: (1) release of the product, (2) removal of solids, (3) capture, (4) purification, (5) polishing, and (6) product conditioning/packing. Figure 1.3 summarizes the purposes of these steps, as well as the units that are commonly used. Of course, the sequence of steps presented in Figure 1.3 must be used only as a general guideline. Chapter 11 gives more details about the design of bioseparation cascades.

Further knowledge of the chemical physical properties of the contaminants, a greater awareness of which contaminants are "allowed" and which contaminants are "forbidden," and an improved level of productivity will define the final purification train. Characterization of the impurities is necessary to design rationally the separation of any single protein from a complex mixture. Two physical properties of contaminant proteins that are particularly valuable for the design of a separation protocol are their molecular weight and isoelectric point, which may be determined by using gel permeation chromatography (HPLC-GPC), isoelectric focusing (IEF), polyacrylamide electrophoresis (PAGE), and sodium dodecyl sulfate PAGE (SDS-PAGE).

Although the philosophy of any separation train is to remove as many impurities as possible, the addition of stabilizers is often necessary not only at the end of the process (just before storage), but also at intermediate steps. Typical examples of molecules added early in the purification cascade is the addition of protease inhibitors, which are needed to minimize/eliminate protein degradation, and DNAse, which is added to degrade DNA and thus decrease the viscosity of the solution. The addition of stabilizers or excipients is not per se a separation step. However, because protein therapeutics are currently injectable and long-term dosing of these protein therapeutics is required, the final preparation of the product has a large economical and medical impact. In addition to improving the stability of the injectable protein, current research has been focused on emerging delivery routes like transdermal, inhalation, and oral/nasal routes.

The key problem faced in bioseparations is the isolation of a component present at dilute concentration in a very complex mixture. Table 1.3 shows the approximate composition of a commonly found host, *Escherichia coli*. There is a total of about four thousand different compounds ranging in size from a few angstroms to a few nanometers.

Although not all these products are released from the cell, it should be obvious for the reader that the isolation of a particular protein is a quite challenging problem. All types of cells are complex entities, but the degree of complexity varies from organism to organism. The consequence of that is a need to follow different approaches depending on the species that is chosen to produce a particular biomacromolecule. We can illustrate this with a very simple example. Suppose that the processing plant is divided into two distinct sections: bioreactor and bioseparation. The bioseparation section receives a material from which the product of interest needs to be isolated and purified. Table 1.4 shows possible raw materials, only differences in size and shape are highlighted. The bioseparation engineer will have to make important decisions at the beginning of the separation train based only on the characteristics of the material that she receives. For example, the optimum disruption for filamentous and bacillus are quite different. Subtle differences in the disruption method will be necessary because of differences in the size and shape of the microorganisms. Even

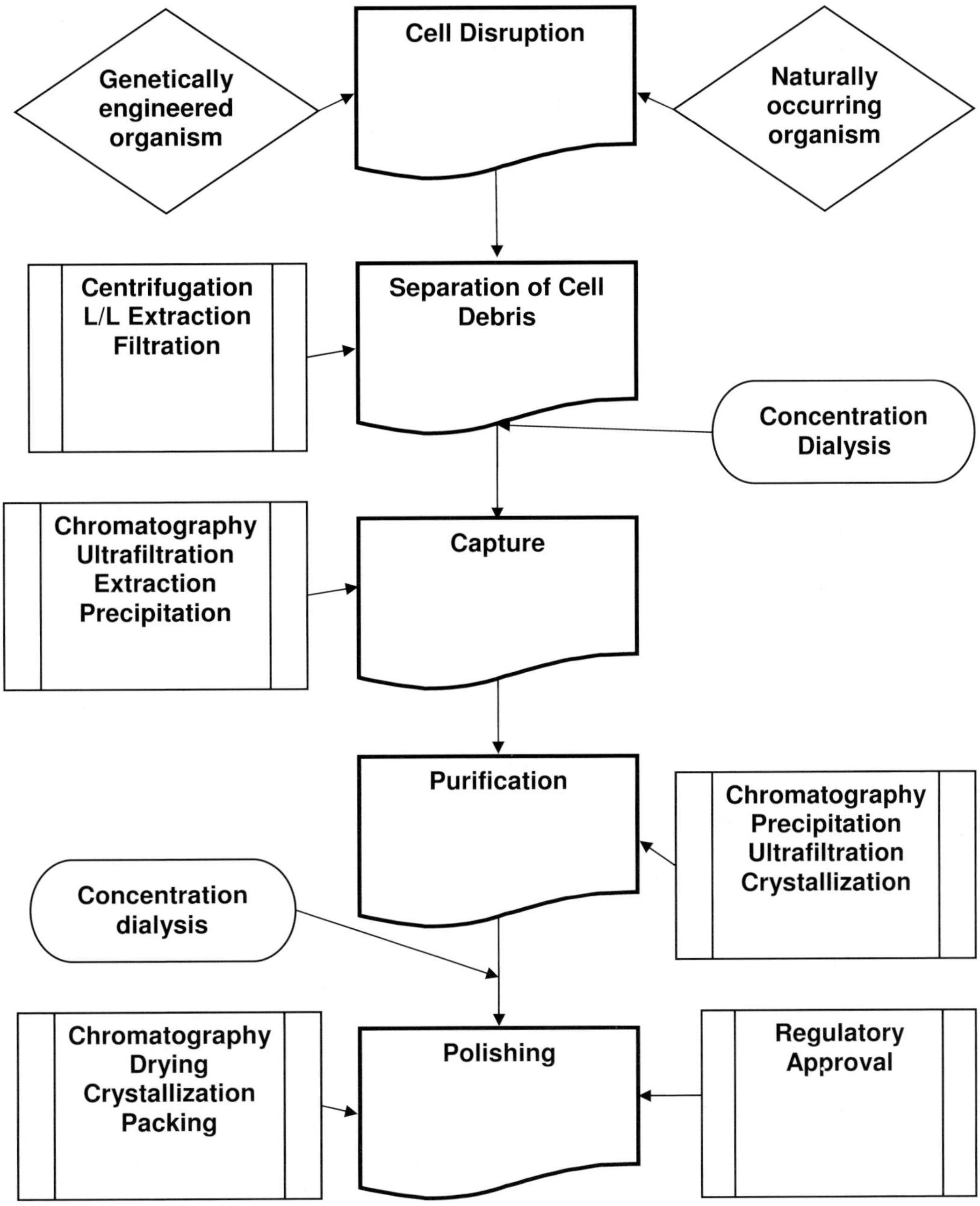

Figure 1.3. Generic bioseparations scheme.

Table 1.3 Approximate Composition of *Escherichia coli*.[3] Bailey, J. E. and D. F. Ollis, 1986. *Biochemical Engineering Fundamentals.* N.Y. McGraw Hill. Reproduced with permission by the McGraw Hill Publishing Company.

Component	Percent of Total Cell Weight	Average Molecular Weight	Approximate Number Per Cell	Number of Different Kinds
Water	70	18	4×10^{10}	1
Inorganic ions (Na^+, K^+, Mg^{2+}, Fe^{3+}, PO_4^{4-}, etc.)	1	40	2.5×10^{8}	20
Carbohydrates and precursors	3	150	2×10^{8}	200
Amino acids and precursors	0.4	120	3×10^{7}	100
Nucleotides and precursors	0.4	300	1.2×10^{7}	200
Lipids and precursors	2	750	2.5×10^{7}	50
Other small molecules	0.2	150	1.5×10^{7}	200
Proteins	15	40,000	1×10^{6}	2000-3000
Nucleic acids				
DNA	1	2.5×10^{9}	4	1
RNA	6			
16SrRNA		500,000	3×10^{4}	1
23S rRNA		1,000,000	3×10^{4}	1
tRNA		25,000	4×10^{5}	40
mRNA		1,000,000	1×10^{3}	1000

Table 1.4. Percent of Cell Dry Weight of Proteins, Nucleic Acids, Polysaccharides and Lipids for Different Types of Cells.

Type	Protein	Nucleic Acids	Polysaccharides	Lipids
Bacteria	40-70	13-34	2-10	10-15
Yeast	40-50	4-10	<15	1-6
Fungi	10-25	1-3	<10	2-9
Algae	10-60	1-5	<15	4-80
Animal	<10	<50	<50	<50
Plant	<70	<30	<90	<80

if the product is extracellular, considerations about the shape and size of the microorganism are important in selecting the sedimentation conditions needed to separate the biomass from the product of interest. Table 1.5 summarizes the differences in size and morphology of various commonly found materials in bioprocessing.

The differences become more important as we look beyond the obvious differences in size and shape. For example, the amount of total protein in fungi and animal cells is typically less than in other organisms. Therefore, if the target product is a protein naturally occurring in fungi or animal cells, it will be at a lower concentration in these cells than if it were produced by bacteria. On the other hand, the amount of protein impurities will be less, thus, separation of said product is facilitated. Finally, high levels of lipids and polysaccharides found in animal and plant cells may complicate the separation of the product. Some specific contaminants are unique for a given host system. For example, tannins and other phenolic compounds present in leaves and plant cells are a major contaminant when antibodies are expressed in tobacco plants.

The location of the protein in the microorganism or in the tissue determines the choices for the

Table 1.5. Size Variability According to Cell Type.

Cell Type or Cellular Material	Dimensions (μm)	Morphology
Cell debris	<0.4	Irregular
Bacterial cells	1×2	Spherical
Yeast cells	7×10	Spherical
Mammalian cells	40×10	Spherical
Plant cells	100×100	Spherical
Filamentous fungi	10×500	filaments
Corn meal	>10μm	irregular

first stages in the separation cascade. Different locations of the target protein represent different challenges. The protein may be secreted by the microorganism, which leads to straightforward separation of the cell mass and the supernatant by centrifugation or filtration. Moreover, the amount of impurities is largely reduced because most of the cells remain intact. Extracellular proteins are easier to purify not only because the cell disruption step is not necessary, but also because the initial purity (defined as the ratio between the amount of the target protein to the total amount of protein) is normally quite high. Intracellular proteins are more difficult to recover and purify. The first step is, obviously, to disrupt the cells, which unfortunately liberates not only the target protein, but also all the other proteins and metabolites (e.g., nucleic acids or polysaccharides) present in the cytoplasm. Furthermore, proteases may still be active after disruption; therefore, it is absolutely necessary to inhibit them. Proteins that are present as inclusion or refractile bodies inside the cytoplasm are easier to isolate. However, the recovery of an active product from the inclusion bodies has proven quite challenging, because the proteins must be solubilized first using astringent conditions and then refolded, which results in lower yields. There is also a substantial increase in the cost associated with the refolding process. Finally, one of the most difficult tasks that one may encounter is the isolation and purification of a membrane-bounded protein. These proteins are highly hydrophobic and hence insoluble in water. A common approach is to isolate pieces of the membrane that contain the protein. Then, the protein is kept in solution by surfactants and it is normally commercialized in that form.

In any bioseparation process, several process-independent problems associated with a bioseparation cascade must be taken into account. Although it is not a bioseparations problem, it is noteworthy to notice that the quantity of a given compound produced by the cell depends upon the point on the growth curve where the product is collected. Therefore, variations upstream may produce changes not only in the amount of product expressed, but also in the kind and concentration of impurities. Product stability (or the lack of) is another problem that must be seriously considered (Chapter 4 deals with this in more detail). A few examples are: 1) exposure of the product to interfaces and to high temperatures may caused denaturation; 2) prolonged exposure to various buffer conditions may cause chemical degradation or racemization; 3) extended periods of time in the presence of impurities (for example, other proteins) may cause degradation by lipases or proteases; and 4) exposure of the product to the environment may cause microbial degradation and contamination.

Two constraints that need to be satisfied during the design of a separation cascade are product purity and yield. Product purity is usually determined by the final use of the product, whereas product yield is determined by economic considerations. For example, a purity level in between 20% and 80% is found in industrial enzymes, whereas the purity of a therapeutic drug is usually close to 100%. The purity level needed in diagnostic drugs is in between 80% to 100%.

Because of the high return in the commercialization of bioproducts, there is no monetary constraint during process development. Still, it is always appropriate to keep in mind the marginal gain produced by a marginal increase in yield. Engineers are used to the concept of scalability, whereas scientists (for the most part) are not. Therefore, we must be sure that whatever technique we propose to use in the lab is scalable to the scale required by upper management. Because high yields and high purity are relatively easy to obtain in laboratory scale using specialized equipment, there is always the risk of proposing a purification cascade that it is of no use in industrial scale. Finally, the process must be reproducible and robust. Many aspects of the production process are regulated by Good Laboratory Practices (GLP) and by current Good Manufacturing Practice (cGMP). It is a very good idea to become very familiar with those terms even during the development stages. This normally eliminates frustrations and saves time in the later stages of product development.

The progress of the separation must be followed through the cascade; a comprehensive characterization and documentation of the final product is needed. During the development of recombinant insulin, scientists were overcautious about the characterization of the final product and thus used a large number of identification and characterization techniques. Today, although not standardized, the progress of the separation may be followed by HPLC and electrophoresis (either in plates or in capillaries). If the target has biological activity, biological activity assays are necessary too. The final characterization of the product may consist of, for example, HPLC-MS, peptide mapping, determination of the amino acid sequence, and a bioassay. It is also necessary to determine the presence of multimeric forms (dynamic light scattering or analytical centrifugation can be used for this), residual organic solvents, trace metals, host cell proteins, recombinant deoxyribonucleic acid (rDNA), endotoxins, moisture, sterility, pH, solubility, and so on.

Advances in Laboratory Scale Approaches Versus Process Development

Chemists and biochemists have done an excellent job in developing powerful analytical techniques for the separation and identification of proteins. Typical examples are liquid chromatography-mass spectroscopy (LC-MS), two-dimensional electrophoresis, and capillary electrophoresis. The genomic revolution has promoted new developments in bioseparations, caused by the need for the isolation of hundreds of proteins in a very short period of time. Although the scaling-up of those techniques is highly questionable and there are even problems in using these techniques as online sensors for process monitoring, we will describe a few here because they represent the state of the art in bioseparations and because their use is widespread in laboratories.

Electrophoresis is an electrically driven separations method based on the transport of a species through a conductive medium. Electrophoresis has been traditionally done on paper or gels (2D) but it is much more powerful in capillary mode (CE) because it is quicker, has better resolution, the analytes are identified by an ultraviolet (UV) detector (so, no staining of the gel is needed) and it is straightforward to collect the samples after separation. In free-zone capillary electrophoresis, the analytes are separated because they have different mobility under the action of an electric field; cations move toward the cathode and anions toward the anode. The rate of transport depends on many factors, including the charge on the ions, the size and mass of the ions, and the conditions of the medium. The flux of a solute under an electric field is the result of two contributions, a concentration-dependent flux (Chapter 3 has more details about this) and an electrostatic contribution.

The flux of species i is given by

$$J_i = -D_{ij}\nabla c_i + \mu_{ele} c_i E \quad 1.2$$

in which D_{ij} is the diffusivity of solute i in solvent j, c_i is the concentration of the solute i, E is the magnitude of the electric field, and μ_{ele} is the electrophoretic mobility of the analyte (solute) given by

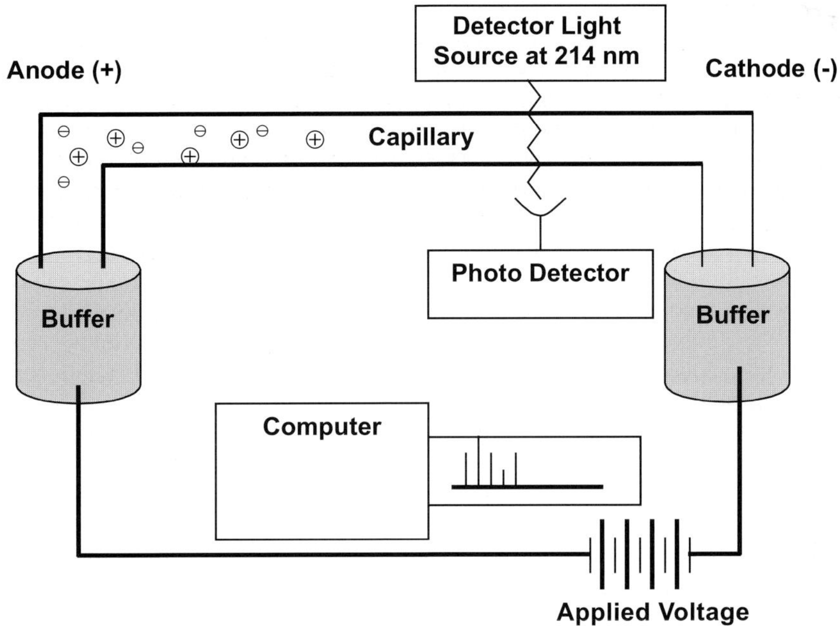

Figure 1.4. Schematic of a capillary electrophoresis instrument. The ends of the capillary are immersed into two buffer reservoirs to which a potential difference is applied.

$$\mu_{ele} = \frac{z}{6\mu\pi a} \qquad 1.3$$

in which z is the effective charge of the analyte, μ is the viscosity of the medium, and a is the radius of the analyte. This expression for the flux can be used in a differential mass balance whose solution can be found (in a circular capillary) by Laplace transformation. The main conclusion reached from that differential equation solution is that (ideally) the analyte moves through the capillary as a Gaussian peak whose height decreases over time and whose standard deviation increases as a consequence of diffusion (diffusion broadens the peaks and therefore, impairs the separation).

Figure 1.4 shows a typical capillary electrophoresis set up. The two ends of the capillary are immersed in two buffer reservoirs and a potential difference is applied between the two ends of the capillary. A typical capillary is made of silica coated with a polymer layer to improve its mechanical properties. A small window (1 or 2 millimeters in length) is "burned" on the coating and a detector is used to monitor the movement of the analytes.

One of the problems intrinsic to electrophoresis is Joule heating. In CE the problem of Joule heating is reduced by using a thick tube wall and a small diameter inner bore (20 to 100 μm). For the same potential drop (applied voltage) the current flowing through the sample is reduced as the diameter is reduced. Moreover, the larger outer diameter also allows for more heat dissipation. Because the geometry of the capillary allows larger voltages to be applied than in two-dimensional (2D) electrophoresis the migration rate is increased and the analysis time is reduced leading to better separations and higher resolutions. The sample can be introduced by electrokinetic injection, in which the sample is introduced by applying an electric field to cause the sample to migrate into the tube, or by hydrodynamic injection, in which the sample is introduced using a pressure drop or siphoning. Injection of a consistent amount of sample is a historical problem of CE, the small internal diameter of the capillary creates interfacial anomalies, and therefore, variations in the sample volume. Detection of the analytes can be done using fluorescence and laser fluorescence, UV absorbance, mass spectroscopy, conductivity tests, or radiometric tests.

Another problem that must be addressed is the generation of a "secondary" flow that may deteriorate the quality of the electropherogram or make it difficult to obtain reproducible results. Species inside the capillary move either because of the force as a result of the applied electric field or because they are dragged along with the buffer solution, which is responding to the electric field. This "dragging" is known as electro-osmotic flow.

The electrophoretic mobility is the mobility of an ion in an electric field, whereas the electro-osmotic mobility is the mobility as the result of a species being swept along in a flow arising from a buffer solution's response to an applied electric field. In CE, negative charges on the capillary wall create an electrostatic double layer. The outer layer is rich in mobile, solvated cations that move toward the cathode and drag along anions and neutral molecules. The electro-osmotic mobility, μ_{eof}, is given by

$$\mu_{eof} = \frac{\varepsilon\xi}{4\mu\pi} \qquad 1.4$$

in which ε is the electric permittivity of the solvent and ξ is the zeta potential, which is, approximately, the electrostatic potential at the wall of the capillary. The zeta potential increases proportionally with the charge on the capillary walls. The condition of the buffer, such as the pH can change the wall charges. The zeta potential increases as the thickness of the double layer (Chapter 3 gives more details about this) decreases. It is known that if the ionic strength of the buffer increases, the increased number of cations will reduce the thickness of the double layer; this is called double-layer compression and Chapter 3 explains it in greater detail. The total mobility of a solute, μ_{total}, is the sum of the electrophoretic, μ_{ele}, and the electro-osmotic mobilities:

$$\mu_{total} = \mu_{ele} + \mu_{eof} \qquad 1.5$$

In capillary electrophoresis, the electrophoretic mobility depends on the properties of the buffer and the analyte as in the general electrophoresis case.

Besides free-zone capillary electrophoresis, the technique can be used in electrofocusing and in micellar modes. The first technique is used to determine isoelectric points, whereas the second is used to separate neutral analytes. Capillary isoelectric focusing uses a pH gradient for separating amphoteric species. This pH gradient is created by the effect of the electric field on the ampholytes dispersed in a gel that fills the capillary. The ampholytes are zwitterions, which are molecules with both a positive and negative charge. In the presence of the electric field, H^+ and OH^- separate and bind to the zwitterions to create a pH gradient. Amphoteric species, such as proteins, migrate against diffusion to the optimum pH position, and thus, they are focused to one position along the tube corresponding to the pH at which they have zero charge. Capillary isoelectric focusing is an extremely sensitive technique. For example, it is able to distinguish between two proteins whose differences are a single charge (something that may be originated by the de-amidation of a single residue).

There are some success stories of laboratory techniques or approaches that have been implemented at process scale. The most notorious one is the use of disposable devices, which were (and are) common sights in any bioprocessing laboratory. Interestingly enough, the two most common disposable devices are ultrafiltration cartridges and chromatography columns that are the two most commonly used units in process scale separations of proteins.

Disposable chromatography systems have been implemented in a number of processes. A 0.5-L

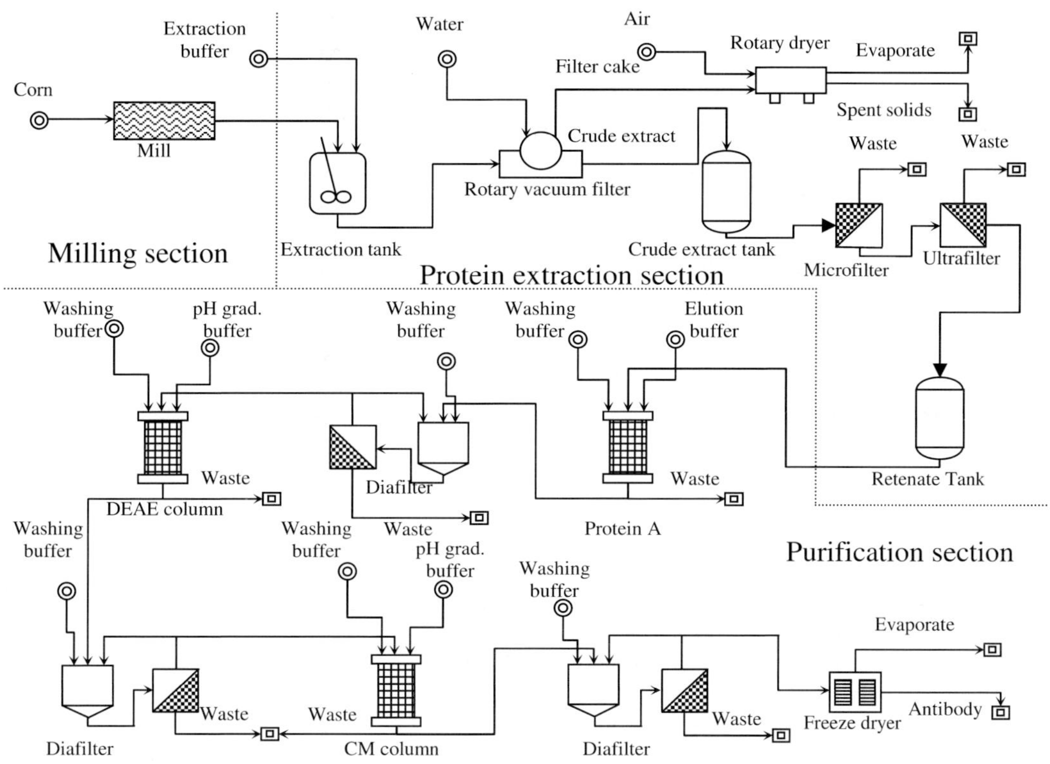

Figure 1.5. Flow sheet of a transgenic antibody purification. Modified from Evangelista, R. L., A. R. Kusnadi, J. A. Howard, and Z. L. Nikolov. 1998. Process and economic evaluation of the extraction and purification of recombinant ß-glucuronidase from transgenic corn. *Biotechnology Progress* 14:607-14. Copyright American Chemical Society; reproduced with permission.

membrane chromatography device can remove process-derived impurities, such as DNA, viruses, and endotoxins, from a 1000-L feed stream just as well as a stainless steel column that is a hundred times larger (Gottschalk 2006). The economics of such a system demonstrated that disposable equipment can provide a cost saving of up to 70% in MAb polishing (Sinclair 2005). This is because disposable options are cheaper to manufacture and significantly reduce material even as they consume 95% less buffer. Disposable chromatography columns offer additional advantages because they eliminate the need to carry out virus carryover and cleaning validation. This is of considerable economic impact, because virus filtration (the other main method for virus removal) is the most expensive downstream step, accounting for up to 40% of costs. The use of disposable devices for virus removal illustrates the viability of using such a device in very specific areas of the process.

Current Approaches to Bioseparations

Figure 1.5 shows a typical flow chart for antibody production (Kusnadi, Nikolov, and Howard 1997). The three sections are milling, protein extraction, and purification. There are three different chromatographic columns (one affinity column and two ion-exchange columns), a diafiltration unit, and a freeze-drying unit in the purification section. Either a protein A or a protein G column can be used as the affinity column. Protein A is a cell wall protein (42 kD), which is normally found in *Staphylococcus aureus*. It binds to immunoglobulins by interacting with the Fc fragment of the antibody. Protein A binds with high affinity to not only human IgG1 and IgG2, but also murine IgG2a and IgG2b. It also binds to human IgM, IgA, and IgE with moderate affinity. On the other hand, protein A does not bind to human IgG3 and IgD. Protein A chromatography is a fast and powerful technique to obtain a pure antibody in high yield. Unfortunately, its price is extremely high, because it is packed with recombinant protein, which makes its use in large-scale equipment too expensive. For example, chromatography columns, including a protein A column, accounted for 49% of the total equipment costs in the purification of ß-glucuronidase (Evangelista et al. 1998).

The other two columns are an anion-exchange diethylaminoethyl (DEAE) column and a cation exchange carboxymethyl (CM) column. The DEAE column is used to remove impurities and DNA residues. The CM column also removes impurities and filters protein A ligands leaking from the affinity column.

The approach followed to purify other antibodies, like immunoglobulin M (IgM) is slightly different. IgM purification is challenging because this antibody is less stable in solution (it slowly precipitates out of aqueous solutions) than immunoglobulin G (IgG). Although an affinity gel is available to purify this antibody by affinity chromatography, it is normally purified by a judicious choice of ion exchange and gel permeation chromatography. For example, IgM COU-1 has been purified from hybridoma cell supernatant by chromatography with hydroxylapatite, hydrophobic interaction, cation exchange, and anion exchange.

Proteins can also be extracted from tissue cultures. An example of this is shown in Figure 1.6, which is a flow sheet for the isolation of Factor XII (a coagulation factor) from human plasma. A simple ammonium precipitation step is followed by two zinc chelate chromatography columns, which are the centerpiece of the separation cascade. The separation in a zinc chelate column is based on the differential binding of proteins containing a different number of histidine residues. The last column is a gel permeation column that is used to eliminate traces of macroglobulin. In between each chromatography step, filtration units operating in diafiltration mode are used to change buffers. Diafiltration consists of eliminating the old buffer through dialysis (diffusion through the membrane) while the new buffer is pumped downstream of the membrane.

A common theme in the examples described above is the use of filtration and chromatography. Although there are other ancillary operations, including but not limited to cell/tissue disruption, centrifugation, and precipitation, it is obvious that both filtration and chromatography are the main unit operations in any bioseparations process. An important difference between the two techniques is that although ultrafiltration is a continuous technique, chromatography is typically a batch technique. This feature complicates the design of separation cascades, because continuous and batch operations must be cleverly combined to optimize the process. Filtration is a well-established approach that has reliable models suitable to scale up. Because it is a mature technique, its evolution has reached a plateau. Most of the recent advances are in the development of new membranes that improve separations and minimize fouling. Although chromatography is also a mature technique, its modeling is less reliable. Still, simple scale-up arguments are normally used and they are quite successful. Recent advances include the development of monolithic "gels," membrane-based systems, and new modes of operation. For example, a simulated moving bed (SMB), common in other industrial sectors, has been explored in bioseparations to eliminate the problem of combining continuous and batch processes (SMB makes chromatography a continuous process). Still, it is questionable whether there is a need to make chromatography a continuous process. The production volume in the biotechnology world is relatively low, so continuous processes may not be necessary.

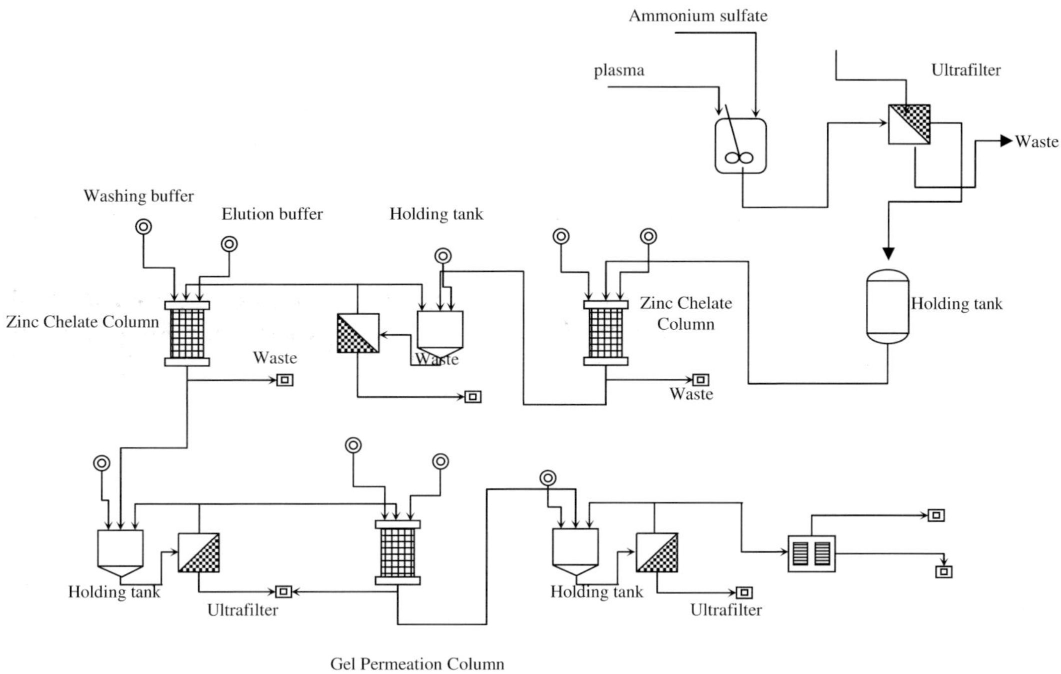

Figure 1.6. Flow sheet for the purification of coagulation Factor XII from human plasma.

Summary

Biotechnology is a growing field that needs expertise from the basic sciences and from engineering. This is most apparent in the branch of biotechnology called bioseparations. Although it is not possible to make an accurate forecast of this dynamic field, it is possible to anticipate that the production of therapeutical proteins or nucleic acids will be of great importance in the future, particularly with an aging population.

Biotechnology has shown in the past that it can affect humanity in an explosive manner. The example that comes to mind is the industrialization of penicillin. For years after its discovery, antibiotic therapy was too expensive for most people. The development of a process by Pfizer to mass produce penicillin has had immediate consequences in the extension of life expectancy for most of the western world.

Nomenclature

- a radius of the analyte (m)
- C_i concentration of the solute (kmol/m^3)
- D_{ij} diffusion coefficient (m^2/sec)
- E magnitude of the electric field (V/m)
- J_i flux of species i (kmol/m^2 sec)
- z effective charge of the analyte (C)

Greek Letters

ε permittivity of the medium (J/V^2 m)
μ viscosity of the medium (kg/m sec^2)
μ_{ele} electrophoretic mobility of the analyte (solute) (m^2/V sec)
μ_{eof} electro-osmotic mobility of the analyte (solute) (m^2/V sec)
μ_{total} total electrophoretic mobility of the analyte (m^2/V sec)
ξ zeta potential (V)

References

Ainsworth, S. J. 2005. Biopharmaceuticals. *Chemical and Engineering News* 83:21-9.

Bailey, J. E., and D. F. Ollis. 1986. *Biochemical Engineering Fundamentals.* New York: McGraw-Hill.

Evangelista, R. L., A. R. Kusnadi, J. A. Howard, and Z. L. Nikolov. 1998. Process and economic evaluation of the extraction and purification of recombinant ß-glucuronidase from transgenic corn. *Biotechnology Progress* 14:607-14.

Gottschalk, U., ed. 2006. The renaissance of protein purification. *BioPharm International* 12:42.

Kohler, G., and C. Milstein. 1976. Derivation of specific antibody-producing tissue culture and tumor lines by cell fusion. *European Journal of Immunology* 6(7):511-9.

Kusnadi, A. R., Z. L. Nikolov, and J. A. Howard. 1997. Production of recombinant proteins in transgenic plants: practical considerations. *Biotechnology and Bioengineering* 56:473-84.

Prouty, W. F. 1991. Production-scale purification processes. In *Drug Biotechnology Regulation 13,* edited by Y.-Y. H. Chen and J. L. Gueriguian, pp. 221-62. New York: Marcel Dekker.

Sinclair, A. 2005. *Disposable Technologies: Impact on Biomanufacturing.* London: Institution of Chemical Engineers.

Chapter 2
Thermodynamics in Bioprocessing

Introduction

Most scientists and engineers are exposed to the principles of thermodynamics to some extent. Yet, there is a lack of connection between the study of the science and the application of the concepts of thermodynamics during their professional life. Rather than attempt to provide a comprehensive review of thermodynamics, this chapter reviews some basic concepts and how they may be used in the design of various separation processes.

The Basic Laws

The science of thermodynamics is based on four postulates (or laws), which provide the foundation for the development of a relatively complex network of properties. It is in that simplicity that one finds its beauty. In Chapter 1, we concluded that the key in any separation process is the uneven distribution of the target molecule and the impurities between two or more phases. Ideally, that distribution is limited only by the laws of thermodynamics. This is true, because it establishes a boundary for the distribution of target and impurities. That boundary is reached when the system is at thermodynamic equilibrium. Therefore, it is in our own interest to know what that boundary is and how we can alter it.

The four fundamental laws of thermodynamics are the Zero Law (Thermal Equilibrium), the First Law (Conservation of Energy), the Second Law (Spontaneity of Processes), and the Third Law (Nerst Postulated). All thermodynamic variables and relationships, as well as the fundamental principles that rule equilibrium between phases, can be derived from these four fundamental laws.

Some definitions are needed before one can begin developing this science. In order to avoid making naïve mistakes, a clear understanding of this "vocabulary" is important.

The first step in applying the laws of thermodynamics is to define the **system**, as the part of the **universe** that we are studying. Systems can be **closed** (no mass is allowed in or out), **open** (mass and energy may flow in and out), **isothermal** (the temperature of the system is constant), **adiabatic** (no heat may come in or out), or **isolated** (no mass or energy can flow in or out of the system).

Theoretical "**walls**" separate the system from the **surroundings** (the universe minus the system). In thermodynamics, we study properties of matter. These properties can be **intensive**, which are independent of the size of the system (will be denoted in small caps), **extensive**, which depend on the size of the system (specific and molar properties will be denoted in small caps and capital letters, respectively), **path variables** (their value depends on the path to reach that particular state), or **state variables** (their value does not depend on the path followed to reach that particular state). There is a special group of properties called "**partial molar properties**" that are defined by

$$\xi_i = \left(\frac{\partial M}{\partial N_i} \right)_{T,P,Nj \neq i} \qquad 2.1$$

in which ξ_i is the partial molar property of M (an extensive property), N_i is the number of moles of species i, T is the absolute temperature, and P is the pressure. In simple words, a partial molar property measures the changes in a total property as the number of moles of one species, i, is changed while keeping constant temperature, pressure, and the number of moles of the remaining species. Because of that, they measure the behavior of a particular species in a mixture. Therefore, they are very useful in the thermodynamics analysis of any separation process because mixtures consist of a large number of species and only one is of interest. Other important concepts are equilibrium, which is defined as the state of a closed system in which there is no apparent changes with time, internal energy, which is equal to the sum of potential and kinetics energy of atoms and molecules ($U=U_{potential}+U_{kinetic}$) work, which is defined as the product of the applied force times the distance traveled ($W = \int_A^B Fdx$), and pressure, which is defined as the force applied per unit area (P=F/A).

Zero Law

Two systems in thermal contact are in thermal equilibrium when the temperature, an intensive property, is the same for both systems.

Rate of accumulation of Energy within the system boundaries	=	Rate of input of Energy into the system throughout its boundaries	-	Rate of output of Energy out of the system throughout its boundaries	2.2

First Law

Sometimes the First Law is stated as the Law of Conservation of Energy. However, the actual statement of the First Law is that the difference between heat and work of a system, both path variables, is defined as a state variable of that system called the internal energy (U). The energy balance then becomes a consequence of the First Law. It is convenient to state this principle in words to clarify concepts.

Unfortunately, "boxes" are not sufficient to solve problems. For example, how can we calculate the power needed by a pump to move a fluid through a chromatography column, or how much energy is needed to sublimate one kilogram of water. Therefore, if we want to benefit from this, we need to translate the boxes into equations.

The meaning of energy needs to be clarified before the first box is replaced by symbols. Energy can be classified into potential, kinetic, and internal energy. In the context for this textbook, we associate potential energy with heights, the higher a body is the more potential energy, E_p, the body has, in symbols,

$$E_p = gL \tag{2.3}$$

in which g is the acceleration of gravity and L is the distance of the body from some reference point. The kinetic energy, E_k, is associated with movement. We are all familiar with this concept,

$$E_k = \frac{1}{2} m v_s^2 \tag{2.4}$$

m is the mass of the body and v_s its velocity. The last "type" of energy, internal (U), is a little more difficult to conceptualize. One relatively simple way of understanding the concept is to associate it with the intrinsic energetic content of matter (including the energy associated with the movement of atoms, the energy associated with the relative position of atoms (or molecules) in the system, and the energy stored in the atoms themselves). Therefore, the total energy, E, is,

$$E = U + E_k + E_p \tag{2.5}$$

The first box is then equal to the change of E with respect to time as a process occurs. The kinetics energy and potential energy terms discussed above are defined for the entire system and not for whatever is inside the walls of the system (i.e., molecules, fluid). The first box is:

$$\frac{dE}{dt} = \frac{d[E_p + E_k + U]}{dt} \tag{2.6}$$

To replace the boxes on the right-hand side by equations we need to recognize the various ways by which energy can enter or leave the system. For example, energy may enter or leave the system accompanying mass flow. An element of fluid entering or leaving with a flow-rate m_k has associated with it an internal energy, u_k per unit mass, a kinetics energy, $v^2/2$, and a potential energy per unit mass, ς. For K ports of entry or exit,

$$\sum_{k=1}^{K} \dot{m}_k (u + v_s^2 / 2 + \varsigma)_k \tag{2.7}$$

Energy may also enter or exit in the form of heat, Q, for J ports of entry or exit,

$$\dot{Q} = \sum_{j=1}^{J} \dot{Q}_j \tag{2.8}$$

According to Joule's work (1845), heat and work are equivalent; therefore, energy may enter or exit the system as work. There are three types of work that must be considered. First, there is work associated with moving parts (an agitator, a pump); this type of work is called shaft work, W_{shaft}. Second, there is work associated with the movement of the boundaries (wall) of the system; this type of work in three dimensions is $-P\ dV/dt$,

$$\dot{W} = \dot{W}_{shaft} - P\frac{dV}{dt} \qquad 2.9$$

in which P is the pressure on the system and V its volume. There is a third kind of work associated with fluid flow. Consider that a stream entering the system is composed by a series of volume elements (like the wagons in a train). Assume that the fluid element J is just inside the system. Then, the fluid element J-1 (upstream) will be pushing element J deeper inside the system as it tries to enter too. Therefore, fluid element J-1 is doing work on element J. A similar reasoning can be made for fluid elements leaving the system. The net work is,

$$= \sum_{k=1}^{K} \dot{m}_k (P\,v)_k \qquad 2.10$$

in which v is the specific volume. Combining Equations 2.5 through 2.10,

$$= \sum_{k=1}^{K} \dot{m}_k (P\,v)_k \qquad 2.11$$

Using H=H+PV (and calling this new thermodynamic, variable enthalpy), we obtain,

$$\frac{d[E_p + E_k + U]}{dt} = \sum_{k=1}^{K} \dot{m}_k (h + v^2/2 + \varsigma)_k + \dot{Q} + \dot{W} \qquad 2.12$$

Equations 2.11 and 2.12 are two equivalent versions of the energy balance. A number of simplifications are possible depending upon the system under study. For a closed system, all the mass fluxes are equal to zero, then,

$$\frac{d[E_p + E_k + U]}{dt} = \dot{Q} + \dot{W} \qquad 2.13$$

For a closed system with negligible kinetic and potential energy contribution to the total energy (E_p=E_k=0),

$$\frac{dU}{dt} = \dot{Q} + \dot{W} \qquad 2.14$$

For a closed system with negligible kinetic and potential energy contribution to the total energy and no shaft,

$$\frac{dU}{dt} = \dot{Q} - P\frac{dV}{dt} \qquad 2.15$$

Second Law

A consequence of the First Law of Thermodynamics is that energy must be conserved. However, the First Law cannot be used to determine whether a process is feasible or not, or in which direction a process will occur. Also, the First Law will allow us to calculate, for example, the amount of work needed to pump a fluid, but it will not allow us to calculate what will be the minimum work that needs to be done to pump the fluid. The Second Law of Thermodynamics gives this type of information. Although there are several ways that the Second Law of Thermodynamics can be stated, it is convenient to state it as "There is a property of isolated systems called entropy, S, that always increases in a spontaneous non-adiabatic and reversible process. When the system reaches equilibrium, changes in this property are equal to zero." Therefore, from these two sentences it is easy to conclude that the entropy of a system at equilibrium is at its maximum. For a reversible process, the change in entropy from state A to state B can be calculated from,

$$\Delta S = \int_A^B \frac{\delta Q}{T} \qquad 2.16$$

in which T is the temperature at which the transfer of heat is taking place. The concept of entropy can be used to define temperature, rather than the intuitive sensation of hotness or coolness that is actually a consequence of heat transfer. Consider the entropy of matter to be a function of the extensive variables N, V, and E. Then,

$$dS = \left(\frac{\partial S}{\partial N}\right)_{V,E} dN + \left(\frac{\partial S}{\partial V}\right)_{N,E} dV + \left(\frac{\partial S}{\partial E}\right)_{N,V} dE \qquad 2.17$$

Let's define:

$$\left(\frac{\partial S}{\partial E}\right)_{N,V} = \beta \qquad 2.18$$

Considering two solid blocks with energy E_1 and E_2 in thermal contact, the total entropy of the system is,

$$S(E_1) = S_1(E_1) + S_2(E_1 - E_1) \qquad 2.19$$

in which $E_T=E_1^0+E_2^0$. At equilibrium, $S'(E_1)=0$, so, $S'_1(E_1)=S'_2(E_2)$. However, according to Equation 2.18, $\beta_1=\beta_2$. Therefore, β has the meaning of temperature.

It is relatively simple to write the Second Law for open systems. However, it is not sufficient to recognize the means by which entropy can enter or leave the system because entropy is not a conserved quantity. Therefore, it is necessary to add a term to the balance that accounts for the creation of entropy (it is useful to contrast this with the energy balance). Calling the generation of entropy term S_g, the open systems version of the entropy balance is,

$$\Delta S = \sum_{i=1}^{N} s_i \dot{m}_i - \sum_{i=1}^{M} s_i \dot{m}_i + \int \frac{\delta Q}{T} + S_g \qquad 2.20$$

The analysis of the generation of entropy during a process is a valuable tool to evaluate its efficiency. This is discussed in the last section of this chapter.

The Third Law of Thermodynamics has a marginal use in bioseparations. Just for completeness, we state it as "the entropy of a perfect crystal at the absolute zero is equal to zero."

Phase Equilibrium

Understanding the basic thermodynamic postulates is a good starting point, but it is not enough to fully benefit from it. There is one additional thermodynamics potential that plays a key role in the analysis of equilibrium between phases and in the thermodynamic analysis of processes. That quantity is the Gibbs free energy (beautifully introduced by Gibbs [1993] in his exceptional work), which is defined by,

$$G = H - TS \qquad 2.21$$

A related quantity was defined by Helmholtz (Sandler 2006) as,

$$A = U - TS \qquad 2.22$$

A is called the Helmholtz free energy. The meaning of both quantities is the amount of work that a system potentially can do during a process at constant pressure and temperature (G) or at constant volume and temperature (A). Besides, they are convenient quantities to characterize the equilibrium between phases.

At this point, the concept of equilibrium can be formalized. We defined it before as a time invariant state of a system. It can be rigorously proven that at equilibrium either the entropy is a maximum, or the free energy (including Gibbs') is a minimum. Applying the principle of maximum entropy, one can show that when two phases, α and β, are at equilibrium the following conditions must be satisfied,

$$T_\alpha = T_\beta \qquad 2.23$$

$$P_\alpha = P_\beta \qquad 2.24$$

$$\mu_\alpha^i = \mu_\beta^i \qquad 2.25$$

in which,

$$\left(\frac{\partial G}{\partial N_i}\right)_{T,P,Nj} = \mu_i \qquad 2.26$$

is called the chemical potential of species i, which is a function of T, P, and concentration. Equation 2.23 is the statement of thermal equilibrium, Equation 2.24 is the statement of mechanical equilibrium, and Equation 2.25 is the statement of chemical equilibrium. Equation 2.26 is written for all the species in the mixture. μ'_α is called the chemical potential of species i in phase α and it is the partial molar property of the Gibbs free energy. After a reliable model for the chemical potential is used in the last set of equalities, these equations can be solved to calculate, for example, the distribution of the species in a mixture between two or more phases. It is standard practice to write,

$$\mu_i = \mu_i^0 + RT \ln \gamma_i(x_i, T, P)x_i \qquad 2.27$$

in which μ_i^o is the chemical potential at some convenient reference or standard state, x_i is the mole fraction of species i and γ_i is its activity coefficient. By using Equation 2.27, the problem of developing a model for the chemical potential is replaced by the problem of developing a model for the activity coefficient. If the activity coefficient is equal to one, one says that the solution is ideal. Thus, the activity coefficient measures deviation from that ideal behavior. Analogous to the activity coefficient is the fugacity coefficient, which is treated in many thermodynamic textbooks (see Further Reading suggestion at the end of this chapter); however, in this book we will deal solely with activity coefficients. Advances in the developing of models for activity coefficients have been mostly driven by the need of the petrochemical companies. Therefore, it is not surprising that there are several reliable models for activity coefficients of small molecules or for non-polar ones but that the menu is much less abundant for proteins. One of the most popular models is universal quasi-chemical activity coefficient (UNIQUAC) group-functional activity coefficient (UNIFAC) that can be found in most process simulators. The idea of the UNIFAC approach is that the thermodynamic properties of a complex molecule are given by the contribution of the groups (e.g., carbonyl groups, amino groups, methylene groups, phenol groups) that make the molecule. The key in the application of this approach is the availability of experimental data for the contribution of a particular group and the judicious combination of the contribution for each group in a molecule to obtain the property of that molecule.

We emphasized in Chapter 1 that the "secret" of a successful separations process is to achieve an unequal distribution of the product and impurities between two different phases. These phases may have different states (gas, liquid, or solid) of aggregation and they may be separated by an interface or by a membrane. In the following sections, we examine solid/liquid, solid/vapor, liquid/liquid, and liquid/solid equilibrium.

Solid/Liquid Phase Equilibrium

Crystallization, precipitation, and chromatography are three unit operations commonly used in bioseparations that involve equilibrium between solid and liquid phases. Protein solubility plays a key role in the first two techniques, whereas the affinity of a protein for a solid surface is key in chromatography.

Protein solubility is affected by a variety of factors, including pH, type and concentration of salts, presence of non-ionic or ionic polymers, and temperature. A salt commonly used for the precipitation of proteins is ammonium sulfate. Triangular diagrams can be used to illustrate this. Figure 2.1 shows one such diagram for the precipitation of bovine serum albumin (BSA) by ammonium sulfate (Elysee-Collen and Lencki 1999). The upper vertex represents 100% BSA, the lower left vertex represents 100% water and the lower right vertex represents 100% ammonium sulfate. Lines parallel to the side opposite to the 100% vertex represent decreasing quantities of the component whose concentration in that vertex is 100%. The concentration of that component on the opposite side is zero; therefore, the sides of the triangle represent binary mixtures, whereas the plane of the triangle represents ternary mixtures. The curve lines inside the triangle divide it into regions that represent different phases. If a system falls on any of those lines, the system consists of two phases at equilibrium. Region 1 corresponds to a transparent liquid, clear and colorless at low BSA concentrations; it becomes increasingly yellow at concentrations above 15%. Region 2 corresponds to a particulate structure with no free liquid present. It is mostly a dry powder, with small quantities of solid salt sometimes present. Region 3 corresponds to a liquid that contains a precipitate, which redissolves upon dilution. Finally, region 4 corresponds to a liquid that contains a protein precipitate and salt crystals, which redissolve upon dilution. For example, point A represents a system consisting of two phases at equilibrium. One is a particulate with no free liquid (region 2) and the other is a liquid that contains a precipitate (region 4).

If the phase diagram is not available experimentally, it may be generated by solving the systems of equations that define the equilibrium between phases. It is of course necessary to have a reliable model for the activity coefficient of the protein and in the case of precipitation by salts, for the activity coefficient of the various ions.

It has been argued that one of the problems with classical thermodynamics is that it is not "predictive." This may be misleading. It is true that its models rely on experimental data of some kind, but most simulators will use well-established equations of state or expressions for activity coefficients to "predict" the behavior of the plant. If experimental data is not available, then one needs to move away from the realm of classical thermodynamics and into the realm of statistical thermodynamics. An advantage of this approach, in a scientific world obsessed with microscopic scales, is

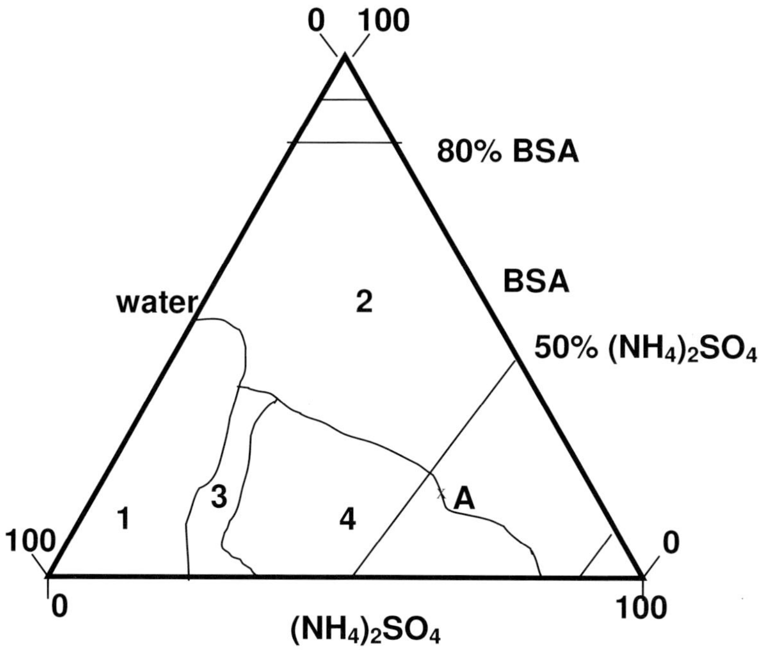

Figure 2.1. Triangular diagram for phase equilibrium of mixtures of bovine serum albumin (BSA), ammonium sulfate, and water. Adapted from Elysee-Collen, B., and R. W. Lencki. 1999. Influence of protein surface morphology on the ultrafiltration flux resistance of bovine serum albumin. *Biotechnology Progress* 15:732-9.

that it yields information at the molecular level that can be used to manipulate the process at a macroscopic scale. For example C. K. Hall's group has used statistical thermodynamics to study the precipitation of proteins by neutral polymers (Mahadevan and Hall 1990; Mahadevan and Hall 1992). The method is complex yet simple to follow if we keep in mind that the quantity of interest is the Gibbs free energy. To determine the coexistence points, they calculate the Gibbs free energy of each phase, and plot them versus the pressure, which is a function of protein concentration, for a given polymer volume fraction. Figure 2.2 shows this.

The intersection point between the solid and fluid Gibbs free energy lines defines the conditions for equilibrium between the two phases. This process can be repeated for several polymer concentrations to obtain a solid/liquid phase diagram similar to the one shown in Figure 2.3.

Although true equilibrium is rarely reached in chromatography, it is common to assume that equilibrium is reached for a particular solute between the liquid phase (mobile) and the solid phase (stationary). This equilibrium may be represented by,

$$C + S \xleftrightarrow{Keq} CS \qquad 2.28$$

in which C is the dissolved chemical species, S is the adsorption site, and CS is the adsorbed species. Assuming that the adsorption is reversible, only one site is involved in the binding of a solute, and that all the sites are equivalents, we can write,

$$K_{eq} = \frac{[CS]}{[C][S]} \qquad 2.29$$

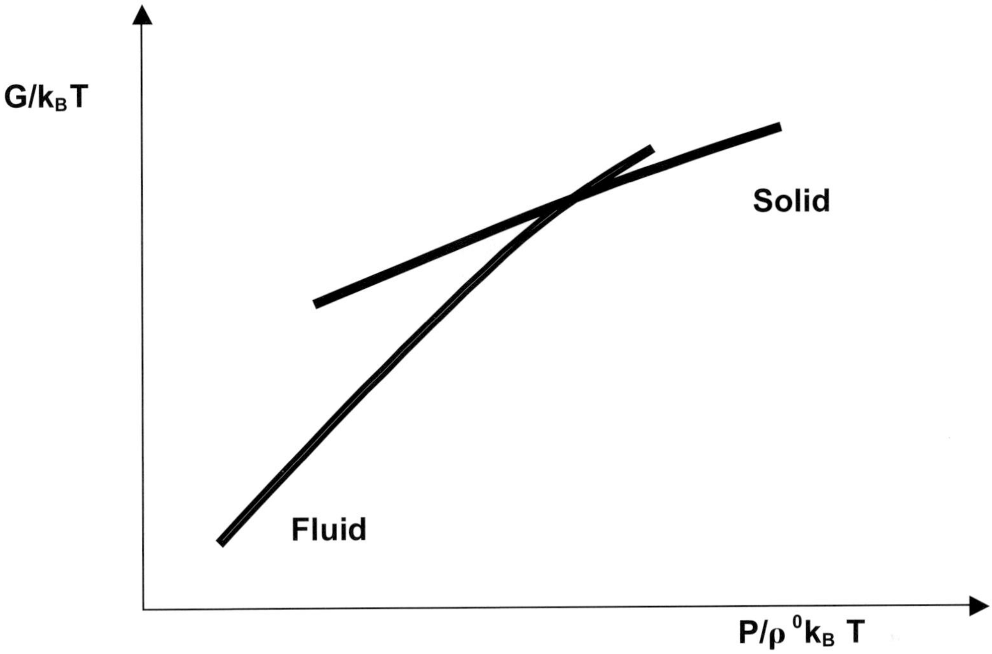

Figure 2.2. Normalized Gibbs free energy versus normalized pressure for two condensed phases. The point of intercept corresponds to equilibrium between a solid and a liquid phases. Adapted from Mahadevan, H., and C. K. Hall, 1990. Statistical-mechanical model of protein precipitation by nonionic polymer. *AIChE Journal* 36(10):1517-28. Copyright American Institute of Chemical Engineers; reproduced with permission.

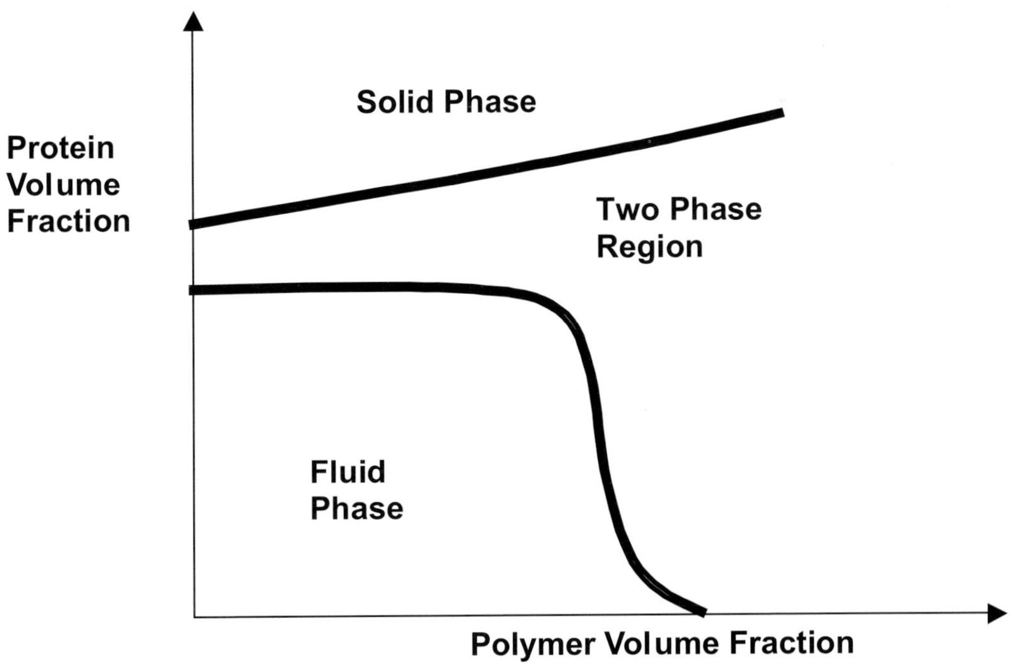

Figure 2.3. Phase diagram for a protein/neutral polymer mixture. Mahadevan, H., and C. K. Hall, 1990. Statistical-mechanical model of protein precipitation by nonionic polymer. *AIChE Journal* 36(10):1417-28. Copyright American Institute of Chemical Engineers; reproduced with permission.

Mass balances for the dissolved species and the sites are:

$$[C_0] = [C] + [CS] \tag{2.30}$$

$$[S_T] = [S] + [CS] \tag{2.31}$$

in which $[C_0]$ is the total concentration of the dissolved species and $[S_T]$ is the total concentration of sites. Therefore,

$$[CS] = K_{eq}[C]([S_T] - [CS]) \tag{2.32}$$

and

$$[CS] = K_{eq}[C]([S_T] - K_{eq}[C][CS] \tag{2.33}$$

or

$$[CS] = \frac{Keq[C][S_T]}{1 + K_{eq}[C]} \tag{2.34}$$

If the amount adsorbed, [CS], follows the dependence on [C] indicated by this expression at constant temperature, one says that the system follows a Langmuir isotherm. Equation 2.34 may be simplified. Two particular cases are of interest. Case 1 occurs if K_{eq} [C] <<1 then $[CS]=K_{eq}[C]$ $[S_T]$. Therefore, the "isotherm" is linear. The concentration of the adsorbed species is a multiple of [C] and thus, there is no saturation limit. Case 2 occurs if K_{eq} [C] >>1 then [CS]=[ST] and the adsorption sites are saturated. The connection between absorption and thermodynamic potentials like H or S is given by the relationship between K_{eq} and ΔG,

$$\Delta \underline{G}_{ad} = -RT \ln K_{eq} \tag{2.35}$$

Furthermore,

$$\Delta \underline{G}_{ad} = \Delta \underline{H}_{ad} - T\Delta \underline{S}_{ad} \tag{2.36}$$

Therefore, the adsorption process is driven by an enthalpy contribution and an entropic contribution. If the adsorption process is athermal then only changes in entropy are driving it. The change in entropy is for the entire system, which includes protein, water, and surface. Normally, the entropy of the surface does not change. Therefore, the entropy contribution is given by either a gain in entropy of the solvent or a loss of entropy by the adsorbed species because it loses degrees of freedom in the adsorbed state. There are several more adsorption models. One used as frequently as Langmuir is the Freundlich Isotherm (empirical),

$$\Delta \underline{G}_{ad} = \Delta \underline{H}_{ad} - T\Delta \underline{S}_{ad} \tag{2.37}$$

in which n is greater than 1. "Energetic heterogeneity" of the surface has been invoked to explain Freundlich. Figure 2.4 shows the different cases for adsorption isotherms discussed here.

Solid/Vapor Phase Equilibrium

Figure 2.5 is a typical phase diagram for a fluid. The plane is divided by coexistence curves into three regions: vapor, liquid, and solid. CP is the critical point and TP is the triple point (coexistence

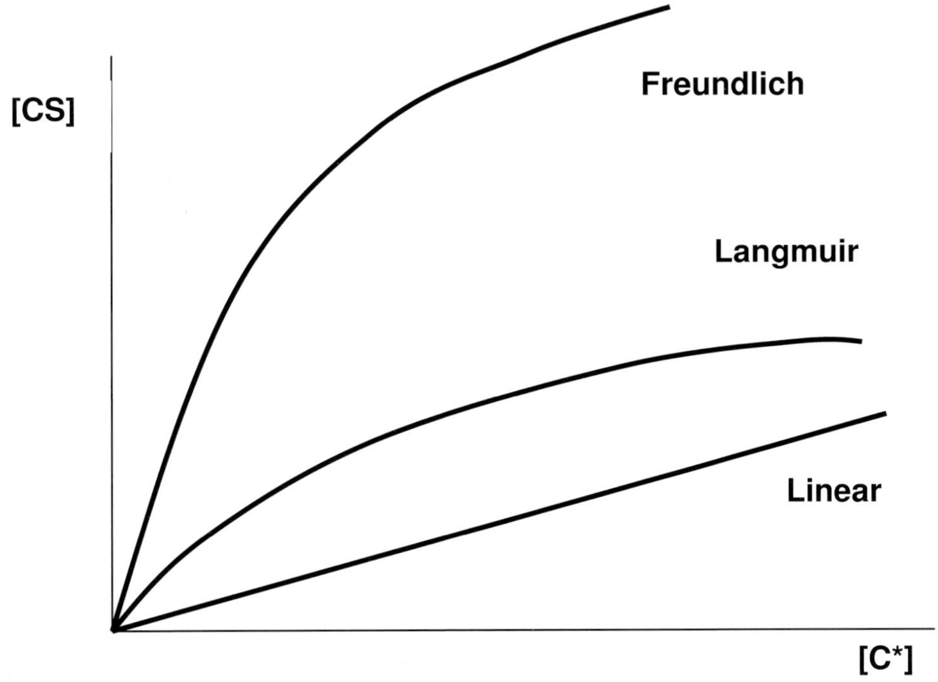

Figure 2.4. Different adsorption isotherms. [CS] is the concentration in the adsorbed state at equilibrium with a bulk concentration equal to [C*].

between three phases). The arrow indicates a phase transition called sublimation, in which a solid becomes a vapor directly—without melting first. For sublimation to take place the equilibrium line needs to be crossed at very low pressures ($<10^{-3}$ mbar in the case of pure water) and low temperatures (<-50°C, again in the case of water). The arrow represents the typical transition that occurs during freeze drying that is done at very low temperatures and high vacuum. Most of the time, a protein solution in water will contain salts and their presence decrease the freezing point of water.

Liquid/Liquid Phase Equilibrium

Liquid/liquid phase equilibrium may be depicted in triangular diagrams or in standard Cartesian coordinates. A typical liquid/liquid phase equilibrium diagram in an x-y plane is shown in Figure 2.6. Concentrations of the phase-forming species (the components that are used to create the two liquid phases) are plotted in the x and y axes respectively. The solid curve line, called the coexistence line, divides the plane into two regions. Two phases are present to the right of the coexistence line, whereas only one phase exists under the curve. The straight line joining two points of the coexistence line is called a tie-line and the two points a and b represent the composition of each phase. The composition of each phase is obtained by reading the concentration values where the dashed line intercepts the axes. Point c represents the overall composition of the system. The volume ratio between the two phases is given by:

$$\frac{V_{top}}{V_{bottom}} = \frac{\overline{CB}}{\overline{AC}} \tag{2.38}$$

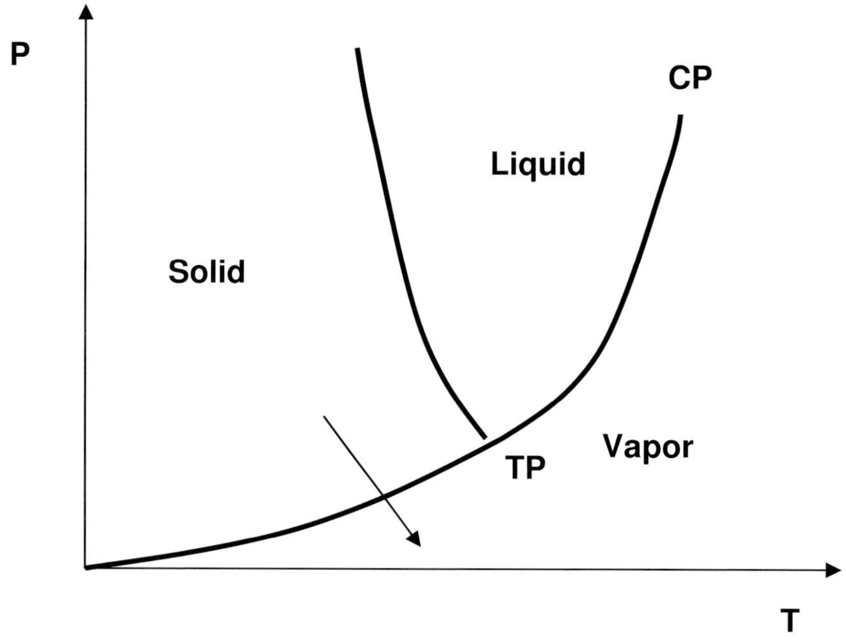

Figure 2.5. Phase diagram for a pure fluid. CP is the critical point and TP is the triple point.

When solutes are added to two phases in equilibrium, they will distribute between the phases according to:

$$K = \frac{[A]_{top}}{[A]_{bottom}} = \left(\frac{\gamma_A^{bottom}}{\gamma_A^{top}}\right) \exp \frac{\left[\mu_0^{A,top} - \mu_0^{A,bottom}\right]}{RT} \qquad 2.39$$

in which γ_A is the activity coefficient of species A in either the top or bottom phase and $\mu_0{}^A$ is the standard chemical potential of species A in either the top or bottom phases.

Thermodynamics of Electrolytes

Most large biomolecules are not stable in organic solvents. Therefore, the most common solvent used in bioseparations is water. This may sound trivial, but water is actually one of the most complex fluids that we may encounter. The fact that water is a polar molecule makes well-known approaches used in the petrochemical sector to calculate phase equilibrium useless. Besides, the buffers used and the biomolecules themselves may contain charges depending on the pH of the medium. The presence of a polar solvent and ions, such as salts, proteins, or nucleic acids, make the calculation of phase equilibrium in bioseparations quite challenging.

Many approaches used to calculate chemical potentials or activity coefficients of "simpler" molecules such as hydrocarbons fail for electrolytes because even at very dilute conditions electrostatic interactions are so long ranged that one needs to account for multiple body interactions. Electrolytes

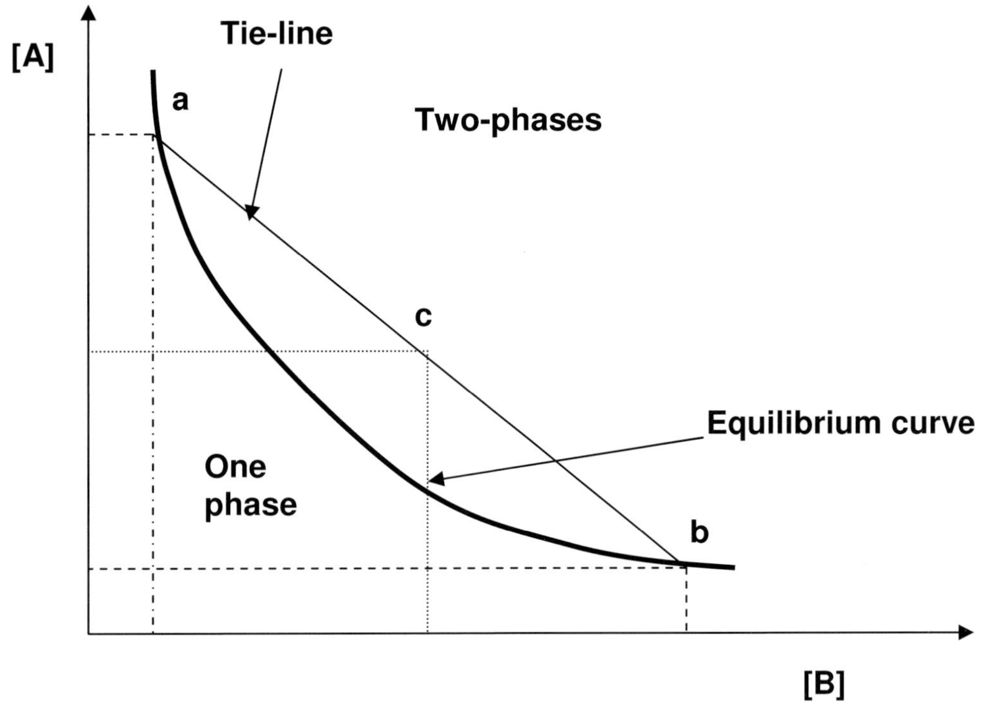

Figure 2.6. Phase diagram for a liquid/liquid two-phase system. The tie line joins the composition of the top phase (point a) with the composition of the bottom phase (point b) and the overall composition of the system (point c). The segments ac and cb are proportional to the volumes of the bottom and top phase respectively.

are distinctively different from neutral molecules in that they dissociate into two or more species in solution. A simple example is a solution of sodium chloride that dissociates into sodium and chloride ions. The treatment of electrolyte solutions starts with the recognition that more than one species will be present. The common approach is to define a mean activity coefficient, $\gamma_{\pm}$, and a mean ionic molality, m± by,

$$\gamma_{\pm}^{\nu} = \left[\gamma_A\right]^{\nu+} \left[\gamma_B\right]^{\nu-} \tag{2.40}$$

$$m_{\pm}^{\nu} = \left(m_A\right)^{\nu+} \left(m_B\right)^{\nu-} \tag{2.41}$$

in which γ_A is the activity coefficient of cation, γ_B is the activity coefficient of the anion, ν^+ and ν^- are the stoichiometric coefficients of the cation and anion as the salt dissociates (1:1 in the case of NaCl) and ν is equal to the sum of stoichiometric coefficients of the cation and anion. Notice that because an isolated charge cannot exist, the activity coefficients of the individual ions do not have a clear physical meaning. Figure 2.7 shows the dependence of the activity coefficient with the molality of a simple electrolyte (Goldberg, 1981). Observe that at concentrations normally found for salts in biosystems, the activity coefficient deviates appreciably from one. On the other hand, because of the low concentration of proteins, their activity coefficient is normally close to one.

The simplest approach to calculate chemical potentials of electrolytes was developed by Debye

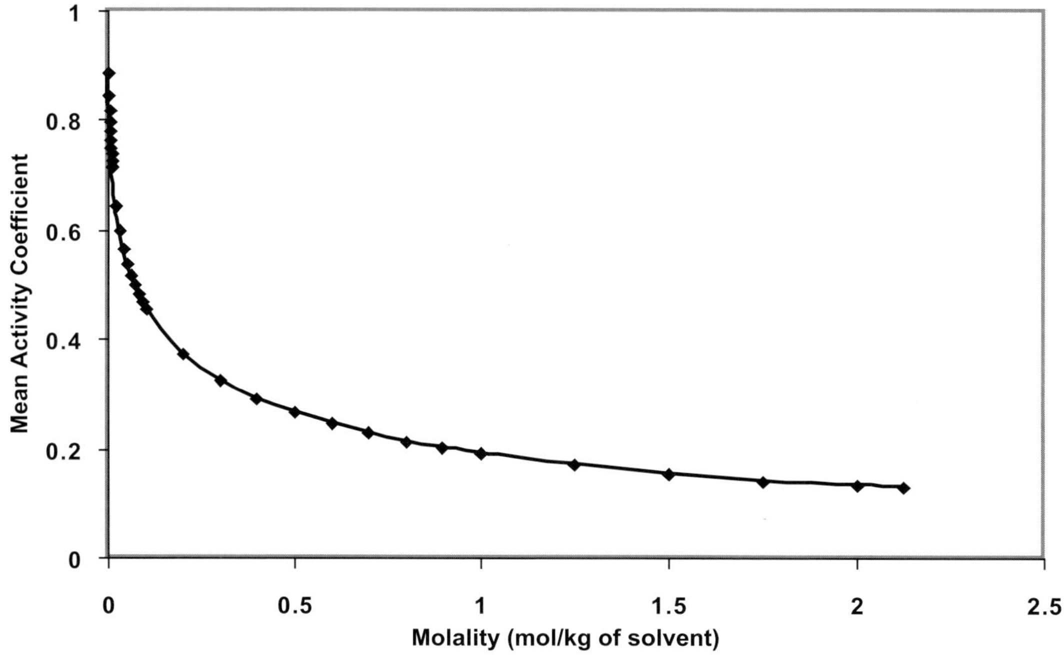

Figure 2.7. Mean activity coefficient versus molality for a simple electrolyte. Data plotted from Goldberg, R. N. 1981. Evaluated activity and osmotic coefficients for aqueous solutions: thirty-six uni-bivalent electrolytes. *J Phys Chem Ref Data* 10(3):671-764.

and Huckel (1923) many years ago. Using statistical mechanical arguments, they found the following expression for the mean activity coefficient,

$$\ln \gamma_{\pm} = -a|z_+ z_-|\sqrt{I} \qquad 2.42$$

in which I is the ionic strength, given by,

$$I = \frac{1}{2}\sum_{i=ion} z_i^2 m_i \qquad 2.43$$

and a is a constant that is a function of temperature and solvent type, z_+ and z_- are the charges of the positive and negative ions, and m_i is the molality of ion i. The Debye-Huckel model is normally called a limiting law because it is valid at very low ionic strengths. Some corrections to the Debye-Huckel law have been proposed (Grenthe and Plyasunov 1997; Lim 1993). For example, the addition of a linear term in ionic strength improves the correlation at less dilute conditions,

$$\ln \gamma_{\pm} = -\frac{a|z_+ z_- \sqrt{I}|}{1 + ba\sqrt{I}} + dI \qquad 2.44$$

in which a is a constant related to the hydrated radius of the ions and d is an adjustable parameter. This equation can follow experimentally determined activity coefficients quite well as long as no constraints are put on the parameter d. Pitzer (1991) proposed to use

$$\ln \gamma_{\pm} = -b\kappa(1 - \frac{2}{3}\kappa r_0 + \frac{5}{9}\kappa^2 r_0^2 - ....) \qquad 2.45$$

with,

$$b = \frac{|Z_i Z_j| e^2}{8\pi\varepsilon_0 D_s k_B T} \qquad 2.46$$

$$\kappa = \left[\left(\frac{2000 e^2 N_A \rho_s}{\varepsilon k_B T}\right) I\right]^{1/2} \qquad 2.47$$

in which r_0 is the radii of the ion, e is the charge of an electron, ε_0 is the permittivity of empty space, D_s is the static dielectric constant of the solvent, N_A is the Avogadro's number, ε is the permittivity of the solvent, and k_B is the Boltzmann's constant. An alternative expression is,

$$\ln \gamma_{\pm} = -b\kappa(1 - \kappa r_0 + \kappa^2 r_0^2 - ....) \qquad 2.48$$

in which b and κ have the same meaning as before. One of the best models for activity coefficients of electrolytes is the so-called Pitzer ion interaction model,

$$\ln \gamma_{\pm} = |Z_+ Z_-| f^{\gamma} + m\left(\frac{2\nu_+\nu_-}{\nu}\right) B_{\pm}^{\gamma} + m^2\left(\frac{2(\nu_+\nu_-)^{3/2}}{\nu}\right) C_{\pm}^{\gamma} \qquad 2.49$$

in which f is the Debye-Huckel term equal to,

$$f^{\gamma} = -3A_{\phi}\left[\frac{I^{1/2}}{1 + bI^{1/2}} + \frac{2}{b}\ln(1 + bI^{1/2})\right] \qquad 2.50$$

and,

$$B_{\pm}^{\gamma} = 2\beta_0 + \frac{2\beta_1}{\alpha^2 I}\left[1 - (1 + \alpha I^{1/2} - 0.5\alpha^2 I)\exp(-\alpha I^{1/2})\right] \qquad 2.51$$

for a single electrolyte,

$$C_{\pm}^{\gamma} = \frac{3}{2} C_{\pm}^{\phi} \qquad 2.52$$

$$A_{\phi} = \frac{1}{3}(2\pi(1000) N_A \rho_s)^{1/2}\left(\frac{e^2}{4\pi\varepsilon_0 D_s kT}\right)^{3/2} \qquad 2.53$$

In the above equations m is the molality of the salt, ρ_s is the density of the solvent, α=2.0, b=1.2 $(kg/mol)^{1/2}$. β_0, β_1, and $C_{\pm}^{\gamma}$ have to be fitted for each i,j electrolyte. The Pitzer model has been extended to mixtures of electrolytes and his equation is an integral part of many process simulation packages. A much simpler approach was developed by Meissner and Tester (1972) who extended to electrolytes the idea of corresponding states, commonly used to organize or collapse PVT data for neutral fluids. The Meissner model is,

$$\Gamma_{ij}^{0} = \left(\gamma_{\pm}\right)^{1/|Z+Z-|} = \left(1 + B(1+0.1I)^{qij} - B\right)\Gamma_{ij}^{DH} \quad 2.54$$

with,

$$\log \Gamma_{ij}^{DH} = \frac{-0.5107 I^{1/2}}{1 + C I^{1/2}} \quad 2.55$$

$$B = 0.75 - 0.065 q_{ij}^{0} \quad 2.56$$

$$C = 1 + 0.055 q_{ij}^{0} \exp\left(-0.023 I^{3}\right) \quad 2.57$$

Finally, it is worthwhile to mention Chen's local composition model (Chen et al. 1982; Chen and Evans 1986). Again here, he followed the idea of the UNIFAC model to develop a local composition model for electrolytes. Chen's model considers three contributions to the mean activity coefficient. They are a Debye-Huckel contribution—properly adjusted according to Pitzer, a Born contribution, which accounts for changes in the dielectric constant of the solvent versus that of water, and a contribution by short range forces properly accounted for by an approach similar to non-random two liquid theory (NRTL) model used for non-ionic species,

$$\ln \gamma_i = \ln \gamma_i^{PDH} + \ln \gamma_i^{Born} + \ln \gamma_i^{LC} \quad 2.58$$

in which γ^{PDH} is the long-range contribution according to Pitzer's modification of the Debye-Huckel approximation, γ^{Born} is,

$$\ln \gamma_i^{Born} = \frac{e^2 Z_i^2}{800\pi k T \varepsilon_0 r_B}\left[\frac{1}{D_s^{mix}} - \frac{1}{D_s^{w}}\right] \quad 2.59$$

in which D_s^{mix} , D_s^{w} is the dielectric constant of the mixed solvent and water respectively. As an example, the short-range contribution for the cation is,

$$\frac{1}{Z_C}\ln \gamma_C^{LC} = \frac{X_s^2 \tau_{CS} G_{CS}}{\left(X_C G_{CS} + X_a G_{as} + X_s\right)^2} - \frac{X_a \tau_{sa} X_s G_{sa}}{\left(X_c + X_s G_{sa}\right)^2} + \frac{X_s \tau_{sc} G_{sc}}{\left(X_a + X_s G_{sc}\right)} - \tau_{sc} - G_{cs}\tau_{cs} \quad 2.60$$

with

$$G_{ij} = \exp(-\alpha_{ij}\tau_{ij}) \quad 2.61$$

in which $\alpha_{ij} = 0.2$, X_i stands for local mol fractions, τ_{ij} are binary parameters that need to be fitted to experimental data, and the sub indexes are s:solvent, c:cation, and a=anion.

Osmosis and Reverse Osmosis

The analysis of osmosis or reverse osmosis is a good example of the use of thermodynamic principles to understand separation processes. Suppose that one has one cell divided into two compartments by a membrane. In one compartment (I) we place the pure solvent and in the other compartment (II) a solution of a solute that cannot cross the membrane. At equilibrium,

$$P^{II} - P^{I} = \Pi = -\frac{RT}{\underline{V}_{solvent}} \ln(x^{II}_{solvent}\gamma^{II}_{solvent}) \qquad 2.62$$

in which Π is the osmotic pressure. In such a cell, the solvent will flow from the solvent compartment into the solute compartment until the pressures in both compartments are equal. In other words, Π is the pressure that needs to be applied to the solute compartment not to allow the solvent to flow into it. In reverse osmosis, pressure is applied to the solute compartment to move the solvent into the solvent compartment. This is commonly used to desalinate seawater. Osmometry may be used to determine the molecular weight, M_w, of macromolecules. At dilute conditions, the activity coefficient is equal to one and thus,

$$M_W = \frac{RTC_{solute}}{\Pi} \qquad 2.63$$

Therefore, measurement of the osmotic pressure for a known concentration of solute at a fixed temperature will yield the molecular weight.

Heat and Work Requirements

The principles of thermodynamics can be used to estimate the heating/cooling requirements in different portions of a separation process and to estimate the power requirements for pumping fluids in a particular unit or in between units. The methods are simple and the information needed to obtain good estimates is readily available.

Heating and Cooling

Heating and cooling have a moderate impact in bioseparations. Cooling is necessary in freeze-drying and many chromatographic separations are performed below room temperature (5°C). Cooling is achieved by circulating cold water through a jacket that covers, for example, the chromatography column. This configuration is similar to the traditional tube heat exchanger utilized in other industries. In both cases, two streams (cooling water and the mobile phase inside the column) are placed into contact through a metal surface. Two contact modes are commonly encountered: co-current and counter-current. Figure 2.8 shows the temperature evolution for the cold (the stream whose temperature increases) and the hot (the stream whose temperature decreases) the streams along the heat exchangers. An energy balance is,

$$n_H(\Delta\underline{H})_H + n_C(\Delta\underline{H})_C = 0 \qquad 2.64$$

or

$$n_H\underline{C}_p(T_{H2} - T_{H1}) + n_c\underline{C}_p(T_{C2} - T_{C1}) = 0 \qquad 2.65$$

in which $\underline{C}_p$ is the molar heat capacity of the fluid, n_H and n_C are the molar flows of the hot and cold streams, T_{H1} is the entrance temperature of the hot stream, T_{H2} is its exit temperature, and T_{C1} and T_{C2} are the entrance and exit temperatures of the cold stream. This energy balance can be used to

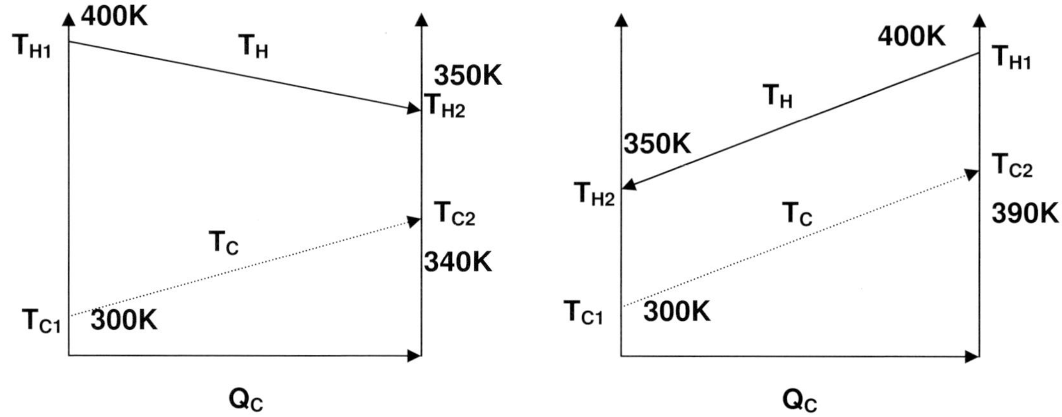

Figure 2.8. Schematic of two heat exchangers configurations. Co-current (left panel) and counter-current (right panel). The arrows indicate the direction of the fluid flow for the cold and hot streams.

determine the amount of cooling water that is needed to keep the temperature of the other stream at a prescribed value. Using the Second Law, more information for this system can be determined. Consider an entropy balance for a heat exchanger,

$$\Delta(\underline{S}n) = n_H(\Delta\underline{S})_H + n_C(\Delta\underline{S})_H \tag{2.66}$$

$$\Delta(\underline{S}n) = n_H \underline{C}_p \left(\ln\frac{T_{H2}}{T_{H1}} + \frac{n_C}{n_H}\ln\frac{T_{C2}}{T_{C1}} \right) \tag{2.67}$$

It is relatively straightforward to prove that because of the heat exchanged, the system has lost its ability to do work in the amount of,

$$W_{lost} = T\Delta(\underline{S}n) \tag{2.68}$$

Simple calculations show that the amount of lost work is larger for the co-current mode. Therefore, a counter current heat exchanger is more efficient from a thermodynamic viewpoint. This idea of "lost work" is explored further in the last section of this chapter.

Pumps

A pump is a device used to increase the potential energy of a fluid. An ideal pump is adiabatic and reversible. An easy way to describe thermodynamically how a pump works is to consider the "system" a pack of fluid moving through the pump. Then, its change in enthalpy is

$$\Delta h = \int_{in}^{out} v dP \tag{2.69}$$

Furthermore, because the process is adiabatic (isoentropic) we can write,

$$W_s = \int_{in}^{out} v dP \tag{2.70}$$

Therefore, to pump a fluid, we need to use an amount of work equal to the change in enthalpy of the fluid. The actual work will be equal to the calculated value using Equation 2.70 divided by the efficiency of the pump. If the fluid is incompressible, the volume is constant, and Equation 2.69 can be integrated to obtain,

$$\Delta h = v(P_2 - P_1) \qquad 2.71$$

If there is a temperature change across the pump, the enthalpy change can be calculated from,

$$\Delta h = c_p \Delta T + v(1 - \beta T)\Delta P \qquad 2.72$$

in which β is the coefficient of thermal expansion. For water at 20°C $\beta = 2.07 \times 10^{-4}\ K^{-1}$.

Thermodynamics Analysis of Processes

In the previous section we introduced the idea of "lost work." The quantification of "lost work" can be used for the thermodynamic analysis of a process's components to find the most energy-efficient method of running such a process. The concept of lost work reached its popularity peak in the energy crisis of the seventies, and for obvious reasons has regained its popularity in the twenty-first century. Different formalisms can be used. Gibbs and Helmholtz free energies analyses are limited to constant pressure and temperature or constant volume and temperature processes. There is an additional thermodynamic potential called exergy that can be used for the analysis of any arbitrary process. The exergy or availability is a state function that is defined by,

$$\Delta B = \Delta H - T_0 \Delta S \qquad 2.73$$

in which T_0 is the temperature of an infinitely large reservoir with which the system exchanges heat. Notice that the only difference between exergy and Gibbs free energy is that in the definition of exergy we use a reference temperature T_0, whereas in Gibbs free energy we use the temperature at which the process takes place. An exergy balance at stead state for an open system is,

$$B_{in} = B_{out} + B_Q + W + B_{loss} \qquad 2.74$$

$$B_Q = \sum \frac{Q^*(T_r - T_0)}{T_r} \qquad 2.75$$

Q* is the amount of heat transferred from the system at a temperature T_r. W is the exergy transfer as a result of shaft work done by the system and B_{loss} is the exergy loss because of irreversibilities within the system. The exergies associated with mass flow are given by,

$$B_{in} = \sum m_i b_i \qquad 2.76$$

with

$$b_i = (h_i - h_{o,i}) - T_0(s_i - s_{0,1}) \qquad 2.77$$

or,

$$b_i = b_{chem} + b_{phys} + \Delta b_{mix} \qquad 2.78$$

with,

$$b_{phys} = \Delta_{actual \to 0}(\sum x_i h_i - T_0 \sum x_i s_s) \qquad 2.79$$

and,

$$\Delta b_{mix} = \Delta h_{mix} - T_0 \Delta s_{mix} \qquad 2.80$$

If the mixing properties of the different streams are known, B_{loss} can be calculated from Equation 2.74. Repeating these calculations for different process units allows identification of the less-efficient ones. Moreover, parametric studies can be done to identify the process conditions leading to the most efficient process. Unfortunately, exergy balances are not common in bioseparation. The interested reader may want to read the article by Maia, Ottens, and van der Wielen (2004), who used exergy analysis to optimize chromatographic separations in fixed beds. There is a particular exergy analysis of processes called the "pinch method." This method allows the identification of the bottleneck in a process that impedes further heat integration. The pinch method is standard in the heat integration of energy demanding industries like petrochemical and chemical but it has yet to find its niche in bio-related industries.

Summary

Thermodynamics is important to bioprocesses. Sometimes we find thermodynamic statements in the literature that we overlook because of our lack of knowledge of the science. Other times, we fail in making a connection between the pure science and its applications. Of course, this chapter is not meant to be the only source for the use of thermodynamics in the analysis of processes. The reader is referred to excellent textbooks that deal with the problem in detail. This chapter can be used, however, to decide what to read and what not to read.

Nomenclature

A	Helmholtz free energy (J)
B	exergy (availability) (J)
$\underline{C}_p$	molar heat capacity (J/mol K)
c_p	specific heat capacity (J/kg K)
D_s	static dielectric constant of the solvent
e	charge of an electron (1.6022×10^{-19} C)
E	total energy (J)
E_k	kinetic energy (J)
E_p	potential energy (J)
F	force (N)
g	acceleration of gravity (m/s^2)
G	Gibbs free energy (J)

G_{ad}	Gibbs free energy of adsorption (J/kmol)
H	molar enthalpy (J/mol)
h	specific enthalpy (J/kg)
I	ionic strength (M)
K	distribution coefficient
k_B	Boltzmann's constant ($1.3807 x 10^{-23}$ J/K)
K_{eq}	equilibrium constant
L	distance (m)
m	mass (kg)
m_i	molarity of ion i
$m\pm$	mean molality (kmol/kg of solvent)
$\dot{m}_k$	mass flow-rate of stream k (kg/sec)
M	extensive property
M_w	molecular weight (kg/kmol)
N_i	number of moles of species i
n_i	molar flow rate of stream (kmol/sec)
N_A	Avogadro's number ($6.02 x 10^{23}$ molecules/mol)
P	pressure (Pa)
$\dot{Q}$	heat flow-rate (J/sec)
R	ideal gas constant (1.98 J/mol K)
r_0	radii of the ion (cm)
S	molar entropy (J/kmol K)
S_g	generation of entropy (J/K)
s	specific entropy (J/K kg)
t	time (sec)
T	absolute temperature (K)
U	internal energy (J)
u	specific internal energy (J/kg)
v_s	velocity (m/s)
v	specific volume (m^3/kg)
V	molar volume (m^3/kmol)
W	work (J)
$\dot{W}$	rate of work (J/sec)
z+ and z-	charges of the positive and negative ions

Greek Letters

α	coefficient of thermal expansion (1/K)
ξ_i	partial molar property
ρ	density (kg/m^3)
ε_0	permittivity of empty space (C^2/J m)
ε	permittivity of the solvent (C^2/J m)
ζ	potential energy per unit mass (J/kg)
γ_i	activity coefficient of species i
$\gamma_{\pm}$,	mean activity coefficient
Π	osmotic pressure (Pa)
μ_i	chemical potential of species i (J/mol)

References

Chen, C. C., H. I. Britt, J. F. Boston, and L. B. Evans. 1982. Local composition model for excess Gibbs energy of electrolyte solutions. Part 1: Single solvent, single completely dissociated electrolyte systems. *AIChE Journal* 28(4):588-96.

Chen, C. C., and L. B. Evans. 1986. A local composition model for the excess Gibbs energy of aqueous electrolyte systems. *AIChE Journal* 32(3):444-54.

Debye, P., and E. Huckel. 1923. The theory of electrolytes. I. Lowering of freezing point and related phenomena. *Physik Z* 24:185-206.

Elysee-Collen, B., and R. W. Lencki. 1999. Influence of protein surface morphology on the ultrafiltration flux resistance of bovine serum albumin. *Biotechnology Progress* 15:732-9.

Gibbs, J. W. 1993. *The Scientific Papers of J. Willard Gibbs, Vol 1: Thermodynamics.* Woodbridge, CT: Ox Bow Press.

Goldberg, R. N. 1981. Evaluated activity and osmotic coefficients for aqueous solutions: thirty-six uni-bivalent electrolytes. *J Phys Chem Ref Data* 10(3):671-764.

Grenthe, I., and A. Plyasunov. 1997. On the use of semiempirical electrolyte theories for the modeling of solution chemical data. *Pure and Applied Chemistry* 69(5):951-8.

Joule, J. P. 1845. On the existence of an equivalent relation between heat and the ordinary forms of mechanical power. *Philosophical Magazine* 27(3):205.

Lim, T.-K. 1993. The calculation of activity coefficients of binary mixed electrolytes. *Pure and Applied Chemistry* 65(12):2561-72.

Mahadevan, H., and C. K. Hall. 1990. Statistical-mechanical model of protein precipitation by nonionic polymer. *AIChE Journal* 36(10):1517-28.

Mahadevan, H., and C. K. Hall. 1992. Theory of precipitation of protein mixtures by nonionic polymer. *AIChE Journal* 38(4):573-91.

Maia, M. L. O., M. Ottens, and L. A. M. van der Wielen. 2004. Exergy analysis of chromatographic separations in a fixed-bed column. *Ind Eng Chem Res* 43:3183-93.

Meissner, H. P., and J. W. Tester. 1972. Activity coefficients of strong electrolytes in aqueous solutions. *Industrial and Engineering Chemistry Research* 11(1):128-33.

Pitzer, K. S., ed. 1991. *Ion interaction approach: theory and data correlation. In Activity Coefficients in Electrolyte Solutions*, 2nd ed, pp. 75-173. Boca Raton, FL: CRC Press.

Sandler, S. I. 2006. *Chemical, Biochemical and Engineering Thermodynamics*, 4th ed. New York: John Wiley & Sons.

Further Reading

Tester, J. W., and M. Modell. 1997. *Thermodynamics and its Applications,* 3rd edition. Upper Saddle River, NJ: Prentice Hall Inc.

Chapter 3
Mass Transfer in Bioprocessing

Introduction

The objective of a separation process is to place the compound of interest and the impurities in different phases, which may be separated by a physical constraint like a membrane or by an interface. The mixture to be separated is in a single phase and is either put in contact with a second phase or the second phase is induced by changing the temperature, the temperature and the pressure, or by adding one or more additional components. Therefore, molecules have to move from one phase into the other. The result of that movement is the transfer of mass from one phase to another. If that transfer can be affected any particular unit can be manipulated to achieve a separation.

Mass can be transferred by two distinct mechanisms: convection and diffusion. Mass is transferred by convection because of the bulk movement of a fluid; for example, the transfer of a fluid from a reservoir into a chromatography column actuated by a pump. Diffusion is a more complex phenomenon that is explained in detail in the following paragraphs because it plays a central role in most separation units.

Diffusion

Let us consider the following simple experiment. Place 1 mL of a 1 mg/mL hemoglobin solution into a rectangular cuvette. Place the cuvette in a spectrophotometer so that the beam does not reach the solution. Carefully (an automatic syringe is ideal) place 1 mL of buffer on top of the hemoglobin solution; the fresh buffer will be in the path of the light beam. If we monitor the absorbance at 420 nm we will notice a steady increase in optical density. This increase in the optical density takes place because hemoglobin is migrating into the freshly deposited buffer layer. If the layer of fresh buffer was carefully deposited, the solution of hemoglobin should not have been disturbed. Therefore, mechanical mixing is not the reason for the migration of the protein. It is obvious that the protein is moving from a phase in which it is very concentrated (crowded!) to the phase with more "empty" space. The rate at which the protein is moving is determined by the concentration difference between the phases and by an intrinsic property of molecular motion called diffusion coefficient.

Table 3.1. Different Types of Diffusion.

Types	Driving Force
Ordinary Diffusion	ΔC_A
Pressure Diffusion	ΔP_A
Thermal Diffusion	ΔT
Force Diffusion	ΔF

In the above example, diffusion is induced by a concentration gradient. We can generalize that result by stating that the movement of a solute from one place (phase) to another is induced by a force. Table 3.1 presents different types of these driving forces that are used to classify diffusion into various categories.

Concentration driving forces are by far the most common driving force in bioprocessing. Therefore, the discussion below is focused on ordinary, or concentration difference-driven, diffusion. At a molecular level, the movement of molecules by Brownian motion was first successfully described in an equation by Albert Einstein in 1906. The units of the diffusion coefficient are length^2/time; usually cm^2/sec. The order of magnitude difference between diffusion coefficients for gases, D ~ 1 cm^2/sec, and that for liquids, D ~ 10^{-5} cm^2/sec, suggests that in liquids molecules find more resistance to their movement compared to gases because they encounter other molecules more frequently than do gases. These other molecules may be either the same liquid or a solute dissolved in a solvent. It is important at this point to differentiate between the self-diffusion coefficient and the mutual diffusion coefficient. Self-diffusion coefficients refer to systems consisting of a single species, whereas mutual diffusion coefficients refer to binary systems (e.g., a protein dissolved in a solvent). The self and mutual diffusion coefficients are the same for gases at low pressure but they are quite different for liquids. The diffusion coefficient is, in general, a function of temperature, pressure, and composition. For binary gases at low pressures, the diffusivity is inversely proportional to the pressure, directly proportional to the temperature and independent of concentration. For gases at high pressures, the diffusivity is independent of pressure, proportional to the temperature, and a function of concentration.

The study of diffusion coefficients started many years ago. By then, most of the efforts were put on the diffusion of gases into gases. Later on, studies were done on the diffusion of molecules in liquid phases. Kinetic theory has been widely used for the study of the diffusion of gases. For example, the theory yields for a gas (consisting of rigid spheres) at low density (Slattery and Bird 1958),

$$D_{AB} = \frac{2}{3}\left(\frac{K^3}{\pi^3}\right)^{1/2}\left(\frac{1}{2M_A}+\frac{1}{2M_B}\right)^{1/2}\frac{T^{3/2}}{p\left(\frac{d_A+d_B}{2}\right)^2} \tag{3.1}$$

in which d is the diameter of molecule A or B, M is the molecular mass of molecules A or B, T is the absolute temperature, p is the pressure, and K is a constant.

The study of mass transfer in bioseparations is complicated because of the presence of condensed phases, and because the species that diffuses, such as proteins and nucleic acids, is normally charged. Approaches developed to calculate diffusion coefficients in liquid phases for neutral particles are valid only for charged molecules at low concentrations and/or in the presence of high concentrations of salts, where Columbic interactions are screened out.

In the absence of salts, Columbic interactions cannot be ignored even at low concentrations. It is convenient to separate our studies of diffusivities in liquids into infinite dilution diffusivities and then incorporate concentration effects as correction factors.

The starting point for one popular approach known as the Stokes-Einstein formulae, used to estimate diffusion coefficient at infinite dilution, is the Nerst-Einstein equation, which gives the diffusion coefficient of a molecule in a stationary fluid by,

$$D_{AB} = 10^{-4} k_B T \frac{v_A}{F_A} \quad 3.2$$

in which k_B is the Boltzmann's constant, v_A is the velocity of the particle, and F_A is the applied force. If the fluid does not slip at the surface of a spherical particle, the Stokes-Einstein formulae is obtained

$$R_H = \frac{k_B T}{6\pi\mu D_{ps}} \quad 3.3$$

in which R_H is the hydrated radius of the protein, D_{ps} is the diffusion coefficient of a protein, p, in solvent, s, and μ is the viscosity of the solvent. At high protein concentrations and dilute salts this expression fails.

Other models for the diffusion coefficient of proteins are Polson models, Svedberg Method, Flory-Mandelkern-Scheraga Method, and Young-Carroad-Bell Method. One of the Polson models states a simple relationship between the diffusivity and the molecular weight (Polson 1950),

$$D_{ij} = \frac{A}{M^{1/3}} \quad 3.4$$

in which A = 2.85×10^{-5} $cm^2 sec^{-1} g^{1/3} mol^{-1/3}$ or (Polson 1967),

$$D_{ij} = \frac{B}{(M\eta)^{1/3}} \quad 3.5$$

in which B = 4.185×10^{-5} $cm^3 sec^{-1} mol^{-1/3}$ and η is the intrinsic viscosity. The model proposed by Svedberg is (Svedberg 1925),

$$D_{ij} = \frac{sRT}{M(1 - v\rho)} \quad 3.6$$

in which s is the Svedberg constant in sec^{-1}, v is the partial specific volume in cm^3/g and ρ is the density of water at 20°C in g/cm^3. Considering that a protein, fairly rigid in its natural form, is flexible (polymer like) Flory-Mandelkern-Scheraga proposed (Scheraga and Mandelkern 1953),

$$D_{ij} = \frac{\beta kT}{\mu(M\eta)^{1/3}} \quad 3.7$$

in which β ranges from 2.12×10^6 to 2.5×10^6 $cm^2 g\ sec^{-2} erg^{-1} mol^{-1/3}$. Another correlation proposed by Young-Carroad-Bell is (Young, Carroad, and Bell 1980) is,

$$D_{ij} = \frac{8.34 \times 10^{-6} T}{\eta M^{1/3}} \quad 3.8$$

Table 3.2. Diffusion Coefficients of Some Proteins.

Protein	Diffusion Coefficient $X10^7$ cm^2 sec^{-1}
Lysozyme (egg white)	11.2-11.8
Chymotrypsin B	9.9
Hemoglobin (human)	6.3-6.9
BSA	5.81-6.15
IgG	3.8-4
γ-Globulin	2.03-4.0
Fibrinogen	1.50-2.02

A careful reader would notice that this is the same as Polson's model with $A=8.34x10^{-6}T/\eta_0$. A different approach was followed by Tyn and Gusek (1990) who correlated diffusion coefficients at infinite dilution of 86 proteins with the inverse of the radius of gyration and found,

$$D_{ij} = \left(1.69x10^{-6} \frac{cm^2 \overset{o}{A}}{\sec} \right) \frac{1}{R_g} \tag{3.9}$$

and

$$D_{ij} = \left(5.78x10^{-8} \frac{cm \overset{o}{A} g}{\sec K} \right) \frac{T}{\mu_o} \frac{1}{R_g} \tag{3.10}$$

in which μ_0 is the viscosity of water at standard conditions, equal to 1.002 cP, and R_g is the radius of gyration defined by,

$$R_g^2 = \frac{1}{2N^2} \sum_{i,j} \left(r_i - r_j \right)^2 \tag{3.11}$$

in which r_i and r_j are vectors defining the position of atoms i and j in the molecule and N is the total number of atoms. Still, these corrolations (Equations 3.9 and 3.10) fail for rod-like molecules like fibrinogen. With the exception of fibrinogen, a good correlation for rod-like proteins was found by replacing the R_g in the above equation with one-tenth of the length of the particle.

Any of the correlations described above can be used to estimate the diffusion of a given protein in water or in another solvent in the absence of experimental data. A recent compilation of diffusion coefficients of proteins and nucleic acids can be found in Tyn and Gusek (1990). Table 3.2 shows a collection of diffusion coefficients for a few proteins of various sizes and shapes. At room temperature, diffusion coefficients of proteins are of the order of 10^{-7} cm^2/sec.

As the concentration of the protein increases, the diffusion coefficient becomes a function of protein concentration. A careful analysis of the hydrodynamic interactions encountered as a large molecule diffuses through a continuous medium leads to,

$$\frac{D_{ij}}{D_0} = 1 + \Gamma\phi + o(\phi^2) \tag{3.12}$$

in which D_0 is the "infinite dilution diffusion coefficient", ϕ is the volume fraction of the large molecule and Γ is a parameter that accounts for direct long-range forces between particles and for hydrodynamic interactions between particles. For example, the diffusion coefficient of bovine serum albumin (BSA) at two ionic strengths is (Anderson and Reed, 1976)

$$\frac{D_{ij}}{D_0} = 1 - 1.5\phi \qquad 0.1M \qquad salt \tag{3.13}$$

$$\frac{D_{ij}}{D_0} = 1 + 75\phi \qquad 0.001M \qquad salt \tag{3.14}$$

Therefore, at a volume fraction of 1%, D_{ij} is 2% lower than its infinite dilution value at 0.1M ionic strength but 75% higher at 0.001 M.

Diffusion coefficients in liquids may be experimentally determined using a variety of techniques; for example, sedimentation centrifugation, NMR, and dynamic light scattering. The discussion that follows focuses on the latter technique. In a typical light scattering experiment, a laser beam illuminates the sample and the scattered light is measured by a photodiode at one or several angles. Because particles in a liquid are subjected to Brownian motion, they scatter light that travels different distances. This scattered light is a function of time and is quantified by a random fluctuation in intensity. The fundamental quantity measured in a dynamic light scattering experiment is the time correlation function. The understanding of this property is important for the interpretation of dynamic light scattering experiments. Although it can be obtained in a very elegant manner by using statistical mechanics techniques, some intuitive arguments may be used to understand its meaning. Suppose that a laser beam illuminates a small volume of the sample. Also assume that there are molecules in this volume and that they are large enough to scatter the light. Those molecules are randomly moving according to their Brownian motion. The intensity of the scattered light will fluctuate because of the random motion of the scatters. A correlator calculates the product of the intensity at time t and time t+δt, where δ is a very small number. The intensity at time t and time t+δt will be very similar (the signals will be highly correlated). In other words, knowledge of the signal at time t will allow to estimate the signal at t+δt. As signals are separated by longer and longer times, they will become more dissimilar (less correlation). Finally, as the signals are infinitely separated there will be no correlation between them. Figure 3.1 shows a typical correlation function for a sample with only one type of scatter.

The scattered light intensity pattern may be fitted by an equation of the form,

$$C(t) = Ae^{-2\Gamma t} + B \tag{3.15}$$

in which C(t) is the correlation function and $\Gamma = D_{ij}\, q^2$ with q equal to the momentum transfer. If there is more than one size particle scattering the light, the curve may be fitted by a family of exponentials of the form,

$$g(t) = \int G(\Gamma)e^{-\Gamma t}d\Gamma \tag{3.16}$$

Therefore, by fitting the experimental data, the diffusion coefficient can be obtained.

In the two most commonly found bioseparation methods, ultrafiltration and chromatography, diffusion occurs inside pores whose dimensions are of the same order of magnitude as the proteins. The observed diffusion coefficient in the porous media is smaller than the bulk diffusion coefficient. The diffusion coefficient in constraint geometries is affected by hydrodynamic and thermodynamic effects and by the particular structure of the porous media. The effective diffusion coefficient in porous media may be written as (Tallarek, Vergeldt, and Van as 1999),

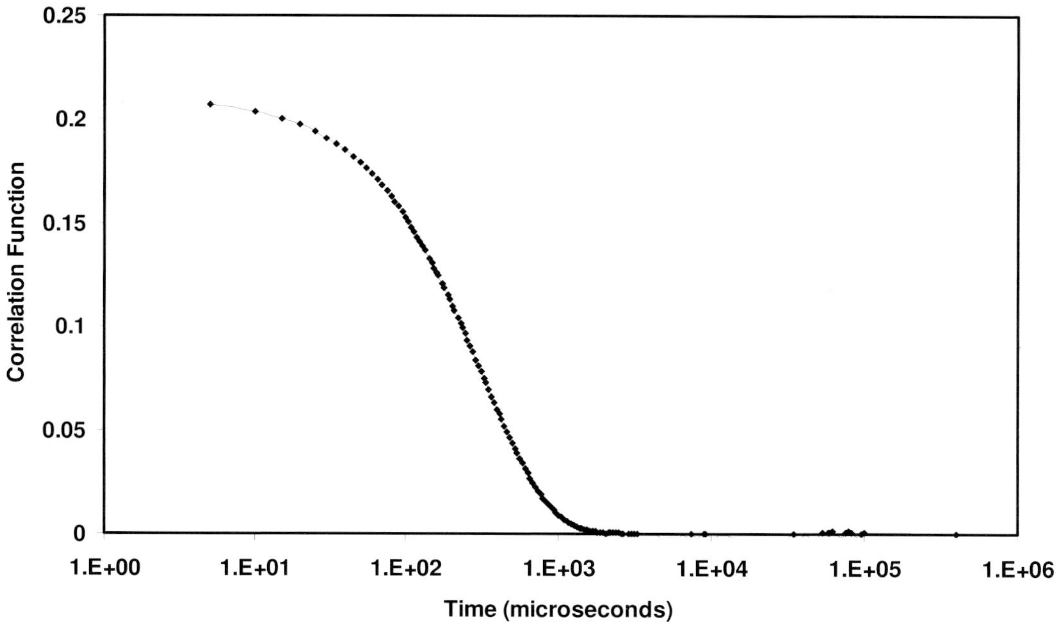

Figure 3.1 Time correlation function for a dilute and monodisperse solution.

$$D_{eff} = D_{AB} \frac{\varepsilon_{intra}}{\tau_{intra}} \frac{K_p}{K_H} \quad 3.17$$

in which ε_{intra} is the porosity of the particle, τ_{intra} is the tortuosity of the particle, K_p is the partition coefficient that represents the ratio of solute concentration inside the pores to the concentration in the bulk, and K_H accounts for the additional drag that the particles "feel" because of the proximity of the pore's wall. K_H and K_p are properties of single pores, whereas the porosity and tortuosity are properties of the pore's network. Hydrodynamic and thermodynamic effects have been studied for many years and the consensus is that both effects depend on the ratio between the size of the molecule and the radius of the pore (λ). For example, Renkin (1954) found that,

$$\frac{K_p}{K_H} = (1-\lambda)^2 (1 - 2.1044\lambda + 2.089\lambda^3 - 0.948\lambda^5) \quad 3.18$$

For small solutes the ratio between K_p and K_H goes to 1 and therefore, the effective diffusivity is a function of the pore morphology only. For proteins, that is not generally true.

The effective diffusivity may be estimated using Equation 3.17 with Equation 3.18, provided that values for the porosity and tortuosity of the particles are available. It can also be measured by a variety of techniques ranging from the fitting of chromatography data with the appropriate model to the direct measurement by, for example, confocal laser scanning microscopy. Figure 3.2 is a plot of the ratio between the effective and bulk diffusivity for several proteins; the effective diffusivity was experimentally measured using confocal microscopy (Schruder, von Lieres, and Hubbuch 2006). The figure shows that the effective diffusivity is about half the value of the bulk one and that the

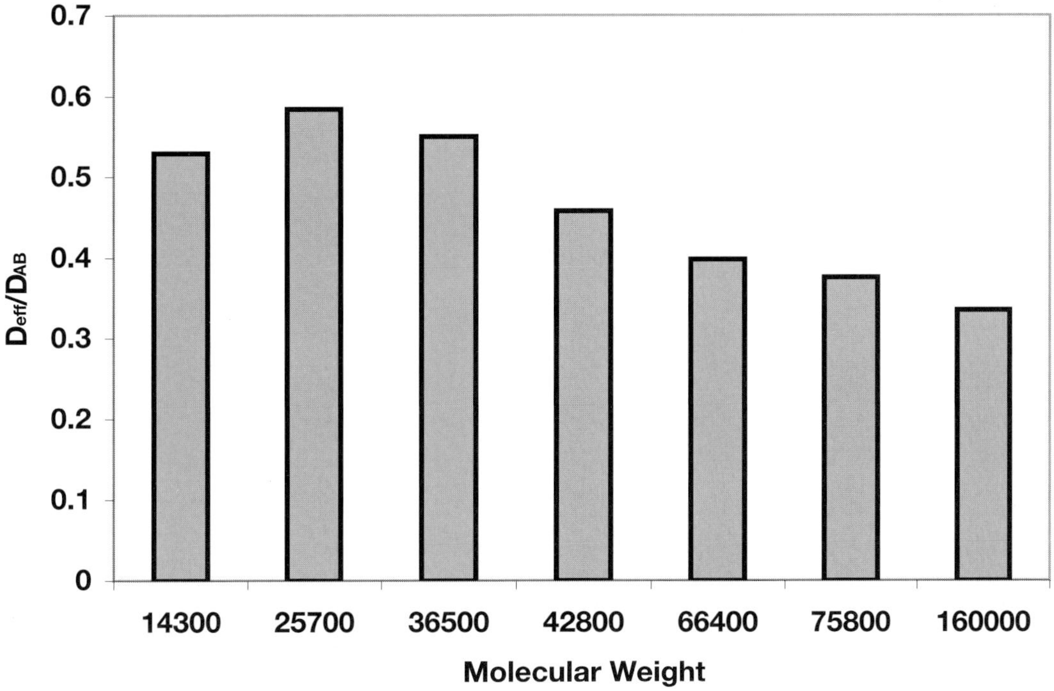

Figure 3.2 Ratio between the effective diffusivity and the bulk diffusivity for a series of proteins versus the protein's molecular weight (data plotted from Schruder et al. 2006)

ratio decreases as the molecular weight increases. This is expected, because the increase in hydrodynamic drag and a decrease in partition coefficients as the molecular weight of the solute increases. For smaller molecules the effective diffusivity is practically independent of the molecular weight because the porosity and tortuosity of the beads are the dominant effects (λ is small).

Some Definitions

Because mass transfer refers to the movement of a fluid (or a species) from one point to another it is necessary to choose a reference point for that movement. The typical analogy is that of a person walking on a moving train. We can measure his velocity with respect to the train or with respect to the train station. Besides needing a point of reference, it is necessary to choose the units for concentration. Although this seems to be trivial, the difficulty of solving a problem may be lessened by a judicious choice of units of concentration. Most of this material has been adapted from Bird, Stewart, and Lightfoot (2000). Here are some definitions (and notation).

The mass concentration, ρ_A in units of g of A/cm^3 of solution is,

$$\rho_A = C_A M_A \tag{3.19}$$

The molar concentration, c_A, in units of moles of A/cm^3 of solution is,

$$C_A = \frac{\rho_A}{M_A} \tag{3.20}$$

The mass fraction, w_A is,

$$w_A = \frac{\rho_A}{\rho_A + \rho_B} \tag{3.21}$$

and the mole fraction, x_A, is,

$$x_A = \frac{C_A}{C} = \frac{C_A}{C_A + C_B} \tag{3.22}$$

We can define different velocities. For example, we can call $\vec{v}_i$ the velocity of the ith species with respect to stationary axes. Thus, the local mass average velocity, $\vec{v}$ (in units of length/ unit time), will be:

$$\vec{v} = \frac{\sum_{i=1}^{n} \rho_i \vec{v}_i}{\sum_{i=1}^{n} \rho_i} \tag{3.23}$$

Similarly, the local molar average velocity, $\vec{v}^*$, is:

$$\vec{v}^* = \frac{\sum_{i=1}^{n} c_i \vec{v}_i}{\sum_{i=1}^{n} c_i} \tag{3.24}$$

Therefore, we can define velocities of a particular species, i, in reference to any of these two average velocities as:

$$\vec{v}_i - \vec{v} = \text{diffusion velocity with respect to } \vec{v} \tag{3.25}$$

$$\vec{v}_i - \vec{v}^* = \text{diffusion velocity with respect to } \vec{v}^* \tag{3.26}$$

The flux of species A is defined as the amount of mass or moles of A that passes a unit area per unit time. In order to define flux we must specify the coordinates (i.e., with respect to what we are measuring in velocities). We have the following options:

A. Flux relative to stationary coordinates

$$n_i = \rho_i \vec{v}_i \qquad \textit{Mass flux} \tag{3.27}$$

$$N_i = c_i \vec{v}_i \qquad \textit{Molar flux} \tag{3.28}$$

B. Flux relative to local mass average velocity

$$j_i = \rho_i(\vec{v}_i - \vec{v}) \qquad \textit{grams of A/ time cm}^2 \qquad 3.29$$

$$J_i = c_i(\vec{v}_i - \vec{v}) \qquad \textit{moles of A/ time cm}^2 \qquad 3.30$$

C. Flux relative to local molar average velocity

$$j_i^* = \rho_i(\vec{v}_i - \vec{v}^*) \qquad 3.31$$

$$J_i^* = c_i(\vec{v}_i - \vec{v}^*) \qquad 3.32$$

All these different definitions may look like a waste of time. However, their use becomes apparent when analyzing different problems, because the governing equations, which we have not introduced yet, take different forms according to the flux definition chosen. It is, therefore, convenient to look at the relationship between these different quantities. For example, it is straightforward to find a relationship between $\vec{J}^*_i$ and $\vec{N}_i$. By definition,

$$J_i^* = c_i(\vec{v}_i - \vec{v}^*) = c_i\vec{v}_i - \frac{c_i}{c}\sum_{j=1}^{n} c_j\vec{v}_j \qquad 3.33$$

but $c_i\vec{v}_i = \vec{N}_i$. and $x_i = c_i/c$. Therefore,

$$J_i^* = \vec{N}_i - x_i\sum_{j=1}^{n}\vec{N}_i \qquad 3.34$$

in which the first term has the meaning of flux as a result of diffusion and the second one flux as a result of convection. Summing from i-1 to i=n,

$$\sum_{i=1}^{n} J_i^* = \sum_{i=1}^{n}\vec{N}_i - \sum_{i=1}^{n} x_i\sum_{j=1}^{n}\vec{N}_i = 0 \qquad 3.35$$

Therefore, the sum of molar diffusion fluxes relative to the molar average velocity is zero in any mixture. For a binary mixture,

$$\vec{J}_A^* = \vec{J}_B^* \qquad 3.36$$

Notice that the same relationship does not hold for fluxes defined with respect to fixed coordinates.

Mass balances are done by considering the fluxes of mass coming in and exiting a previously defined control volume. Therefore, the fluxes need to be specified as a function of the concentration of the species that is moving and as a function of position of the species in the control volume. This relationship between the flux, concentration, and position is called Fick's Law and it is a "mass transfer constitutive equation." It has no theoretical basis, but it is yet to be disproved. In terms of fluxes relative to the movement of the bulk fluid,

$$\vec{J}_A^* = -cD_{AB}\vec{\nabla}x_A \qquad 3.37$$

The flux relative to fix coordinates, $\vec{N}_i$, is:

$$\vec{N}_A = -cD_{AB}\vec{\nabla}x_A + x_A(\vec{N}_A + \vec{N}_B) \qquad 3.38$$

Notice that both equations are different representations of the same fundamental equation.

Rate of transport of mass of A into the system	-	Rate of transport of mass of A out of the system	+	Rate of production of mass of A by homogeneous chemical ` inside the system	=	Rate of accumulation of mass of A within the system

Figure 3.3 "Boxes" mass balance for species A in a control volume.

Law of Conservation of Mass

Consider a piece of the fluid that is enclosed by well-defined borders (boundaries) that define a control volume. Species can move in and out from the control volume by diffusion and/or convection and they may be produced or consumed by a chemical reaction. A mass balance for species A in that control volume is shown in Figure 3.3.

For simplicity, let us consider a binary mixture (A and B) and a cubic volume element (the extension of this to other coordinate systems is relatively straightforward). In the presence of chemical reactions and unsteady state, we obtain the so-called continuity equation. For species A,

$$\frac{\partial c_A}{\partial t} + \left(\overline{\nabla}.\overline{N}_A\right) = R_A \qquad 3.39$$

for species B,

$$\frac{\partial c_B}{\partial t} + \left(\overline{\nabla}.\overline{N}_B\right) = R_B \qquad 3.40$$

Adding the two equations, we obtain the continuity equation for the mixture,

$$\frac{\partial c}{\partial t} + \left(\overline{\nabla}.c\overline{v}^*\right) = (R_A + R_B) \qquad 3.41$$

Replacing fluxes by concentration gradients (Fick's Law),

$$\frac{\partial c_A}{\partial t} - \overline{\nabla}.(cD_{AB}\overline{\nabla}x_A) + \overline{\nabla}.(x_A\overline{N}_A + x_A\overline{N}_B) = R_A \qquad 3.42$$

The quantity within parenthesis in the third term of the left hand side is just $c_A v^*$; therefore,

$$\frac{\partial c_A}{\partial t} - \overline{\nabla}.(cD_{AB}\overline{\nabla}x_A) + \overline{\nabla}.(c_A\overline{v}^*) = R_A \qquad 3.43$$

Several simplifications of Equation 3.42 are possible. For example, in the absence of chemical reactions, when v* is zero, and c is constant, Equation 3.42 becomes,

$$\frac{\partial c_A}{\partial t} = D_{AB}(\nabla^2 c_A) \qquad 3.44$$

which is known as Fick's Second Law or the Diffusion Equation.

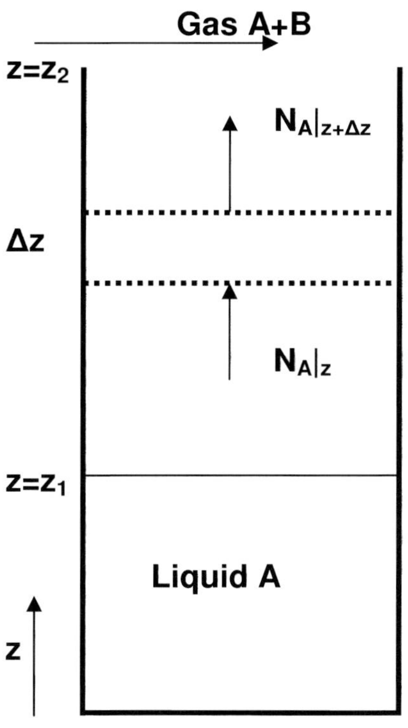

Figure 3.4 Sketch showing the evaporation of a liquid (A) that is swept by a gas flow (B)

In order to obtain valuable information the differential equation needs to be integrated between some limits or initial/boundaries conditions. Commonly found boundary conditions are:

a) The concentration of A at a surface is specified.
b) The mass flux at a surface is specified.
c) If diffusion is occurring in a solid, it may happen that at the solid surface substance A is lost to a surrounding fluid stream according to:

$$N_{A0} = k_c (C_{C0} - C_{Af}) \qquad 3.45$$

in which k_c is a mass transfer coefficient (to be defined!).
d) The rate of a chemical reaction at the surface can be specified.

The most commonly encountered initial condition is:

a) The concentration of A for all relevant length scales (i.e., x, y, and z) is specified at a time equal to zero.

Examples

One of the simplest examples of mass transfer, which appears in many textbooks (Bird, Stewart, and Lightfoot 2000), is the diffusion of one species through a stagnant film of other. This corresponds to the evaporation of one component in a binary mixture of a non-volatile component. Figure 3.4 shows a sketch of this system. The objective of the exercise is to calculate the rate of evaporation of species A.

Some assumptions are needed to obtain a manageable descriptive equation:

i) At $z = z_2$ x_{A2} is constant because the stream of A+B flows very slowly (enough to just remove what evaporates).
ii) $z = z_1$ is constant (we have a means to supply liquid A into the container).
iii) At $z=z_1$: $x_A = x_{A1}$, where x_{A1} corresponds to the equilibrium concentration of A; i.e., $x_{A1} = P_A^s/P$.
iv) At $z=z_1$: $x_B = 0$ (B is insoluble in A).
v) A is an ideal gas. For ideal gases the product cD_{AB} is not a function of pressure.

A steady state mass balance in a shell of thickness Δz is:

$$SN_{Az}|_z - SN_{Az}|_{z+\Delta z} = 0 \qquad 3.46$$

in which S is the area of the container. Dividing by Δz, taking the limit as Δz goes to zero, and using the definition of a derivative one gets,

$$-\frac{\partial N_{Az}}{\partial z} = 0 \qquad 3.47$$

Now we can use Equation 3.38 for N_{Az} (things are "happening" only in the z direction); that is,

$$N_{Az} = -cD_{AB}\frac{\partial x_A}{\partial z} + x_A(N_{Az} + N_{Bz}) \qquad 3.48$$

Because B is not soluble in A, we have N_{Bz}=0. Therefore,

$$N_{Az} = -cD_{AB}\frac{1}{1-x_A}\frac{\partial x_A}{\partial z} \qquad 3.49$$

One can calculate the rate of evaporation using,

$$N_{Az}|_{z=z1} = \left(cD_{AB}\frac{1}{x_B}\frac{\partial x_B}{\partial z}\right)_{z=z1} \qquad 3.50$$

Therefore, the rate of evaporation per unit area is equal to the flux of A at the liquid/vapor interface. In terms of the mole fraction of species B,

$$J_A = \left(cD_{AB}\frac{1}{x_B}\frac{\partial x_B}{\partial z}\right)_{z=z1} \qquad 3.51$$

Thus, to calculate the rate of evaporation we need to determine the concentration gradient from the liquid/vapor interface to the edge of the container. First, we use Equation 3.49 and substitute into Equation 3.47,

$$\frac{d}{dz}\left(cD_{AB}\frac{1}{1-x_A}\frac{\partial x_A}{\partial z}\right) = 0 \qquad 3.52$$

because for an ideal gases cD_{AB} is not a function of pressure (z) one gets,

$$\frac{d}{dz}\left(\frac{1}{1-x_A}\frac{\partial x_A}{\partial z}\right)=0 \tag{3.53}$$

Integrating once,

$$\frac{1}{1-x_A}\frac{dx_A}{dz}=c_1 \tag{3.54}$$

Integrating again,

$$-\ln(1-x_A)=c_1 z+c_2 \tag{3.55}$$

Using the two boundary conditions to determine c_1 and c_2, we obtain,

$$\left(\frac{1-x_A}{1-x_{A1}}\right)=\left(\frac{1-x_{Az}}{1-x_{A1}}\right)^{\frac{z-z1}{z2-z1}} \tag{3.56}$$

This is an equation for the concentration profile of the volatile species through the vapor space in the chamber. This concentration profile may be used in Equation 3.50 to calculate the rate of evaporation. First, Equation 3.50 is rearranged,

$$N_{Az}\,|_{z=z1}=\left(cD_{AB}\frac{1}{x_B}\frac{\partial \ln x_B}{\partial z}\right)_{z=z1} \tag{3.57}$$

Equation 3.55 is solved for the ln x_B,

$$\ln x_B=\ln x_{B1}+\frac{z-z_1}{z_2-z_1}\ln\frac{x_{B2}}{x_{B1}} \tag{3.58}$$

Taking the derivative of the ln x_B with respect to Z one gets,

$$\frac{d\ln x_B}{dz}=\frac{1}{z_2-z_1}\ln\frac{x_{B2}}{x_{B1}} \tag{3.59}$$

Using equation 3.58 and substituting into 3.57 gives,

$$N_{AZ}\,|_{z=z1}=\frac{cD_{AB}}{z_2-z_1}\ln\frac{x_{B2}}{x_{B1}} \tag{3.60}$$

This equation satisfies physical insights one may expect from this type of problem. First, noticing that the flux is defined as the evaporation rate per area one expects the evaporation rate to increase as the cross-sectional area increases. Second, the evaporation rate increases as the vapor head is decreased (z_2-z_1 becomes smaller). Third, the evaporation rate will increase as the diffusion coefficient is increased. The dependence of the evaporation rate on temperature is given by the diffusion coefficient's dependence on temperature and by the partial pressure of A, which is proportional to the concentration of B near the liquid/vapor boundary (x_{B1}). This dependence shows that the evaporation rate does indeed increase as the temperature increases. Finally, the evaporation rate will increase as x_{B2} becomes larger than x_{B1}.

Electrophoresis

Electrophoresis separates molecules because charged molecules move at different velocities in the presence of an electric field. A charged particle of charge Z_p and mass m placed in an electric field, E, is subjected to two forces. An electric force, F_e,

$$F_e = Z_p * E \tag{3.61}$$

and a drag force, F_d,

$$F_d = f * v \tag{3.62}$$

where v is the velocity of the particle and f is the friction factor, f. Stoke's law gives assuming spherical particles and very slow flows (Creeping Flow) the drag force by,

$$f = 6\pi\mu R_H \tag{3.63}$$

The application of Newton's law to such a particle yields,

$$m\left(\frac{dv}{dt}\right) = F_e - fv \tag{3.64}$$

at steady state the left hand side is equal to zero and the equation can be solved for v,

$$v = \frac{Z_p * E}{f} = \mu_{ele} * E \tag{3.65}$$

in which $\mu_{ele}=Z_p/f$ is the electrophoretic mobility. Thus, the electrophoretic mobility is a function of the particle charge and on its size and shape. The dependence of the electrophoretic mobility on the latter two originates from the friction factor. The Creeping Flow assumption made in deriving Equation 3.65 is justified because the Reynolds number found in a typical electrophoretic experiment is well below 1. There are two electrophoresis modes: in free solution and in gels. It is illustrative to consider the differences in the electrophoretic mobility between these two. Because the charge dependence of the mobility is essentially the same in free solution and in gels, it is sufficient to focus on the friction factor. As stated above, the friction factor depends on the mass and shape of the particle. The molecular weight dependence of f for a few commonly found shapes are: $M^{0.33}$ (spheres), $M^{0.5}$ (random coil), $M^{0.8}$ (long rod), and $MW^{1.0}$ (free draining coil). Thus, the electrophoretic mobility becomes independent of the molecular weight for a free draining coil whose charge is proportional to its length (like in the case of DNA). The determination of the friction factor for a molecule moving through a gel is much more complicated because of the complex nature of the gel. There are several models available and only the Biased Reptation Model is briefly described below.

The idea of the Biased Reptation Model is that a polymer moves into a network of tunnels like a snake moves on the ground, head first and deforming its body. By defining a diffusion coefficient associated with the "reptation" movement it is possible to obtain the following relation between the friction factor and the number of segments of the polymer, N (Noolandi et al., 1987; Duke, Semenov, and Viovy 1992),

$$f_{reptation} \propto N^2 \qquad 3.66$$

In this case, the electrophoretic mobility of DNA becomes proportional to 1/N, in contrast with the molecular weight independence of the electrophoretic mobility in free solution for the same polymer. There are deviations from this behavior at high electric fields and for low molecular weight (or single stranded DNA) at low field intensities (Hubert et al. 1998).

Capillary Electrophoresis

Consider a species A at concentration C_A in an electric field. We start with the equation of continuity,

$$\frac{\partial c_A}{\partial t} = -\nabla . N_A \qquad 3.67$$

Both diffusion and the electric field contribute to the flux. Fick's Law gives the diffusive flux by,

$$N_{A-diff} = -D_{AB} \nabla c_A \qquad 3.68$$

the electric flux is given by,

$$N_{A-ele} = \mu_{ele} C_A E \qquad 3.69$$

in which E is the strength of the electric field. Using this in equation 3.67,

$$\frac{\partial c_A}{\partial t} = D_{AB} \nabla^2 c_A - \mu_{ele} E . \nabla c_A \qquad 3.70$$

The radius of the capillary is much smaller than its length. Therefore, we can assume that the only relevant flux takes place in the Z direction, along the axis of the capillary,

$$\frac{\partial c_A}{\partial t} = D_{AB} \frac{\partial^2 c_A}{\partial z^2} - \mu_{ele} E . \frac{\partial c_A}{\partial z} \qquad 3.71$$

It is worth noticing that the second term of the right hand side looks like a convective term with constant velocity $\mu_{ele}E$. To solve this differential equation it is convenient to make it dimensionless by defining a dimensionless length, Z

$$Z = \frac{z}{L} \qquad 3.72$$

a dimensionless concentration, C

$$C = \frac{c}{c_0} \qquad 3.73$$

and a dimensionless time, τ

$$\tau = \frac{t\mu_{ele}E}{L} \quad 3.74$$

$$\left(\frac{\partial C}{\partial \tau}\right) + \left(\frac{\partial C}{\partial Z}\right) = \frac{1}{Pe}\left(\frac{\partial^2 C}{\partial Z^2}\right) \quad 3.75$$

With

$$Pe = \frac{\mu_e EL}{D_{AB}} \quad 3.76$$

Where Pe is the Peclet number and it represents the ratio between the driving force because of the electric field and the driving force of diffusion. The boundary conditions are,

$$C(Z,0) = 0 \qquad Z > 0 \quad 3.77$$

$$C(0,\tau) = 1 \qquad 0 < \tau < \tau_0 \quad 3.78$$

$$C(\infty,\tau) = \textit{finite} \quad 3.79$$

Using Laplace transforms, we obtain,

$$c(Z,t) = \frac{c_0 t_0}{\sqrt{4\pi D_{AB} t^3}} \exp\left\{-\frac{(Z-\mu_{ele}Et)^2}{4D_{AB}t}\right\} \quad 3.80$$

This is the equation of a normal wave moving at a velocity $\mu_{ele}E$ and amplitude $(4D_{AB}t)^{1/2}$. Therefore, the peak moves with a velocity equal to $\mu_{ele}E$. The maximum in concentration occurs when Z is equal to μ_{ele} Et and it is equal to:

$$c(Z,t)_{\max} = \frac{c_0 t_0}{\sqrt{4\pi D_{AB} t^3}} \quad 3.81$$

Therefore, the height of the peak decreases as $t^{3/2}$. On the other hand, the width of the peak is,

$$c(x,t)_{width} = \sqrt{4D_{AB}t} \quad 3.82$$

Therefore, the peak broadens as $t^{1/2}$ and with $D_{AB}^{1/2}$. Thus, the separation will deteriorate at longer times and larger diffusivities. One of the advantages of CE over gel electrophoresis is that the very high voltages applied (O ~ 10,000 V) allows very fast experiments reducing the width of the peaks.

Mass Transfer Coefficients and Transport in Porous Media

The examples described above suggest that we can write the flux of species A as a coefficient multiplied by a driving force,

$$Flux = [Driving\ Force][Coefficient] \tag{3.83}$$

For example, for equimolar diffusion or for diffusion of A through stagnant B, we obtain,

$$N_A = \frac{D_{AB}}{\delta}(C_A^1 - C_A^2) \quad with \quad C_A^1 > C_A^2 \tag{3.84}$$

in which δ is a thin layer of fluid where the mass transfer resistance takes place. In principle the rate of mass transfer can be calculated if D_{AB} and δ are known. D_{AB} can be measured or theoretically calculated. On the contrary δ is much more difficult to determine. Besides, equations like 3.84 apply only to laminar flow, which are rarely found in practice. A practical approach to calculate mass transfer rates consists in writing,

$$N_A = k\Delta C_A \tag{3.85}$$

in which k is a mass transfer coefficient, which depends on the type of fluid, the type of flow, and the geometry of the system. This coefficient will change along the mass transfer unit or over time; therefore, it is a local mass transfer coefficient. An overall mass transfer coefficient can be calculated by integrating the local mass transfer coefficient along the unit or over time and normalizing it. Although the mass transfer coefficient formalism may be used for all types of flows, it becomes more important for turbulent flow because of the lack of reliable models for this type of flow.

A pragmatic approach to calculate mass transfer coefficients is to use dimensional analysis. Experimental data suggests the following functional form for the mass transfer coefficient,

$$k = f(D_{AB}, d, G, \mu, \rho) \tag{3.86}$$

in which d is a characteristic dimension, and G is the mass velocity. Let,

$$k \propto [D_{AB}]^a [d]^b [G]^c [\mu]^d [\rho]^e \tag{3.87}$$

using the appropriate units for each variable,

$$Mt^{-1}L^{-2} = L^{2a}t^{-a}L^b M^c t^{-c} L^{-2c} M^d L^{-d} t^{-d} M^e L^{-3a} \tag{3.88}$$

and equating powers on M, t, and L of the left and right hand sides, we obtain,

$$\begin{aligned} 1 &= d + e + c \\ 1 &= a + d + c \\ 2 &= 3e + d + 2c - b - 2a \end{aligned} \tag{3.89}$$

retaining a and b,

$$\begin{aligned} a &= e \\ d &= -a - b \\ c &= 1 + b \end{aligned} \tag{3.90}$$

Thus,

$$k \propto G\left(\frac{D_{AB}\rho}{\mu}\right)^a \left(\frac{dG}{\mu}\right)^b \tag{3.91}$$

or

$$\frac{kd}{\rho D} = f\left(\frac{dG}{\mu}, \frac{\mu}{\rho D_{AB}}\right) \qquad 3.92$$

Here it is practical to describe the preceding equation in terms of the Sherwood (sh), Reynolds (RE), and Schmidt (SC) numbers,

$$\frac{kd}{\rho D_{AB}} = Sh \qquad 3.93$$

$$\frac{dG}{\mu} = \mathrm{Re} \qquad 3.94$$

$$\frac{\mu}{\rho D_{AB}} = Sc \qquad 3.95$$

The Sherwood number (analogous to the Nusselt number in heat transfer) is equal to the ratio of the mass transfer rate by convection to the mass transfer rate by diffusion. The Reynolds number is equal to the ratio between inertial and viscous forces and the Schmidt number (analogous to the Prandtl number in heat transfer) collects the properties of the solvent and the diffusive particles. It is illustrative to present some correlations for mass transfer coefficients. Because they are widely used, the discussion below will be limited to hollow-fiber membrane modules and to packed beds.

Three approaches have been followed to correlate mass transfer coefficients with process variables in filtration through membranes. One obvious, but sometimes tedious, approach consists of solving the continuity equation with appropriate boundary conditions. A second approach consists of the extrapolation of known correlations for single pores. Finally, the third approach consists of fitting the experimental data with an appropriate functional form.

The solution of the continuity equation depends on the particular geometry. The following expression was found (Zeman and Zidney 1996) for the local mass transfer coefficient for dead-end filtration,

$$k(t) = \left(\frac{D_{AB}}{\pi t}\right)^{1/2} \qquad 3.96$$

Therefore, the mass transfer coefficient decreases with $t^{1/2}$ because of the formation of a diffusion-driving cake at the surface of the membrane. For cross flow devices the following correlation was (Zeman and Zydney 1996) found,

$$k(x) = 0.538\left(\frac{D_{AB}{}^{2}\gamma}{x}\right)^{1/3} \qquad 3.97$$

in which γ is the shear rate at the surface of the membrane. In this case, the mass transfer coefficient is a function of $x^{1/3}$ where x is the tangential distance. Notice that the mass transfer coefficient for dead end membranes depends on the diffusion coefficient to the one-half power, whereas the

Table 3.3. Coefficients in the Correlation for the Mass Transfer Coefficient Under Different Flow Conditions (Zeman and Zydney 1996).

Flow Conditions	a	b	c	d	Reference
Laminar tube	0.33	0.33	0.33	1.62	Leveque (1928)
Laminar \slit	0.33	0.33	0.33	1.86	Leveque (1928)
Developing	0.50	0.33	0.50	0.664	Grober et al. (1961)
Turbulent	0.80	0.33	—	0.023	Gekas and Hallstrom (1987)
Turbulent	0.875	0.25	—	0.023	Deissler (1961)
Spacer	0.50	0.33	0.50	0.664	Da Costa et al. (1994)
Stirred cell	0.567	0.33	—	0.23	Smith et al. (1968)
Rotating	0.50	0.33	0.42	0.75	Holeschovsky and Cooney (1991)

mass transfer coefficient in cross-flow filtration depends on the diffusion coefficient to the two-thirds power. Experimental correlations for the mass transfer coefficient in membranes have the expected dependence on the Reynolds number and the Schmidt number. This dependence is,

$$Sh = \frac{kd_h}{D_{AB}} = \beta \,\mathrm{Re}^a \, Sc^b \left(\frac{d_h}{L} \right)^c \qquad 3.98$$

in which d_h is the hydraulic diameter (=4x the cross-sectional area available to flow divided by the wetted perimeter of the conduct). The excellent monograph by Zeman and Zydney (1996) compiled the values for the coefficients a, b, c, and d shown in Table 3.3.

All the exponents in these correlations were obtained assuming that the density, viscosity and diffusion coefficient remain constant. Experimental evidence suggests that such an assumption may not be valid in many practical cases. For example, in a filtration process it is likely that the viscosity will increase as the process advances because the solution becomes more concentrated. Furthermore, as the solute becomes more concentrated, the diffusion coefficient will increase too. In order to calculate the mass transfer coefficient for a protein in a ultrafiltration membrane we need to use its calculated diffusion coefficient (see above for calculation of diffusion coefficient) and account for its changes.

A typical chromatographic separation consists of the injection of a sample into a column filled with a porous material and subsequent elution by the addition of various buffers. Therefore, the solute moves in the downwards direction through interstices in between particles and through the pores within the particles. Four main mass transfer mechanisms have been recognized: axial dispersion, radial dispersion, film mass transfer and intraparticle diffusion. As we have four distinct mass transfer mechanisms, we will need four distinct mass transfer coefficients.

It is a well-known fact that when a solute moves through a pack column, the solute moves in the axial and radial direction by a combination of molecular diffusion and convective flow. The fluid, then, disperses in both the radial and axial directions. It is customary to introduce a new coefficient, called the dispersion coefficient, D, to account for both convective and diffusive effects. Axial dispersion is caused mainly by eddy diffusion but molecular diffusion, channeling and wall effects also contribute. Dispersion is a function of the length of the bed, the viscosity and density of the fluid, the ratio of column diameter to particle diameter, ratio of column length to particle diameter, particle size distribution, particle size, and solute molecular diffusion coefficient. Following a recent review by Delgado (2006), we will divide this discussion into two parts: Longitudinal (axial) dispersion and radial dispersion. An exhaustive collection of correlations for axial and radial dispersion coefficients can be found in that review; only a few are reproduced below.

Dimensional analysis yields the following dependence of the axial dispersion coefficient with process variables, $Pe_L = Vd/D_L$

$$Pe_L = f\left(\frac{L}{D_{AB}}, \frac{D_{AB}}{d}, Pe_m, Sc\right) \quad 3.99$$

in which Pe_m is the molecular Peclet number ($=Vd/D_{AB}$), L is the length of the bed, D is its diameter, d is the diameter of the particles, V is the velocity of the fluid, D_{AB} is the molecular diffusivity of the fluid, Sc is the Schmidt number ($=\mu/\rho D_m$). The dependence of Pe_L on L/D and D/d is weak in packed beds. Therefore, $D_L/D_m = f(Pe_m, Sc)$. In the absence of convection,

$$D_L = D_m / \sqrt{2} \quad 3.100$$

in which the factor square root of two accounts for the tortuosity.

There are several correlations for Pe_L with fluid velocity and with Sc number (a function of temperature through the temperature dependence of density, viscosity and molecular diffusivity); one of them is the correlation of Chung and Wen (1968),

$$Pe_L = \left(0.2 + 0.11\,\mathrm{Re}^{0.49}\right) L / (2R_p \varepsilon_b) \quad 3.101$$

Here the Reynolds number is defined by,

$$\mathrm{Re} = \frac{2R_p \varepsilon_b v \rho}{\mu} \quad 3.102$$

in which R_p is the particle radius, ε_b is volume fraction of the bed, v is the intersticial velocity, ρ is the density of the mobile phase, and μ is the viscosity of the mobile phase. One of the latest correlations that have appeared is (Delgado 2006),

$$\frac{1}{Pe_L} = \frac{Pe_m}{5}(1-p)^2 + \frac{Pe_m^2}{25} p(1-p)^3 \left\{ \exp\left[-\frac{5}{p(1-p)Pe_m} \right] - 1 \right\} + \frac{1}{\tau Pe_m} \quad 3.103$$

$$p = \frac{0.48}{Sc^{0.15}} + \left(\frac{1}{2} - \frac{0.48}{Sc^{0.15}} \right) \exp\left(-\frac{75Sc}{Pe_m} \right) \quad 3.104$$

in which $\tau = \sqrt{2}$ for spheres and to 1.93 and 1.8 for solid and hollow cylinders respectively. This correlation is recommended for well-packed random packings of isometric spherical particles.

The radial dispersion coefficient is normally less important than the axial one. The following correlation has been proposed recently for Sc numbers larger than 550 (Delgado 2006),

$$\frac{1}{Pe_R} = \frac{1}{\tau} \frac{1}{Pe_m} + \frac{1}{12} - 8.110^{-3} (\tau Pe_m)^{0.268} \quad \textit{for } Pe_m \le 1{,}600 \quad 3.105$$

and,

$$\frac{1}{Pe_R} = 45.9 - 33.9 \exp\left(-\frac{15Sc}{Pe_m} \right) \quad \textit{for } Pe_m > 1{,}600 \quad 3.106$$

The mass transfer coefficient for the film resistance may be estimated by the correlation by Wakao et al. (1958),

$$\frac{2R_p k}{D_{AB}} = 2 + 1.45\,\mathrm{Re}^{0.5}\, Sc^3 \qquad 3.107$$

which is valid for Re numbers less than 100. Finally, the intraparticle diffusion may be estimated by the correlation of Yau et al. (1979),

$$D_{eff} = D_{AB}(1 - 2.104\lambda + 2.09\lambda^3 - 0.95\lambda^5)0.476 \qquad 3.108$$

in which

$$\lambda = \frac{d_i}{d_p} \qquad 3.109$$

with d_p equal to the average diameter of the pores.

Summary

This chapter has tried to review the main concepts of mass transferred. Rather than emphasizing the mathematics, the concepts of diffusive flux and convective flux are highlighted. The two coefficients that determined those fluxes, the diffusion coefficient and the coefficient of mass transfer were introduced and correlations that can be used in different circumstances were explained. Emphasis was put in mass transfer through membranes and in packed beds because of the indisputable importance of ultrafiltration and chromatography in bioseparation. Because capillary electrophoresis is an important analytical tool but it is not relevant in processing, its principles were introduced in this chapter. Besides, its analysis is an excellent example of the use of mass transfer principles in separations.

Nomenclature

C_i	molar concentration (mol/L)
C(t)	correlation function
d	characteristic dimension
d_i	diameter of molecule i (cm)
d_h	hydraulic diameter (cm)
d_p	pore diameter (cm)
D_0	infinite dilution diffusion coefficient (cm^2/sec)
D_{ij}	diffusion coefficient of i in j. (cm^2/sec)
D_L	dispersion coefficient (cm^2/sec)
D_{eff}	Effective diffusivity (cm^2/sec)
E	strength of the electric field (V/m)
f	friction factor (kg/sec)
F_A	applied force (N)
G	mass velocity (g/sec)

j_i mass flux relative to local mass average velocity (g/cm^2 sec)
J_i molar flux relative to local mass average velocity (mol/cm^2 sec)
j_i^* mass flux relative to local molar average velocity (g/cm^2 sec)
J_i^* molar flux relative to local molar average velocity (mol/cm^2 sec)
k mass transfer coefficient (cm/sec)
k_B Boltzmann's constant (1.3807x10^{-23} J/K)
Kh hydrodynamic coefficient
Kp partition coefficient
L length of the capillary (m)
m mass of a particle/molecule (kg)
M molecular mass of molecules (g/mol)
n_i mass flux relative to stationary coordinates (g/cm^2 sec)
N_i molar flux relative to stationary coordinates (mol/cm^2 sec)
p pressure (Pa)
Pe Peclet number
Pe_m molecular Peclet number
Pe_L longitudinal Pe number
Pi^s saturation pressure of species i (Pa)
q momentum transfer (cm^{-1})
R ideal gas constant
R_g radius of gyration (Å)
R_i rate of formation of species i (moles/L sec)
R_p particle radius (cm)
R_H hydrated radius of the protein (cm)
s Svedberg constant (sec)
Sc Schmidt number
T absolute temperature (K)
$\bar{v}$ partial specific volume (cm^3/g)
$\vec{v}$ local mass average velocity (length/ unit time)
$\vec{v}^*$ local molar average velocity (length/unit time)
V velocity (length/unit time)
w_i mass fraction
x_i mol fraction
z distance (m)
z_p charge (C)

Greek Letters

δ boundary layer thickness (cm)
η intrinsic viscosity (cm^3/g)
ε_b volume fraction of the bed
ε_{intra} porosity of the particles
ϕ volume fraction
γ shear rate at the surface of the membrane (1/sec)
λ ratio between pore and molecule radii
μ_e electrophoretic mobility (m^2/V sec)
μ viscosity (g/cm sec)
ρ density or mass concentration (g/cm^3)
τ_{intra} tortuosity of individual porous particles

References

Anderson, J. L., and C. C. Reed. 1976. Diffusion of spherical macromolecules at finite concentrations. *Journal of Chem Physics* 64(8):3240-50.

Chung, S. F., and C. Y. Wen. 1968. Longitudinal dispersion of liquid flowing through fixed and fluidized beds. *AIChE J* 14:857-66.

Da Costa, A. R., A. G. Fane, and D. E. Wiley. 1994. Spacer characterization and pressure drop modeling in spacer-filled channels for ultrafiltration. *Journal of Membrane Science* 87:79-98.

Delgado, J. M. P. Q. 2006. A critical review of dispersion in packed beds. *Heat Mass Transfer* 42:279-310.

Deissler, R. 1961. Analysis of turbulent heat transfer, mass transfer and friction in smooth tubes at high Prandtl and Schmidt numbers. In *Advances in Heat and Mass Transfer*, edited by J. P. Hartnett, p. 243. New York: McGraw-Hill.

Duke, T. A. J., A. N. Semenov, and J. L. Viovy. 1992. Mobility of a reptating polymer. *Physical Review Letters* 69:3260-3.

Gekas, V., and B. Hallstrom. 1987. Mass transfer in the membrane concentration polarization layer under turbulent cross flow. 1. Critical literature review and adaptation of existing Sherwood correlations to membrane operations. *Journal of Membrane Science* 30:153-70.

Grober, H., S. Erk, and V. Grigoll. 1961. *Fundamentals of Heat Transfer.* New York: McGraw Hill.

Holeschovsky, U. B., and C. L. Cooney. 1991. Quantitative description of ultrafiltration in a rotating filtration device. *AIChE J* 37:1219-26.

Hubert, S. J., M. Krzywinsky, I. L'Heureux, and G. W. Slater. 1998. Reptation dynamics with random local interactions. *Macromolecules* 31:181-92.

Leveque, M. A. 1928. Les lois de la transmission de la chaleur par convection. *Annales des Mines ou Recueil de Memoires sur l'Exploitation des Mines et sur les Sciences et les Arts qui s'y Rattachent, Memoires. Tome XIII* 13:201.

Noolandi, J., J. Rousseau, G. W. Slater, C. Turmel, and M. Lalande. 1987. Self trapping and anomalous dispersion of DNA in electrophoresis. *Physical Review Letters* 58:2428-31.

Polson, A. 1950. Some aspects of diffusion in solution and a definition of a colloidal particle. *Journal of Physical and Colloid Chemistry* 54:649-52.

Polson, A. 1967. The calculation of molecular weights from diffusion and viscosity data. *Biochimica et Biophysica Acta* 140:197-200.

Renkin, E. M. 1954. Filtration, diffusion, and molecular sieving through porous cellulose membranes. *Journal of General Physiology* 38:225-46.

Scheraga, H. A., and I. Mandelkern. 1953. Consideration of the hydrodynamic properties of proteins. *Journal of the American Chemical Society* 75:179-84.

Schruder, M., E. von Lieres, and J. Hubbuch. 2006. Direct quantification of intraparticle protein diffusion in chromatographic media. *Journal of Physical Chemistry B* 110:1429-36.

Slattery, J. C., and R. B. Bird. 1958. Calculation of the diffusion coefficient of dilute gases and on the self-diffusion coefficient of dense gases. *AIChE J* 4:137-42.

Smith, K. A., C. K. Colton, E. W. Merrill, and L. B. Evans. 1968. Convective transport in a batch dialyzer: determination of the true membrane permeability from a single measurement. *American Institute of Chemical Engineering Symposium Series* 64:45-58.

Svedberg, T. 1925. Zentrifugierung, diffusion und sedimentationsgleichgewicht von kolloiden und hochmolekularen stiffen. *Colloid and Polymer Science* 36:53-64.

Tallarek, U., F. J. Vergeldt, and H. Van as. 1999. Stagnant mobile phase mass transfer in chromatography media: Intraparticle diffusion and exchange kinetics. *Journal of Physical Chemistry B* 103:7654-64.

Tyn, M. T., and T. W. Gusek. 1990. Prediction of diffusion coefficients of proteins. *Biotechnology and Bioengineering* 35:327-38.

Wakao N., T. Oshima, and S. Yagi. 1958. Mass transfer from packed beds of particles to a fluid. *Kagaku Kogaku* 22:780-5.

Yau, W. W., J. J. Kirkland, and D. D. Bly. 1979. *Modern Size Exclusion Liquid Chromatography.* New York: John Wiley and Sons, Inc.

Young, M. E., P. A. Carroad, and R. L. Bell. 1980. Estimation of diffusion coefficients of proteins. *Biotechnology Bioengineering* 22:947-55.

Zeman, L.J., and A. L. Zydney. 1996. *Microfiltration and Ultrafiltration: Principles and Applications.* New York: Marcel Dekker.

Further Reading

Bird, B. R., W. E. Stewart, and E. N. Lightfoot. 2000. *Transport Phenomena,* 2nd ed. New York: John Wiley and Sons, Inc.

Deen, W. M. 1998. *Analysis of Transport Phenomena.* New York: Oxford University Press.

Cussler, E. L. 1997. *Diffusion: Mass Transfer in Fluid Systems.* New York: Cambridge University Press.

Crank, J. 1980. *The Mathematics of Diffusion.* New York: Oxford University Press.

Chapter 4

Biomacromolecules

Introduction

It is impossible to rationally design a protein purification train without proper knowledge of protein structure and protein behavior in solution or in solid state. Proteins are already in a well-controlled milieu; for example, in the cell cytoplasm or in the cell membrane. As we attempt to purify a protein, a potentially harsh environment replaces this benign one. The new environment may induce undesirable changes in the protein, such as aggregation, lost biological activity, or general degradation. Compared with separation processes that are used in chemical or petrochemical plants the fragility of the proteins makes bioseparations unique. In the following sections, this background is provided and emphasis is put on protein features that may be of use in their separations, as well as potential problems that may arise in a separation process. Later, this chapter discusses the structure of nucleic acids because of the increasing interest in using nucleic acid fragments as beneficial viruses (nucleic acid therapy). This chapter could not be complete without some connection to actual experiments. Therefore, this chapter presents a series of relatively simple experiments to characterize both the target proteins and more importantly the impurities.

Protein Structure

The building blocks of proteins are twenty amino acids. They are analogs to a molecule of methane (where the carbon in a methane molecule becomes the α-carbon) in which three of its hydrogen atoms have been substituted by: 1) a carboxyl group, 2) an amino group, and 3) a "side chain" that gives a particular amino acid its special characteristics. Table 4.1 presents the most abundant amino acids found in proteins. Because it has been customary to classify amino acids into three categories—ionic, polar, and non-polar—those characteristics are included in the table. Amino acids are zwitterions, they can be negatively or positively charged depending on the pH. The last column in the table is the pH that makes the amino acids neutral, also known as their isoelectric point. The protein will also be charged, because several amino acids bear a net charge at physiologic

Table 1. Amino Acids.

Amino Acid	Abbreviation	Side Chain	Characteristics	Isoelectric Point
Glycine	Gly (G)	Alkane	Non-polar	5.97
Valine	Val (V)	Alkane	Non-polar	5.96
Alanine	Ala (A)	Alkane	Non-polar	6.00
Leucine	Leu (L)	Alkane	Non-polar	5.98
Isoleucine	Ile (I)	Alkane	Non-polar	6.02
Methionine	Met (M)	—S—	Non-polar	5.74
Proline	Pro (P)	Cyclic	Non-polar	6.30
Phenylalanine	Phe (F)	Aromatic	Non-polar	5.48
Tryptophan	Trp (W)	Aromatic	Non-polar	5.89
Glutamine	Gln (Q)	Amide	Polar	5.65
Asparagine	Asn (N)	Amide	Polar	5.41
Serine	Ser (S)	Alcohol	Polar	5.68
Threonine	Thr (T)	Alcohol	Polar	5.60
Arginine	Arg (R)	Amine	Ionic	10.76
Histidine	His (H)	Imidasolium	Ionic	7.59
Tyrosine	Tyr (Y)	Phenol	Ionic	5.66
Lysine	Lys (K)	Amine	Ionic	9.74
Glutamic Acid	Glu (E)	Carboxy	Ionic	3.22
Aspartic Acid	Asp (D)	Carboxy	Ionic	2.77
Cysteine	Cys (C)	Sulfur	Ionic	?

pH. The number of positive and negative charges will be equal at a pH called the isoelectric point of the protein.

Each amino acid, or residue, is joined to the next by the formation of an amide bond between the carbonyl group of one amino acid and the amine group of the next one. This kind of bond is called a peptide bond (Figure 4.1). The figure shows that two consecutive α-carbons lie at opposite vertices of a rectangle, whereas the carboxyl group and the amine group occupy the other two corners. Because of the relative rigidity of the C-N bond, the four groups lie on a plane. The particular characteristics of the peptide bond and the nature of the amino acid side chain set the structure. Rotation is allowed along the αC and the carbonyl carbon, with a solid angle ϕ, and the αC and the amide N, with a solid angle ψ, bonds. The rotation is limited by stearic constraints and is slightly relaxed by the presence of solvent. This is nicely demonstrated in Ramachandran Plots (Ramachandran and Sasisekharan 1968), which are torsion angles (ϕ, ψ) maps. Ramachandran, by just playing with "ball and stick" models, determined the pairs of angles that are allowed and which pairs determine the different types of structures. The plane is divided into four different regions (Figure 4.2):

White—disallowed region
Light gray—generously allowed region
Dark gray—allowed region
Black—most favorable region

After this Nobel-Prize winning finding, researchers have shown that the presence of solvent actually slightly relaxes the constraints found by "simple" stearic arguments.

Proteins consist of one or more polypeptide chains folded into a well-defined three-dimensional shape. The arrangement of four different structural levels—primary, secondary, ternary, and

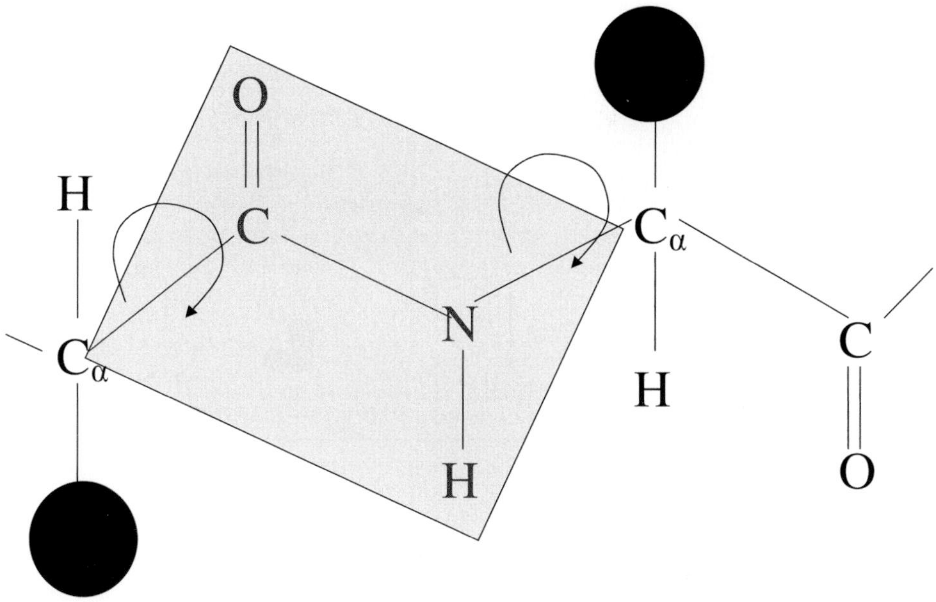

Figure 4.1. Cartoon of the peptide bond. Notice the free rotation along the αC-C bond and along the αC-N bond.

quaternary—defines the three-dimensional shape of a protein. The primary structure is the order in which the amino acids are joined in the polypeptide chain that forms the molecule. Specific spatial organizations between segments of the polypeptide chain constitute the secondary structure. The most frequently found are α-helix, β-sheet, β-turns, and antiparallel β-sheets. Portions of the protein molecule remain disordered in the form of random coils. α-helixes, β-sheets, and disordered structures arrange in space in unique patterns forming the protein's ternary structure. Finally, some proteins are built by more than one polypeptide chain (called a subunit) that cluster together, forming the protein's quaternary structure.

The α-helix is a trademark of most proteins. A delicate arrangement of hydrogen bonds between carbonyl and amino groups determine the structure and its characteristic dimensions. The most common amino acids found in an α-helix are Ala, Leu, Met, Phe, Glu, Gln, Lys, Arg, and His. The α-helix also has a dipolar moment running on the axis, which has been used by some to explain the orientation of proteins in an electric field. Disruption of this hydrogen bond network will "fry" the structure. β-sheet is another secondary structure found in most proteins. The most common amino acids found in β-sheets are Tyr, Trp, Ile, Val, Thr, and Cys. The closest representation of a β-sheet would be that of a piece of paper folded in an accordion pattern.

Various secondary structures arrange in space to define the protein's ternary structure. Motifs are arrangements of secondary structure that are preserved in various proteins. Figure 4.3 shows some simple ones (Creighton 1993). The arrows represent β-sheets and the curved lines disorganized structure. Finally, polypeptides having a well-defined secondary and ternary structure may "associate" to generate the quaternary structure of a protein.

Figure 4.4 shows the structures of four different proteins. The left upper panel corresponds to tubulin (Gigant et al. 2000), which is a typical example of a protein with a well-defined quaternary structure. It shows four different peptides making the individual tubule and an α-helix that keeps the structure together. The upper right panel is a cartoon of glycosyltransferase (Lin et al. 1996).

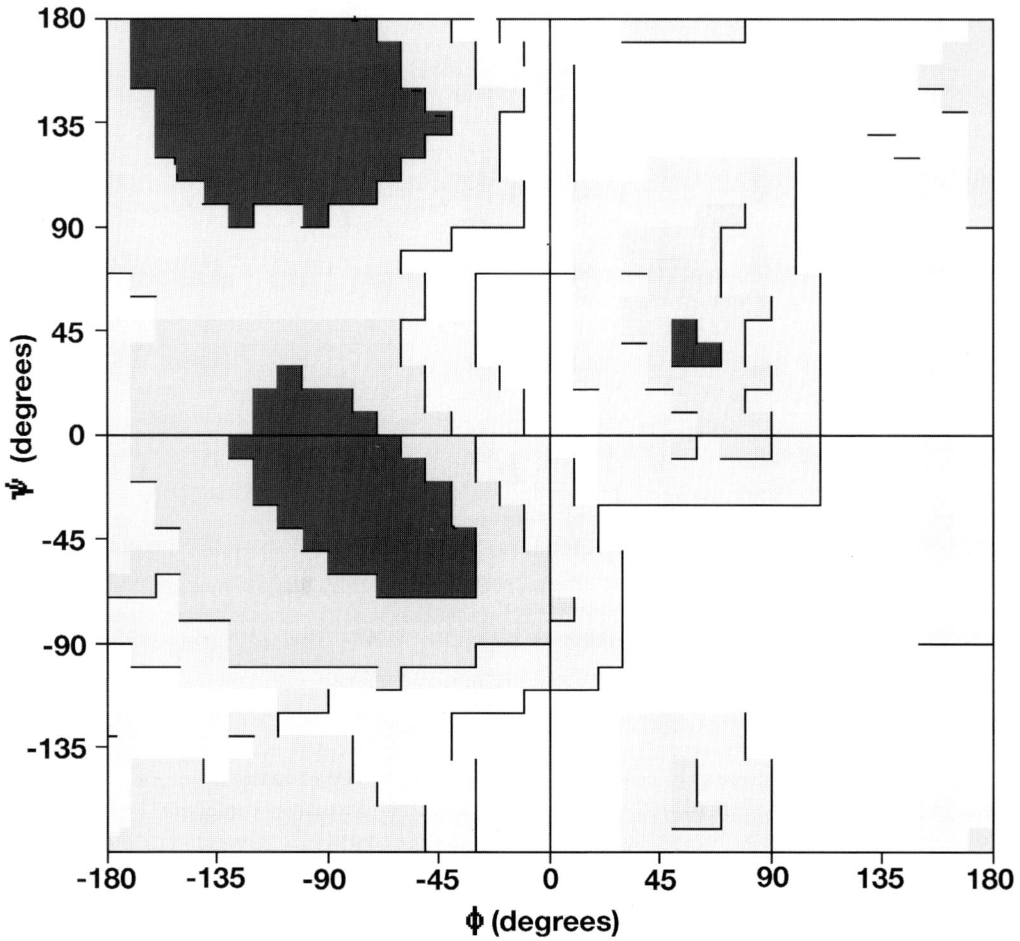

Figure 4.2. Ramachandran plot. White: disallowed region; light gray: generously allowed region; dark gray: allowed region; black: most favorable region.

This protein has four peptide chains and each is very rich in α-helices. The lower left panel is myoglobin (Brunori et al. 2000), which consists almost entirely of α-helices joined by disordered regions. At the other end is prealbumin (Peterson et al. 1998) shown in the right lower corner that consists almost entirely of β-sheets.

Membrane Proteins

A large number of proteins are embedded into the cell membrane and often have unique characteristics. Thus, production of membrane proteins is considered an emergent field in biotechnology. Membrane proteins have a specific function in both mammalian and plant cells. Information exchange, recognition, and sensing are membrane-mediated processes in nature. All membrane receptors are proteins or have protein components in them. They monitor the presence of bacteria in plants and mammals and communicate this information to the appropriate region in the cell.

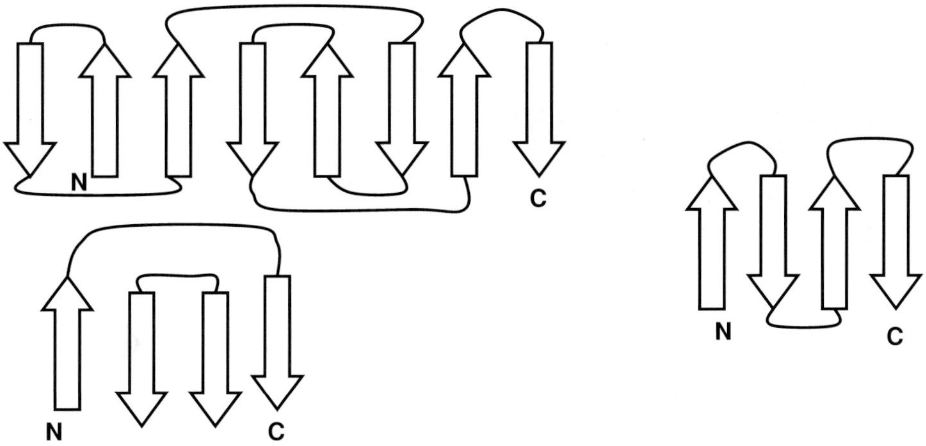

Figure 4.3. Protein ternary structure and motifs. Each block arrow represents a β-sheet and the line's disorganized peptide segments.

Furthermore, membrane proteins have been used in vitro as molecular sieves and/or support devices as components of biosensors.

Membrane proteins must be studied in their natural environment because they are insoluble in aqueous solutions. Therefore, it is common to reconstitute membrane proteins in phospholipid vesicles. Vesicle size, which determines the way in which a protein will fuse into it, is determined by the ratio of phospholipid vesicles. The size is determined by the ratio of phospholipid to protein concentration, transition temperatures, buffer pH, and type and amount of detergent added. The sharp curvature of small vesicles may affect protein packing. Therefore, large vesicles (O~400 nm and higher) are better to use in reconstitution experiments because the curvature of large vesicles resembles the curvature of whole cells.

The phospholipid concentration also plays an important role in the stability of the vesicles and the amount of protein that is incorporated into the vesicle bilayer (Boekema 1990; Jap et al. 1992). For most experiments, the mass ratio of lipid to protein must be 3:1 and higher to form stable proteoliposomes.

Hovmoller et al. (1983) found that varying the pH and temperature during reconstitution contributes to the size and order of the crystals obtained. At low temperatures, lipid bilayers form densely packed crystalline structures (Engelhardt et al. 1994). The incubation at temperatures above the transition temperatures of the phospholipids being used causes the lipid bilayers to become more mobile (almost liquid like) and able to expand to allow the protein to pack in a more ordered array.

Detergent-mediated processes are used in almost all reconstitution procedures. It has been found that the type of detergent added to certain protein-lipid solutions can affect the protein's stability (Rigaud et al. 1997, 1998). It is thought that proteins purified in the presence of non-ionic detergents usually arrange in sheets rather than micelles, which is a better starting point for crystallization (Boekema 1990). Detergent removal from detergent-lipid-protein micellar solutions has also been explored extensively in the literature (Rigaud et al. 1997). Triton X-100 is a non-ionic detergent that has been used in most reconstitution and crystallization experiments. The detergent has a polar head and a long hydrophobic tail that stabilizes the protein in solution. One of the drawbacks of using this detergent is its low critical micelle concentration (CMC), which makes the detergent very difficult to remove. Removal by dialysis can take up to five days with only 90% of the detergent removed. It has been found that polystyrene beads (SM-2 Biobeads) are much more efficient for removing the detergent from the protein-lipid-detergent complex (Jap et al. 1992).

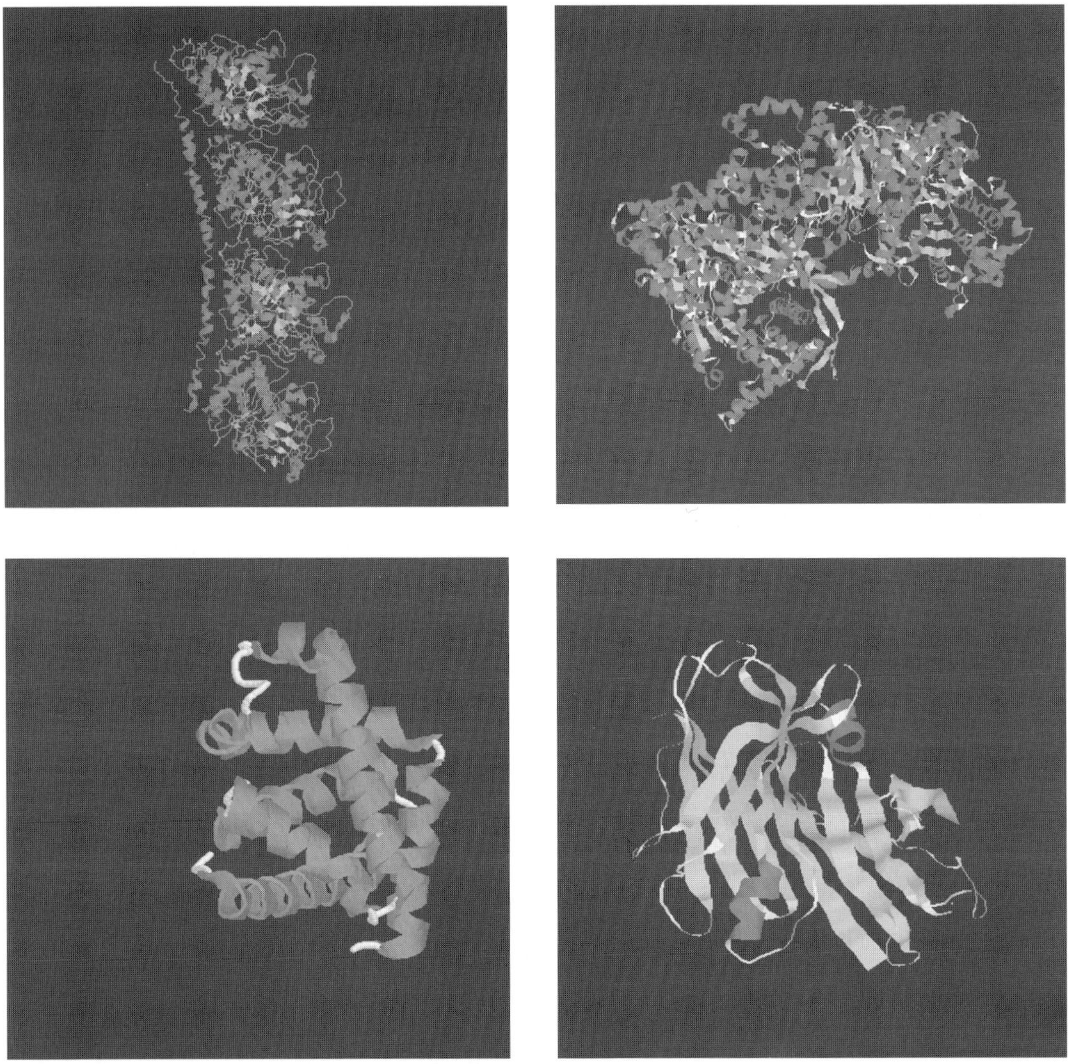

Figure 4.4. Examples of protein quaternary structure. Left upper panel: tubulin; upper right panel: glycosyltransferase; lower left panel: myoglobin; right lower panel: prealbumin. Images are from Brookhaven Protein Data Base and processed with Rasmol.

Cytochrome c reductase is a work-horse in reconstitution experiments that has been reconstituted using several methods in order to produce crystals suitable for electron crystallography studies. Leonard et al. (1981) found that pH and temperature play an important role in the quality of two-dimensional crystals obtained. Initially they crystallized cytochrome c reductase into lipid vesicles by stirring SM-2 biobeads in the protein-lipid-detergent solution at pH 7 and 4°C. They found that crystallizing by this method produced a mixture of sheets, tubes, and small vesicles (Leonard et al. 1981). Two-dimensional crystals mostly in the form of sheets and large vesicles were obtained by lowering the pH to 5.5, raising the temperature to 25°C, and using dialysis to remove the detergent.

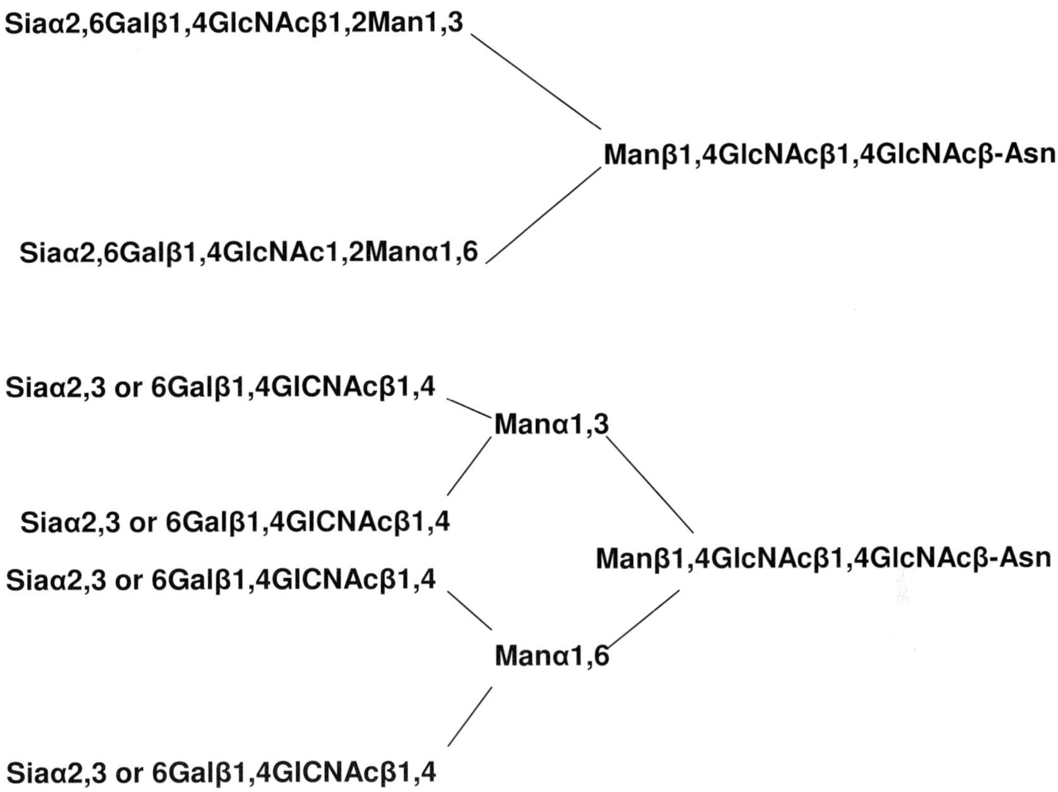

Figure 4.5. Common sugar moieties found in proteins. Abbreviations: Sia: Sialic acid, Gal: Galactose, GlcNAc: N-acetyl-α-D-glycosamine, Man: Mannose,

The techniques used for cytochrome c reductase have provided useful information for crystallizing both membrane and soluble proteins.

Posttranslational Modifications (Glycosylation)

Most proteins are modified posttranslation to some extent. For completeness, translation is defined as the process by which the genetic information is used to build the primary structure of the protein; however, we are interested in this section in the posttranslation modifications. One of those modifications is glycosylation that consists of the addition of various sugars to the polypeptide chain. Although protein glycosylation has been studied for a long time, there is no consensus about the role of glycan groups on protein structure and function. Some think that the glycan group attached to the nascent peptide in the endoplasmic reticulum guarantees the correct folded structure of the peptide. Others think that the glycan groups have a stabilizing role because they increase the hydrophillicity of the peptide surface, which in turn increases its solubility and inhibits aggregation. The fact that glycosylation has an effect on the melting temperature of various proteins supports this view. Both views converge on the fact that glycosylation affects protein stability.

Most sugars present in proteins are complex and very diverse. Figure 4.5 shows a couple of examples. Notice the bifurcation at the mannose residue and the variability in the substitutions. The

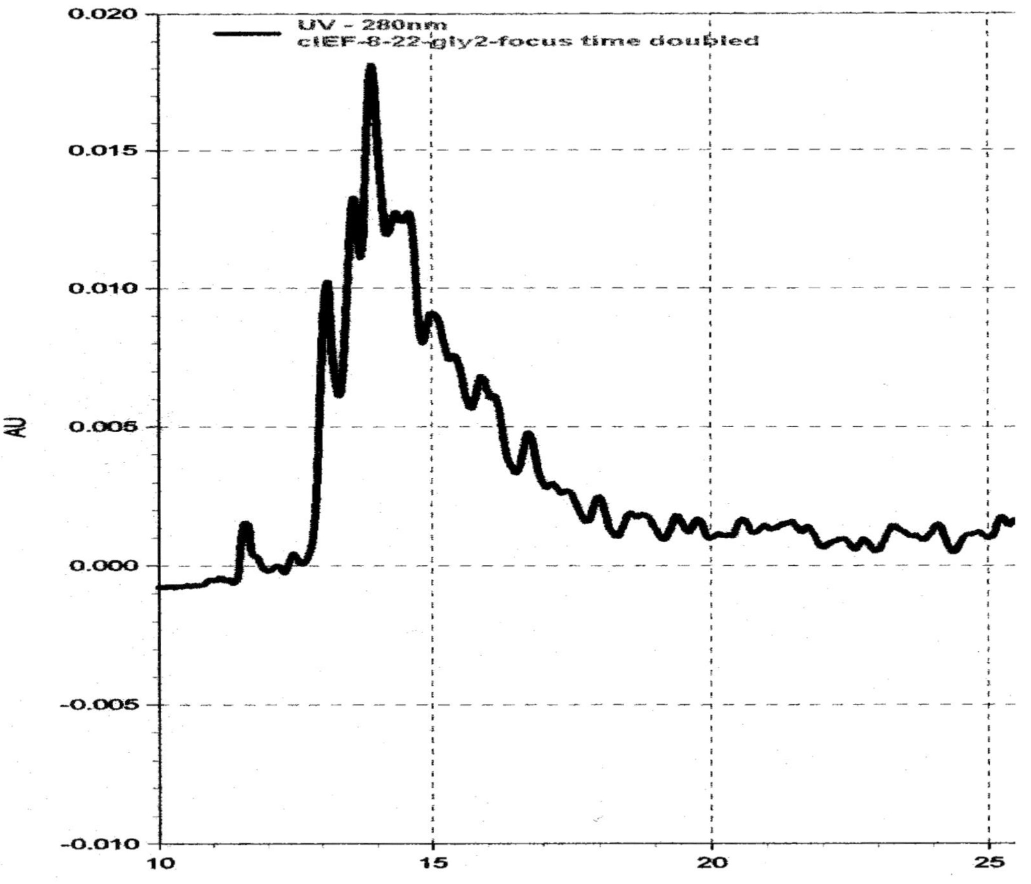

Figure 4.6. Electropherogram of a human antibody showing the resolution of the different isoforms. The absorbance of 280 nm is plotted versus time (in minutes).

oligosaccharides found in proteins may be divided into two groups: high mannose and complex groups. There are four different types of linkages between the oligosaccharide and the polypeptide chain in mammalian cells: 1) N-link to the β-amide nitrogen of asparagine residue, 2) O-link to the β-hydroxyl group of serine, 3) O-link to the β-hydroxyl group of threonine, and 4) O-link to the hydroxyl group of 5-hydroxylysine. The most common structure in N-linked oligosaccharides is a sequence of three sugars: Manβ1-4GlcNAcβ1-4GlcNAcβ-Asn.

After the oligosaccharide is linked to the peptide chain it is trimmed. This trimming produces a family of proteins that differ from each other only in their glycosylation pattern; this is known as microheterogeneity. Normally, there is no need in separating the isomers but variations in the glycosylation pattern may be a problem with regulatory agencies. The isomers can be separated partly or totally by highly sensitive techniques like capillary electrophoresis. This creates a fingerprint of the product. Figure 4.6 shows an electropherogram of a sample of a transgenic human monoclonal antibody showing the separation between various isoforms. If the pattern is reproducible, it can be used to prove the invariability of the product to regulatory agencies.

Glycosylation patterns are different in different types of cells and some are unable to add glycosyl groups. For example, prokaryotes cells do not have the necessary machinery to add glycosyl groups to their proteins; therefore, human proteins expressed in bacteria are not glycosylated. The glycosylation machinery of mammalian and plant cells yields different glycosylation patterns. This is important because recently therapeutic human proteins have been expressed in plants. Although the expression of human antibodies in plants such as tobacco or corn has been quite successful, the plant glycosylation machinery is often turned off because of differences in the glycosylation pattern between the human and the plant variety. Unfortunately, the non-glycosylated antibody is less stable and thus has a shorter shelf life than the native one. The decrease in stability is manifested by the formation of irreversible aggregates, which decreases the solubility of the preparation and impedes their normal function; aggregates are not biologically active even if they do not precipitate.

The presence of oligosaccharides in a protein can be used for their isolation. For example, the presence of sialic acid in the oligosaccharide will change the charge of the protein; therefore, ion exchange chromatography can be used to isolate it. In addition, a group of proteins known as lectins recognizes and binds to sugars. Immobilization of a lectin on a solid support can be used for the affinity chromatography separation of the glycosylated protein.

Protein Modifications to Facilitate Separation

Genetic engineering may be used to modify the target protein to facilitate its purification. Using the appropriate vectors different residues may be added. The most common modification is the addition of a poly(histidine) peptide (Stiborova et al. 2003). Other modifications that have been used include a maltose binding protein (Maina et al. 1988), thioredoxin (Smith et al. 1998), a cellulose-binding domain (Ong et al. 1989), and a streptag (Skerra and Schmidt 1999).

The advantage of using a histidine tag ranging from 6 to 10 residues is that the target protein can be then separated in a single step using metal affinity chromatography, IMAC. The disadvantage is that IMAC is very expensive and prone to fouling. Still, the poly(histidine) tag can be used in less expensive separation approaches. An example of this is the use of a poly(histidine) tag in polymers that may be used to selectively precipitate the target protein.

Protein Solubility

With the exception of membrane-bound proteins, most proteins are soluble in aqueous solutions. Although proteins are not true colloids because the charge is not uniformly distributed on their surface and they are flexible, arguments used to explain the stability of colloids can be used to explain protein stability in solution. Those arguments vary in their degree of complexity, ranging from molecular approaches to more macroscopic ones. Although statistical mechanics has been used to model the interactions between proteins, the models are so complex that they are of little use from a practical point of view. Still, they add some light to our understanding of their behavior in solution and the reader is referred to some of the references at the end of this chapter. It is more convenient to use mean field approaches, because despite their simplicity they capture some of the main features of protein solubility. One common mean field approach based on a fundamental contribution to thermodynamics by Gibbs has been used to understand the stability of colloids and then extended to proteins.

Figure 4.7 shows a cartoon of a typical colloid. They are represented as spherical particles, although they may have any shape, whose surface is decorated by positive and negative charges. To balance the charge, an electrolyte is present in solution (in the figure this is represented by the positive and negative spheres outside the colloidal particle).

If the distance of separation between two spherical particles is small enough, the interaction between them corresponds to the interaction between two flat surfaces bearing an electrical charge. Figure 4.8 shows one such charged surface exposed to a fluid. The surface is positively charged.

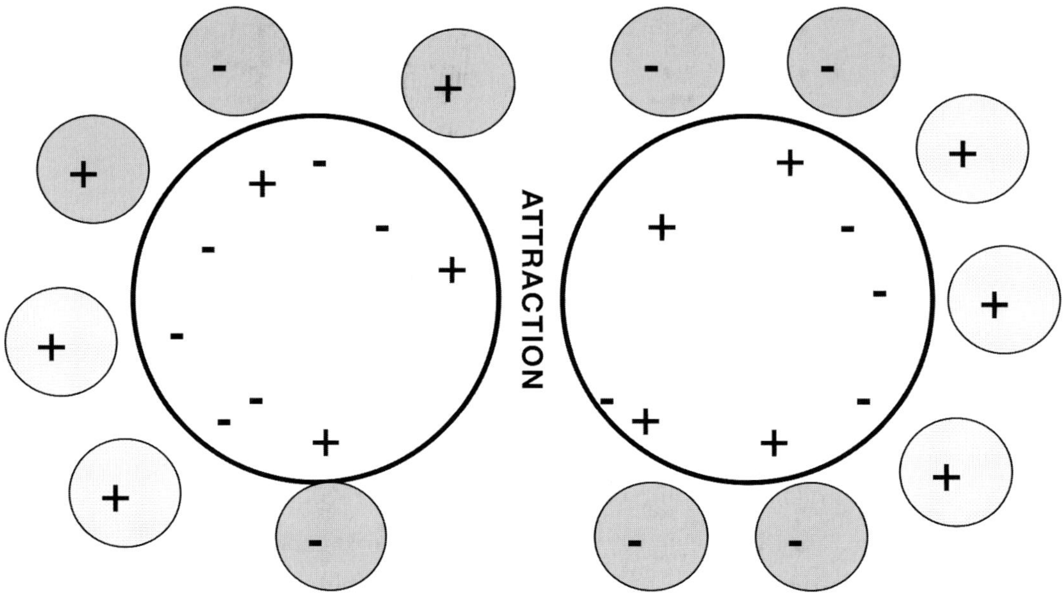

Figure 4.7. Illustration of a colloid. The small gray and white circles represent counterions.

There is a layer of relatively immobile ions of opposite sign next to the surface. The thickness of this layer, which was given the name Stern Layer, is approximately equal to the hydrated volume of the ions. Beyond the Stern Layer, added electrolytes form a diffuse layer of thickness $1/\kappa$. Beyond a distance equal to $1/\kappa$ the structure of the electrolyte corresponds to its bulk distribution. Below the cartoon, we have included a plot of the electrostatic potential as a function of the distance from the solid surface. Notice the discontinuity in the potential caused by the Stern Layer. The potential at the Stern layer is called the zeta-potential. If we neglect that discontinuity, the classical electrostatic potential, ψ_E, relation between two flat surfaces may be used,

$$\Psi_E = \Psi_{E0} \exp(-\kappa r) \qquad 4.1$$

in which

$$\frac{1}{\kappa} = \left(\frac{\varepsilon RT}{16\pi F^2 I} \right)^{1/2} \qquad 4.2$$

ε is the permittivity of the solvent, R is the ideal gas constant, T is the absolute temperature, F is the Faraday's constant and $1/\kappa$ is known as the Debye length. The ionic strength I is given by,

$$I = \frac{1}{2} \sum c_i z_i^2 \qquad 4.3$$

in which c_i is the concentration of ion i, and z_i its charge. Therefore, the thickness of the double layer decreases exponentially with the square root of the ionic strength. When the distance from the surface, r, is equal to $1/\kappa$ the potential is equal to $\psi_0/2.72$ and $1/\kappa$ is approximately equal to 1 nm.

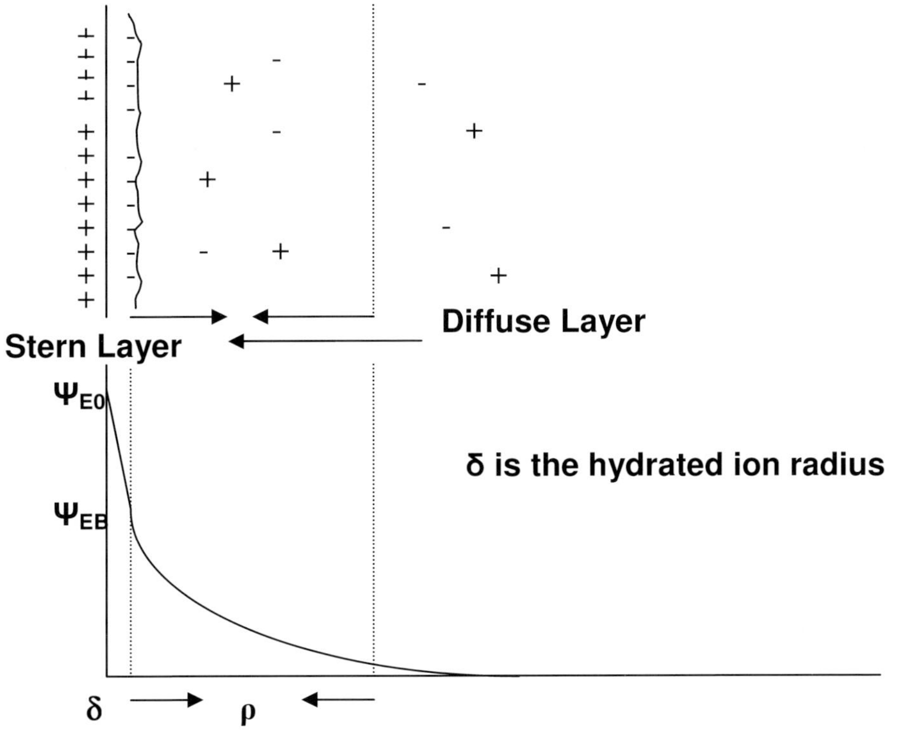

Figure 4.8. Upper panel: electrical double layer. Lower panel: variation in the electrostatic potential with the distance from the charged wall.

Derjaguin, Landau, Verwey, and Overbeek (DVLO) combined this result of electrostatic theory (by assuming that the electrostatic potential originated by a colloidal particle has the same functional form as of a flat wall) with an expression for dispersive forces (wrongly called Van der Waals forces) and developed a potential field for colloid/colloid interactions (DVLO potential). Figure 4.9 shows that potential as a function of distance. DVLO theory predicts that the secondary minimum is located at $r=4/\kappa$. The potential energy barrier represented by the maximum in the potential stabilizes the colloids. If the colloids are brought together at shorter distances the potential becomes quickly negative (attractive) and the colloidal suspension flocculates. This is equivalent to lower the magnitude of the maximum in the energy barrier, whose position and magnitude are a function of the amount of electrolyte added to the colloid.

This is illustrated in Figure 4.10, which shows that the effect of the added electrolyte is to decrease the thickness of the diffuse layer according to Equation 4.2. This is called double layer compression and its consequence is a decrease in the range and magnitude of the repulsive electrostatic potential. Therefore, attractive dispersion forces become dominant and the colloid flocculates.

It is illustrative to consider the value of the Debye Length ($1/\kappa$) for a typical electrolyte concentration. For example, for 0.01 M NaCl at 25°C the Debye length is 0.605 nm. On the other hand, for ammonium sulfate at the same concentration and temperature the Debye length is 0.349 nm. Therefore, the electrostatic double layer compresses as the electrolyte concentration increases. It is clear that an increase in ionic strength decreases the magnitude of the maximum and moves its position to smaller distances. DVLO theory can be used to predict the "critical flocculant concentra-

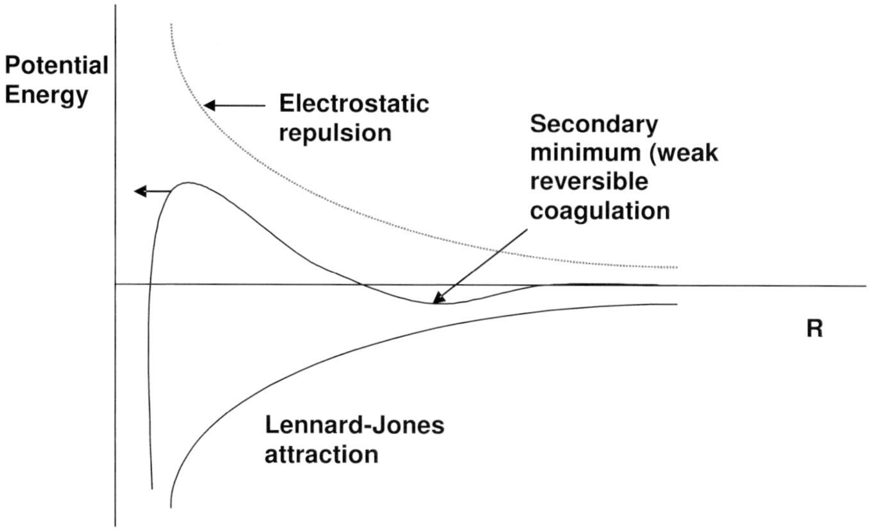

Figure 4.9. Potential energy function versus distance for a colloidal particle according to DVLO theory. The two components of the DVLO potential (electrostatic and dispersion forces) are also shown.

tion" (i.e., the minimum amount of neutral electrolyte needed to initiate coagulation of the colloid). It predicts that the critical flocculant concentration depends on $1/Z^6$. This correlation works relatively well with pure proteins. It, however, breaks down when dealing with protein mixtures as a result of the presence of other proteins indirectly affecting the solubility of the protein of interest.

Protein Denaturation

Denaturation is the loss of part or all of the protein three-dimensional organization because of environmental factors. As mentioned, there are numerous environments where a protein will denature, including exposure to chemical agents such as some organic solvents (methanol), inorganic salts (urea or guanidine hydrochloride), detergents, extreme pHs and temperatures, high pressures, high electrolyte concentrations, physical stresses such as phase changes (e.g., freeze drying and subsequent reconstitution), interfaces, and mechanical shear. Protein denaturation is a consequence of the disruption of intramolecular hydrogen, hydrophobic and/or ionic bonds that define the secondary, tertiary, or quaternary structure of the protein. When the protein is completely denatured, it becomes a random coil. If the denaturation is partial, the quaternary, ternary, or secondary structures may be lost but sometimes only subtle changes in the α-helix or β-sheet content may occur. The consequence of protein denaturation is a decrease in protein solubility and, often, loss of function. Sometimes the protein can be returned to its native state, but often it is rendered inactive.

Modifications in the primary structure are not normally considered denaturation. Even if biological activity is not affected, modification in the primary structure is still important from a processing viewpoint as it will affect the protein's solution behavior and our ability to purify it. Two common alterations of the primary structure are oxidation of cysteine side chains and de-amidation of asparagine (Asn) and glutamine (Gln) side chains (Manning, Patel, and Borchardt 1989). Both of

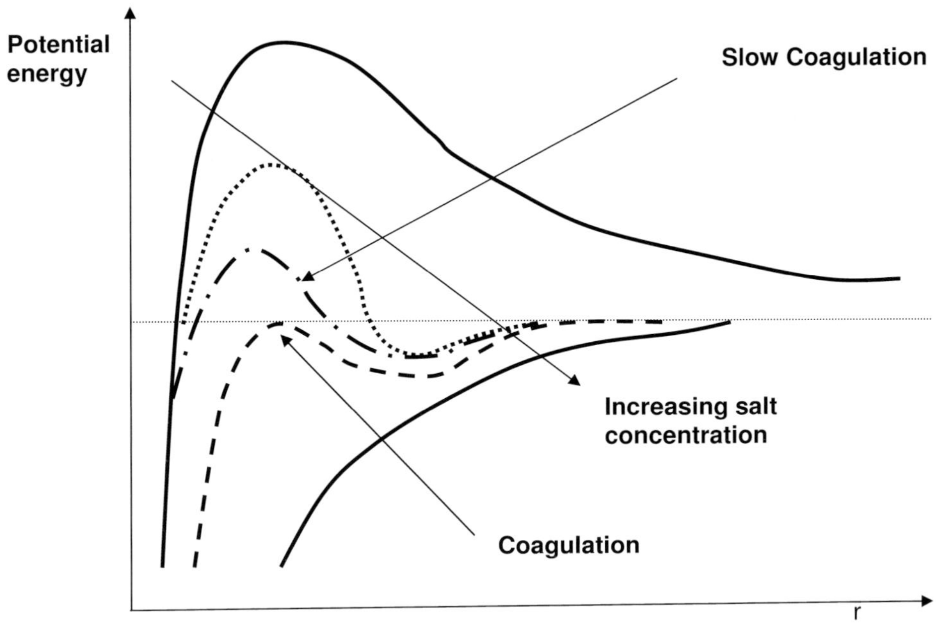

Figure 4.10. Change in the DVLO potential with increasing amounts of salts. When the maximum in the potential (energy barrier) is below zero, complete coagulation of the colloidal suspension occurs.

these alterations may happen during processing or during long-term storage. Less common is the splicing of main subunits destroying the integrity of the protein. Oxidation of cysteine side chains leads to the formation of irreversible aggregates. Some proteins are more prone than others are to oxidation. For example, proteins present in the eye lenses are so prone to oxidation that most of the separation steps leading to the isolation of one particular species are performed under reducing conditions by adding mercaptoethanol. Deamidation of Asn and Gln is a sequence-dependent process that changes the isoelectric point of the protein and it may affect its biological activity. For example, the sequence Asn-Gly is 100-fold more susceptible than the sequence Asn-Pro. De-amidation occurs through the formation of succinimide. This reaction is slow and is followed by the rapid conversion to α or β carbonyl groups. In general, Asn residues are more prone to de-amidation than Gln ones. Human growth hormone and insulin are two commercially important proteins that undergo de-amidation (Brange et al. 1992; Lewis, Cheever, and Hopkins 1970). For proteins that are stable in acidic environments (pH between 3 and 5), a common strategy to prevent de-amidation is to maintain the pH in the acidic range. Other strategies for those proteins that are not stable at these low pHs include having an environment at low ionic strength and low temperatures. Finally, buffer additives that lower the buffer's dielectric constant may help too.

Different experimental techniques, such as photon correlation spectroscopy (dynamic light scattering), ultra violet spectroscopy, circular dichroism, X-ray diffraction, NMR, infrared spectroscopy, differential scanning calorimetry, and capillary electrophoresis can be used to monitor protein denaturation. Photon correlation spectroscopy and ultraviolet spectroscopy are used only to detect the presence of aggregates and gross changes in the shape of protein respectively. Circular dichroism has been traditionally used to detect changes in the secondary structure of proteins by quantifying the amount of α-helix, β-sheet, and random coils. X-ray diffraction yields the complete three-

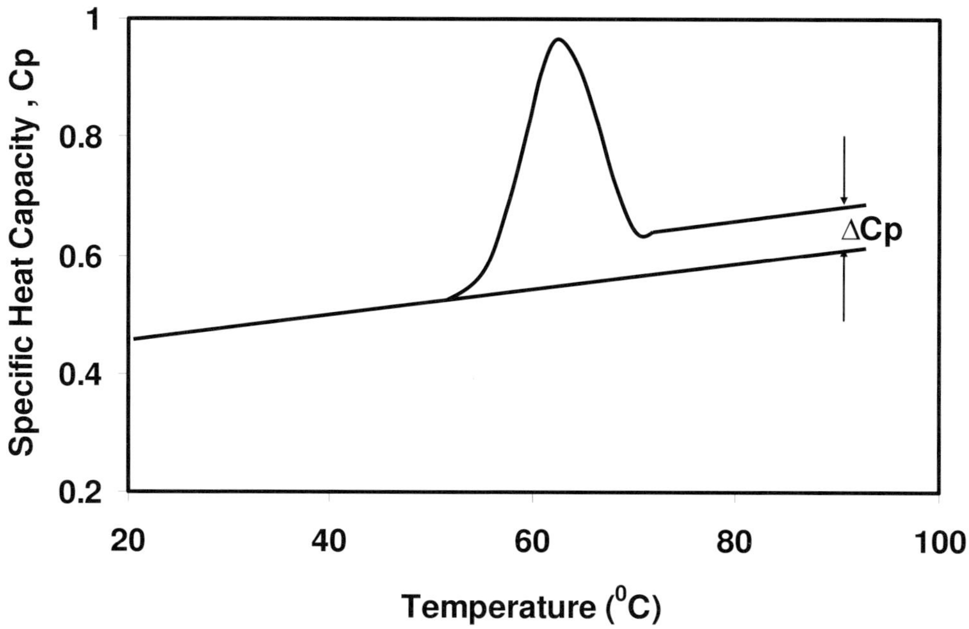

Figure 4.11. Heat capacity versus temperature for a protein solution. The heat capacity increases as denaturation of the protein takes place. ΔC_p represents the difference in heat capacities between the native and denatured protein.

dimensional structure of a protein but requires large crystals that are not always obtainable. It may be used during process development, but it is never used during production. NMR offers molecular details, but it is limited to the study of small proteins (< 20,000 Da). Fourier transform infrared spectroscopy (FT-IR) provides a quick and simple way of determining protein secondary structure. Finally, capillary electrophoresis in electrofocusing mode is a very sensitive method to detect deamidation of glutamine or asparagine side chains. Let's look at some of these techniques in some detail.

Because the denatured and native protein have different heat capacities, calorimetric studies are commonly used to monitor protein denaturation and to characterize the native and denatured states. The heat capacity of a protein measured in a range of temperatures shows a transition between the native and denatured states. This is illustrated in Figure 4.11, which shows the change in heat capacity as a protein solution is heated beyond the melting temperature—the temperature at which the protein becomes a random coil. The figure shows that the heat capacity of the denatured protein is slightly larger than the heat capacity of the native one and that the intermediate species have a much higher heat capacity than the native or denatured protein.

FT-IR (Smith 1996) provides a fingerprint of the protein molecule. This is possible because different functional groups in the polypeptide chain absorb energy at different frequencies depending on their environment. The most important peaks in the infrared region for proteins are the amide I, II, III, and A bands (Dousseau and Pezolet 1990). The amide I band (1690-1600 cm^{-1}) corresponds to C=O stretching, N-H bending, and C-N stretching. The amide II band (1575-1480 cm^{-1}) corresponds to N-H bending and C-N bending. The amide III band (1301-1229 cm^{-1}) corresponds to C-N bending, N-H bending, C=O bending, and O=C-N bending. Finally, the amide A band (3300 cm^{-1}) corresponds to N-H bending. The position of the amide I band depends on the participation of the carbonyl groups in hydrogen bonding. The shift in the position of the amide 1 peak as a result of hydrogen bonding has been confirmed by studies conducted with homopolypeptides and by theoretical calculations. The deconvolution of the amide I peak (i.e., the artificial generation of many peaks from the amide I band) helps to differentiate the different structures. β-sheets, anti-parallel β-sheets, α-helix, and disordered structures are shown by bands at 1620-1640 cm^{-1}, 1670-1695 cm^{-1}, 1648-1657 cm^{-1}, and 1644 cm^{-1} respectively. It is also worthy to note that β-sheets and β-turns adsorb in the same region and therefore, are very difficult to differentiate. The amide II band corresponds to vibrations of N-H groups and is used to study hydrogen/deuterium exchange. The exchange of hydrogen by deuterium moves the amide I and II bands to lower wavelengths.

Studies of the stability of human serum albumin (HSA) in methanol/heavy water solutions at room temperature with a varying concentration of methanol (10% w/w to 40% w/w) can be used to illustrate the use of FTIR to analyze the stability of a protein in solution. The spectra of the native protein show the characteristic amide I and II bands at 1650 and 1575 cm^{-1} respectively.

The amide I peak is deconvoluted into three peaks. The peak at 1653 cm^{-1} corresponds to α-helix structure, the peak at 1630 cm^{-1} to parallel β-sheets, and the peak at 1675 cm^{-1} corresponds to anti-parallel β-sheets. The absence of a peak at 1645 cm^{-1} suggests the absence of disordered structures. Although the extinction coefficients of each peak were not the same, it is possible to conclude that the secondary structure of HSA is dominated by α-helixes-in agreement with crystallographic results. Upon incubation of HSA in 40% methanol for about 30 minutes, the protein precipitates. The precipitate can be mixed with KBr, centrifuged into a pellet, and inspected on the infrared light (IR). The deconvoluted spectrum clearly shows a peak at 1645 cm^{-1}, which corresponds to a disorganized structure.

Ultraviolet (UV) spectroscopy is a cheap and fast method for quantifying aromatic residues (tyrosine, tryptophan, and phenylalanine are chromophores in UV spectroscopy) within proteins and therefore, to determine protein concentration. Monitoring the conformation of a protein is another use of UV spectroscopy that sometimes is neglected. This is possible because the UV spectrum depends on the microenvironment surrounding the chromophores. As the protein denatures, hydrophobic groups are exposed to the solvent. The low resolution of UV spectroscopy is countered by the advantage the technique possesses of being less time-consuming and cheaper than others. Therefore, this method is particularly useful if the goal is not to determine the protein structure, but to monitor conformational changes and possible denaturation in the protein upon changes in composition of the protein solution.

The determination of the degree of exposure of aromatic residues as a result of solvent composition is complicated with ordinary UV spectroscopy because all three chromophores absorb in the same spectral regions with overlapping bands. Hence, contributions from the various chromophores are significantly overlapped in near UV spectra. Derivative spectroscopy is a powerful technique to deconvolute the individual chromophores of a protein from the original spectrum (Balestrieri et al. 1978; Havel 1996). Because of the increased resolution and precision of derivative spectroscopy, even partial denaturation and minor changes in the structure of the protein can be detected. In addition to increasing resolution, another advantage of derivative spectra is that they are unaffected by light scattering that arises from protein aggregation. The broad, featureless background absorption because of scattering is effectively eliminated when the derivative spectrum is calculated. The spectrum is also insensitive to the slowly changing absorption profiles in the 320- to 240-nm region associated with disulfide absorption and light scattering.

Even-order spectra have an advantage over odd-order spectra in that the extrema of the original spectrum will occur at the same position for even-order spectra. Peaks for the zero-order spectrum become minima for second-order spectrum, peaks for the fourth-order, minima for sixth-order spectra, and so on. However, second-order derivative spectroscopy is preferred because, as each additional derivative is calculated, each individual band will exhibit additional negative and positive bands on either side of the central band. These additional bands can interfere with peaks as a result of adjacent principal bands and complicate subsequent analysis.

Because of both the optical activity of the amino acids and the conformation of the polypeptide chain, protein solutions rotate polarized light. As the protein denatures, the rotation becomes more negative. The optical rotation of the protein solution can be measured at various wavelengths and the spectra can be compared with known model compounds. Poly(lysine) has been used for this purpose because it changes its conformation between β-sheet, α-helix, and random coils as the pH is varied. This method known as optical rotatory dispersion is very sensitive to the content of α-helices and β-sheets. Sometimes the results are in very good agreement with crystallographic data, but other times the results can be quite poor. Instead of measuring the optical rotation, one may measure the difference in adsorption of left and right circularly polarized light. The resulting spectra is called the circular dichroism spectra (Greenfield 1996; Bloemendal and Johnson 1995). This spectrum is very sensitive to conformational changes of the protein in the wavelength range of 190 to 300 nm.

Protein Folding

One of the most intriguing and challenging questions faced by scientists today is why proteins fold. To appreciate the magnitude of the challenge, consider the folding of an average size protein consisting of 100 residues (amino acids). For the sake of simplicity, one may ignore the additional conformations that can be generated by the degrees of freedom of the side chains. Thus, all the possible combinations of solid angles ϕ and ψ yields 10^{100} possible conformations. The average time to explore all conformations would be 10^{77} years (assuming 10^{-13} sec to convert from one to the other). However, proteins fold in 0.1 to 1000 seconds! Therefore, the folding process cannot be a stochastic one. The cells must use smart biases to produce these marvels in a reasonable amount of time, and one may think that such an architectural masterpiece would logically be built to resist the passage of time. One would be very wrong in following that line of thought. The following paragraph briefly analyzes the equilibrium and kinetic aspects of protein folding.

From a strictly thermodynamics point of view, the native protein is only marginally stable. Although the enthalpies and entropies of the folded and unfolded proteins are relatively large, their difference is quite small (a few kilocalories per mol). The difference in Gibbs free energy of the folded and unfolded states is a strong function of temperature. This is illustrated in Figure 4.12, which shows a plot of Gibbs free energy difference for small protein.

Protein folding in vivo is catalyzed in some bacteria by enzymes called chaperones, which bind intermediates along the folding pathway, thereby avoiding protein association and probably decreasing the number of conformations explored. Similarly, hydrophobic probes and PEG reduce the formation of protein aggregates during in vitro protein refolding by the same mechanism. The consensus is that unstable intermediates fold slowly and then they quickly explore the final conformation. This idea has developed into the recognition of the existence of a "super" intermediate called the molten globule. The molten globule has some features that are similar to the native protein and others that are similar to the completely denatured one. For example, the molten globule dimensions are less than a random coil and slightly greater than the dimensions of the fully folded protein; however, the secondary structure of the molten globule is similar to that of the completely folded protein. Also, the hydrogen exchange rate of amide groups is faster than in the folded case but slower than the fully unfolded case. The enthalpy of the molten globule is similar to the fully unfolded case. The transition from molten globule to fully unfolded protein is rapid and noncooperative. Therefore, we can represent this transition between unfolded and completely folded protein by,

$$U_s \Leftrightarrow U_F \Leftrightarrow N \qquad 4.4$$

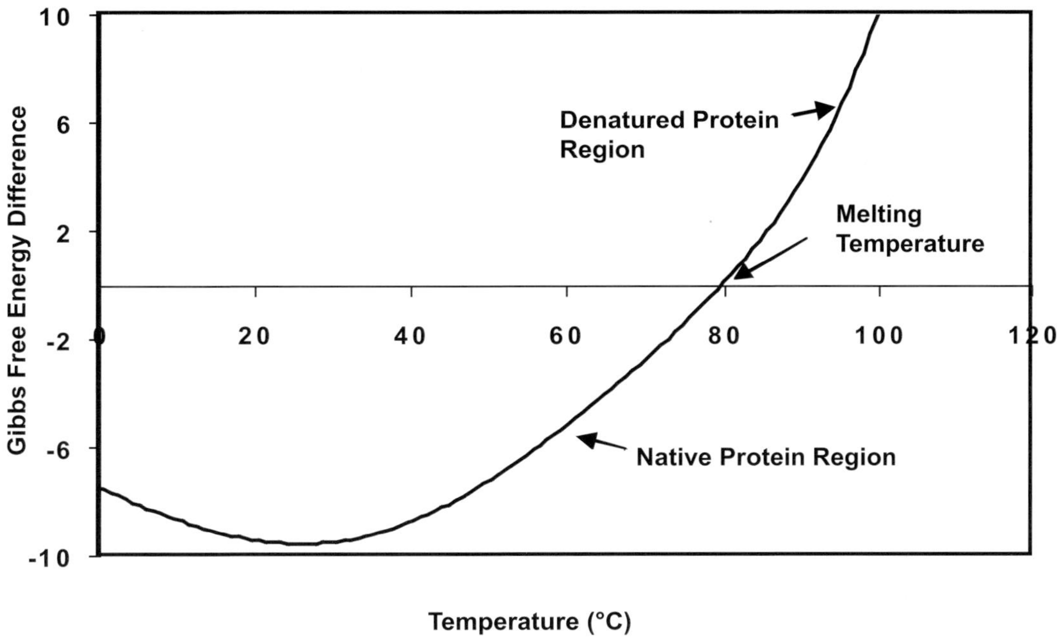

Figure 4.12. Gibbs free energy difference (G_D-G_N) between the denatured and native state for a small protein.

the first reaction is slow, whereas the second is fast. The intermediate has all the right peptide bonds and it is heterogeneous. The heterogeneity arises from cis/trans isomerization of peptide bonds preceding the proline (Pro) residue. An example of this is found in the folding/refolding studies of ribonuclease, which is "cis" in two of the Pro residues. It has three kinetics phases: 15% of molecules fold fast, 65% refold slower, and 20% refold even slower (Creighton 1993).

Other folding pathways have been proposed with different degrees of success. Although it is risky to attempt to reach a unique conclusion, it is reasonable to say that the folding pathway goes through one or more intermediates. Furthermore, it is reasonable to say that not all of them necessarily converge to the final product. Actually, errors in folding are common, but most of the time they are destroyed by proteases—the cell's "garbage family" protein family. Any errors, which happen to survive, may cause severe damage to the organism.

Protein Aggregation

Proteins may lose their biological activity because they are denatured during processing. They can also lose their potency, because both native and denatured proteins may form irreversible aggregates (Cromwell, Hilario, and Jacobson 2006). Aggregation of native proteins is common upon relatively long storage periods. For example, it is a well-known fact that small amounts of dimers and oligomers form when antibodies are stored. Aggregation of native proteins may also occur during their freeze drying, a common polishing step in any bioseparation process. Aggregation of denatured proteins is extremely common and is of particular importance in the recovery of genetically engineered proteins present as inclusion bodies in the cell cytoplasm. Homogenous aggregation can be divided into two groups: insolubility and true aggregation. Insolubility typically results in the formation of precipitates with native protein structures and is reversible upon dilution. For example, ammonium sulfate reduces the solubility and causes native protein precipitation. In true aggregation, we can find protein aggregates with no secondary structure, inclusion bodies, and protein aggregates with secondary structure, for example, non-native β-sheet structure. True aggregation is different than insolubility in that it is irreversible upon dilution. In protein aggregates, there is a specific association of partially folded subunits of the polypeptide that associate preferentially in an intermolecular fashion as opposed to intramolecular association that leads to the native conformation. Inclusion bodies are dense aggregates of protein built up within the cell. The inclusion bodies are solubilized using high concentrations of denaturants and are then diluted to allow the proteins to refold. During the refolding process aggregation occurs, decreasing the recovery of the active protein.

Protein aggregation is effectively a side product of protein folding, because as a protein folds toward its native structure it can also go through alternative folding pathways forming intermediates that aggregate. As in the folding process, the aggregation process is strongly dependent on temperature, pH and salt type, and concentration.

The existence of a molten globule intermediate along the folding pathway may explain the aggregation of proteins and the ability of polyethylene glycol (PEG) and other hydrophobic probes to prevent aggregation. Molten globules are partially folded proteins with large hydrophobic patches that are exposed to the solvent. They are stable at acidic pH, high temperature, and intermediate concentrations of strong denaturants. Molten globules either can evolve toward the native protein or can associate with other molten globules through hydrophobic interactions, thus leading to the formation of aggregates. Hydrophobic probes and PEG may prevent the self-association of molten globules by interacting with them through hydrophobic interactions and thus leading to the native protein.

Two hypotheses have been invoked to explain aggregation of denatured proteins. The first one is that a protein aggregate forms once the protein has completely unfolded and the other, now shown in recent research and thus more accepted, is that aggregation involves interactions between partially unfolded molecules. Specifically, it has been proposed that aggregation occurs in the transition state from folded to unfolded protein and arises from interactions of non-native partially folded conformations, which are driven by interactions between solvent and exposed hydrophobic residues that are normally buried within a protein's interior. The Lumry-Erying model has been used to describe the aggregation mechanism as a result of denaturation. However, this model is only useful for cases in which aggregation occurs from a completely denatured state. The Lumry-Erying model has been recently modified to account for aggregation from intermediates (Kendrick et al. 1998),

$$N \leftrightarrow N^* \rightarrow A \tag{4.5}$$

$$A_m + A \rightarrow A_{m+1} \tag{4.6}$$

N is the native state of the protein, N* is a partially unfolded or a transition state of the protein and is in equilibrium with N, A is the aggregate competent species, and A_m is the aggregated form of the protein. The irreversible reaction of N* to A is the rate-limiting step in the formation of aggregates and, therefore, the aggregation phenomenon becomes a first order process. The aggregate species, A, might be very dilute in solution; therefore, it may not be detected until it grows.

In the biotechnology industry, protein aggregation is encountered routinely during purification, refolding, sterilization, shipping, and storage processes because of the presence of chemical and physical stresses. The presence of aggregates in a protein formulation causes a decrease in the efficiency of the drug, immunogenic reactions, and potential problems in its intravenous administration. Even though the protein preparation may be aggregate free after the last polishing step has been completed, aggregates can form during storage. Aggregation levels as low as 1% over a two-year shelf life can render a product clinically unacceptable (Kendrick et al. 1998). Therefore, monitoring aggregation is an essential quality control practice for any protein-based pharmaceutical. Beyond its implications during processing, protein aggregation is linked to various devastating diseases. Because of their social impact and because of the research opportunities aimed at developing inhibitors of protein aggregation in vivo, the last portion of this section covers this topic.

Aggregation may take place in the absence of interfaces (homogeneous) or in the presence of them (heterogeneous). The interface may be solid/liquid (e.g., the boundary between the surface of a membrane and the bulk fluid), liquid/liquid (e.g., the liquid-liquid boundary observed when two liquid phases are at thermodynamic equilibrium) or liquid/gas (e.g., the interface between gas bubbles and bulk liquid created by stirring a solution). A seemingly harmless interface may subject a protein to extremely large pressures. For example, the air/water interfacial tension is about 1,000 atm. As a protein adsorbs at an interface it must adjust its conformation (unfold) to balance these large tensions. The unfolded protein may desorb and initiate aggregation in the bulk liquid. Shear induced aggregation, which is the result of an increase in interfacial tension because of mixing, is another manifestation of these high pressures. Another heterogeneous mechanism occurs during nucleation. As a precipitant agent is added to a solution, the solution becomes supersaturated and the interaction between molecules is no longer repulsive but slightly attractive. This will cause them to form small clusters. These small clusters begin to arrange themselves in such a way that they are in their most favorable energy state, the crystalline state, and hence a crystal is born.

Inclusion Bodies

The expression of eukaryotic proteins in a foreign host is often accompanied by the formation of aggregates or inclusion bodies of inactive protein that accumulate within the cell. Often this is the result of over-expression of the protein. Although inclusion bodies have been found mainly in *Escherichia coli,* they have also been found in fungi such as *Saccharomyces cerevisiae*, other bacteria, and even in insect cells.

Inclusion bodies are roughly spherical amorphous aggregates of protein that are visible by light and electron microscopy. Typical sizes and densities of inclusion bodies vary from 0.2 to 2 μm in diameter and from 1.1 to 1.3 g/L, respectively. It is generally accepted that these protein aggregates are held together by hydrophobic, ionic, and covalent (disulfide bonds) forces. Because the addition of water-structure-forming salts does not decrease the formation of aggregates, we may conclude that ionic forces are not the primary forces responsible for holding the proteins together in the aggregates. Studies performed on the temperature dependence of protein association (the formation of aggregates) suggest that hydrophobic forces play a dominant rôle in maintaining the proteins in aggregates. This last observation also suggests that the aggregates can be dissolved by chaotropic salts (i.e., salts that disrupt water structure). Indeed, inclusion bodies are solubilized by chaotropic agents such as urea, guanidine hydrochloride (GuClH), detergents, and chaotropic salts. Although urea and GuClH are the most common denaturing agents used, inorganic chaotropic salts are also effective in solubilizing protein aggregates. For example, sodium thiocyanate has been used to solubilize aggregates of tryptophan synthase and carbonic anhydrase (Forciniti 1994). If the aggregate is held together by covalent bonds such as disulfide bonds, it is necessary to add dithiotheitol or β-mercaptoethanol. The addition of thiol reagents can be avoided—although intermolecular and intramolecular disulfide bonds may exist in the inclusion bodies—by solubilizing the aggregates at high temperatures.

The formation of inclusion bodies provides a convenient and clean alternative for protein purification, *if* the active protein can be recovered from the protein aggregates. Under ordinary circumstances, the inclusion bodies formed by the eukaryotic protein contain almost no impurities. Thus, by separating the inclusion bodies from the cell debris by centrifugation or ultrafiltration, it is possible to obtain a highly pure protein in a single purification step. After the inclusion bodies have been isolated, they are suspended in a buffer and exposed to denaturing conditions to solubilize the protein. The denaturing agent is then diluted (either by dialysis or by adding an appropriate buffer) to refold the protein. Unfortunately, during the refolding process the proteins re-aggregate. Thus, only a fraction of the active protein is recovered. The refolding efficiency varies from 94% for human serum albumin to a few percent for prorenin.

Several routes have been explored to prevent the re-aggregation of proteins and thus improve the recovery of the active protein. It has been found that by refolding protein at very low protein concentrations or in the presence of large amounts of denaturant, the recovery of native protein increases. These procedures have two disadvantages: (1) the refolding of the protein occurs at a very low rate, and (2) the refolded protein is obtained at very low concentrations. An alternative is to perform the refolding in a chromatography column (gel permeation, ion exchange, or hydrophobic interaction) by using the appropriate gradient of denaturant in the elution buffer (Lai and Middelberg 2002; Jungbauer, Kaar, and Schlegl 2004) and in the presence of various additives. Inclusion of a hydrophobic probe (e.g., a polypeptide) in the refolding mixture to enhance protein recovery has been investigated. The hydrophobic probe reduces the number of aggregates by interacting with some intermediates (e.g., molten globules) in the folding reaction. Unfortunately, hydrophobic probes are specific for each particular protein; therefore, a more general technique is highly desirable. Other chemicals normally used to improve refolding are EDTA, oxidized glutathione, glutathione, sodium octanoate, and sodium palmitate. Some of these compounds—such as glutathione, oxidized glutathione, or disulfide isomerase—are added to obtain the correct pairs of cysteines. Finally, a series of "hydrophobic" polymers including PEG (Cleland and Wang 1993; Forciniti 1994), Ficoll, dextran (Du et al. 2006), PEO-PPO-PEO (Lee et al. 2006), and PVP (Jiang, Yan, and Zhou 2006) have been used either in batch or as an additive to hydrophobic interaction or gel permeation columns. These polymers act like promiscuous chaperons during the refolding of the protein, minimizing the formation of aggregates and enhancing protein recovery. The inclusion of these probes into chromatography buffers has eliminated the inherent problem of the refolding technique that was the production of a very dilute protein solution contaminated by large amounts of denaturant.

Pathologic Aggregates

Amyloid fibrils are a very special type of protein aggregate with enormous implications to humankind. Amyloid fibrils are proteinaceous deposits associated with a variety of human diseases, such as Alzheimer disease (AD), islet amyloidosis, type II diabetes, systemic amyloidosis, transmissible spongiform encephalopathy, Parkinson disease (PD), familial amyloid polyneuropathy, and so on. The impact of amyloid fibrils on human health is yet to be fully recognized. For example, it has been suggested that amyloid deposits, specifically AD ones, may play a role in age-related macular degeneration of retinas. Amyloid diseases are apparently unconnected; however, amyloid deposits share as a common characteristic the presence of secondary structure (mostly β sheets) in the aggregates. Although the proteins involved in the different types of fibers lack homology, the morphology of the deposits is remarkably similar. Therefore, it is believed that the formation of amyloid fibrils is a universal phenomenon. Still, there are some reports indicating that some peptide chain homologies are necessary for the formation of fibrils.

Amyloid deposits associated with AD are well studied. Amyloid β protein residue 1-42 is the most abundant peptide found in these deposits. This peptide is the result of the cleavage, first by β and then by γ-secretases, of a larger transmembrane protein known as amyloid protein precursor or APP. It has been argued that posttranslational processing (glycosylation) of the precursor determines the production of the amyloid peptide. AD plaque contains advanced glycosylation end products, AGE (AGE is formed by nonenzymatic addition of glucose to protein amino groups, which crosslink the protein deposits increasing their toxicity), cooper iron, Aβ 1-40, τ protein, neuronal thread proteins, apolipoprotein E, glycosylated acetylcholinesterase, serum amyloid P, and so on. For a recent review, see MacRaild et al. (2004). It has been suggested that some of these nonfibrillar components enhance local densities of fibril entanglements.

Detection of Aggregates

Several techniques can be used to detect protein aggregation: light scattering, infrared spectroscopy, circular dichroism spectroscopy, AFM, size exclusion chromatography, X-ray scattering, and centrifugal sedimentation. The structure of protein aggregates (not native) is universally given by intermolecular β-sheets that can be characterized by IR and CD spectroscopy. In IR, β-sheets produce two distinct bands at 1615 and 1695 cm^{-1}.

The most common method to detect aggregates is centrifugal sedimentation. A simple force balance for a centrifuge (for details, see Chapter 5) is,

$$\frac{M(1-\bar{v}\rho)}{Nf} = \frac{v}{\omega^2 r} \equiv s \qquad 4.7$$

in which M is the molecular weight, $\bar{v}$ is the specific volume of the protein, ρ is the density of the solvent, v is the sedimentation velocity, f is the frictional coefficient that depends on the shape and size of the particle, ω is the angular velocity, r is the distance of the particle from the axis of notation, and s is the sedimentation coefficient. Sedimentation coefficients vary from 1 to 100 x 10^{-13} seconds (one Svedberg unit is equal to 10^{-13} seconds). The protein solution is placed in a specially designed centrifuge tube (cell) and the spinning is started. As the protein begins to accumulate at the bottom of the cell, it begins diffusing from the bottom up because of the concentration gradient. At one point (normally in the order of several hours) equilibrium is reached. The key measurement is the concentration at different distances from the center of rotation at any given time. Several methods are available to do just that, including Schlieren optics, interference, photographic absorbance, and photoelectric absorbance. Schlieren optics is less sensitive than interference optics; however, it may be used for proteins at concentrations between 1 and 50 g/L. Interference optics is very accurate but they are restrictive to concentrations between 0.1 to 5 g/L. Newer absorption optical systems have high reproducibility, excellent sensitivity, and operate in a wide wavelength range. An example of such a system is the Optima XL-A developed by Beckman-Coulter (Fullerton, CA), which uses a high-intensity xenon flash lamp that covers wavelengths from 190 to 800 nm. The system can detect protein concentrations at 230 nm as low as several μg/mL.

The two basic types of experiments are sedimentation velocity and sedimentation equilibrium. In a sedimentation velocity experiment, a uniform solution is placed in the cell and spun. The rapid sedimentation of the solute creates a depletion zone near the meniscus and a sharp boundary between the depleted region and the region of uniform concentration. The rate of movement of the boundary is used to calculate the sedimentation coefficient. Measurement of the spreading of the boundary is used to determine the diffusion coefficient. Then, the molecular weight is determined by,

$$M = \frac{s^0 RT}{D^0(1-\bar{v}\rho)} \qquad 4.8$$

in which D^0 is the diffusion coefficient at infinite dilution. The presence of various boundaries is taken as an indication of aggregation. In a sedimentation equilibrium experiment the sample is spun for several hours, until a concentration profile is developed along the cell. Measurement of the concentration profile leads to the determination of the molecular weight. For a single size,

$$M = \frac{2RT}{(1-\bar{v}\rho)\omega^2} x \frac{d^2(\ln c)}{dr^2} \tag{4.9}$$

Sedimentation equilibrium experiments yields more accurate results than sedimentation velocity ones. Sedimentation equilibrium may be used to determine the number-average molecular weight and the weight average molecular weight. For a homogeneous sample, the ratio between M_w and M_n is equal to one. The greater the heterogeneity, the greater the ratio between M_w and M_n. Readers familiar with polymer science will note that this ratio is a measure of the polydispersity of the sample. One of the potential problems in sedimentation experiments is that the high centrifugal force may disrupt aggregates.

Another method that may be used for the rapid monitoring of protein aggregates is dynamic light scattering, DLS. The technique was described in Chapter 3 as one of the means of determining diffusion coefficients. Although there are some publications where DLS was used for the detection of aggregates, it is not commonly used. This is because of some ambiguity in the interpretation of the results. Examples can be found in the works of Rosenqvist, Joessang, and Feder (1987) who used DLS to study the aggregation kinetics of human IgG, Joessang, Feder, and Rosenqvist (1985) who used DLS to study the kinetics of heat-induced aggregation of human IgG, and Bermudez and Forciniti (2004) who used DLS in combination with capillary electrophoresis to study the unfolding and aggregation of human IgG.

As stated previously, the quantity measured in a dynamic light scattering experiment is the autocorrelation function. When the sample is polydisperse—true for a sample containing protein aggregates—the autocorrelation function can be deconvoluted into a family of exponential functions, each corresponding to a different size. Therefore, even minute amounts of aggregates can be detected because large scatters dominate the scattering of light. However, the inexperienced operator may confuse the presence of dust with aggregates. In the experimental section of this chapter, there is a simple recipe that can be used for the detection of aggregates in a protein preparation.

Nucleic Acids

Nucleotides are the building blocks of nucleic acids. Their structure consists of a phosphate group linked to a pentose, which is then linked to a ring structure called a nucleic base. The bases adenine, guanine, and cytosine are found in both DNA and RNA, uracil is found only in RNA and thymine is only found in DNA. Figure 4.13 shows two nucleotides joined by a di-phosphate bond.

DNA and RNA are both linear polymers of nucleotides; however, DNA is much larger than RNA. RNA size varies from less than one hundred to many thousands, whereas cellular DNAs are several hundred million nucleotides in length. The backbone of the acid (to make a similarity with proteins) is repeated units of phosphate-pentose and the base makes the "side chain." A phosphodiester bond that joins a 5′ carbon of one pentose with the 3′ carbon of the other joins nucleotides together.

DNA consists of two anti-parallel polynucleotide strands that wind together to form a double helix. The bases stack into the interior of the helix. The strands are perfectly aligned by base pairs; adenine pairs with thymine by two hydrogen bonds, guanine is paired with cytosine through three hydrogen bonds. A reader should note that the difference in the number of hydrogen bonds between these base pairs will yield different melting temperatures (i.e., the temperature at which the two strands fall apart) of differently composed DNA double helix; thus, giving us a way to characterize the composition of DNA molecules simply. There are four different three-dimensional structures for

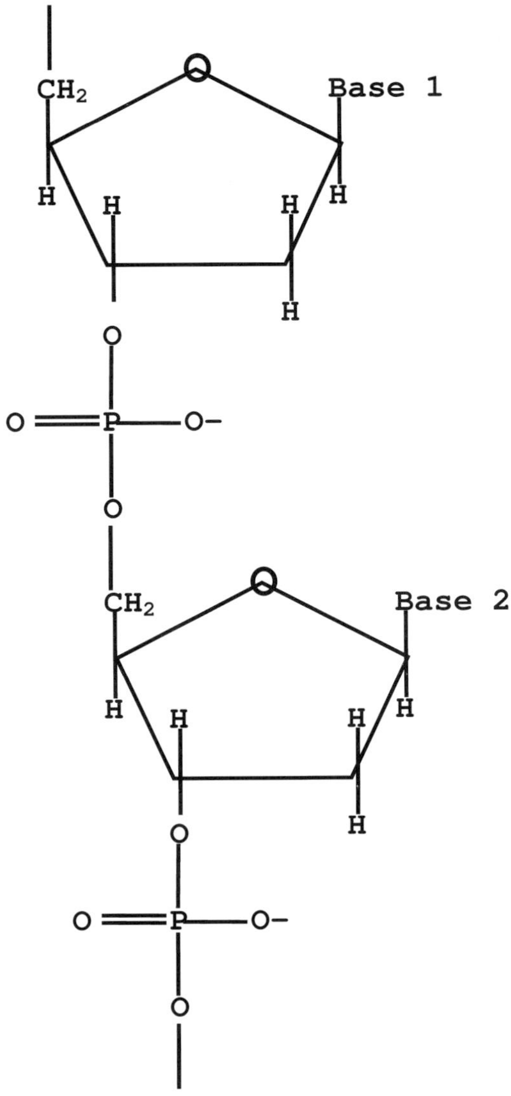

Figure 4.13. Chemical structure of nucleic acids.

DNA, the B-form, the A-form, and the Z-form. The B-form is the normal form of DNA. It consists of a right-handed helix in which the stacked bases are spaced 0.36 nm apart. The helix makes a complete turn every 3.6 nm and it has 10.5 pairs per turn. The A-form is obtained when the B-form is exposed to low humidity. This form is more compact and it has 11 tilted base pairs per turn. The Z-form is a left-handed helix. Finally, a triple-stranded DNA can be obtained in the laboratory by mixing poly(A) and poly(U) polymers. As stated previously, the two strands of DNA can be separated by increasing the temperature above the melting temperature. DNA can be also denatured by common agents such as urea and extreme pHs. When the native conditions are restored, the single strands wind together to recover the original double-stranded molecule.

The two differences in the primary structure between RNA and DNA are that the sugar ribose has

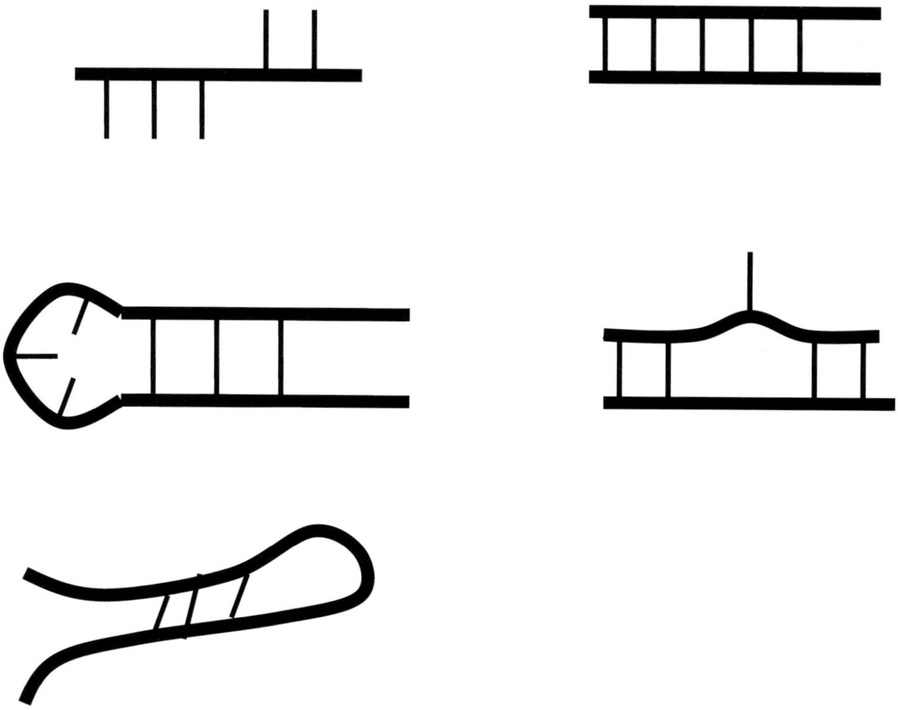

Figure 4.14. Several ternary structures of RNA. The thick solid lines represent phosphate–pentose repeating units and the thin solid lines represent bases.

a hydroxyl group at the 2' position and the thymine in DNA is replaced by uracil. RNA is less stable under alkaline solutions than DNA because of the modification in the sugar. There are several types of RNA. The most important ones in eukaryotes and prokaryotes cells are transfer RNA (t-RNA), messenger RNA (mRNA), and ribosomal RNA (rRNA). Other types of RNA include snRNA (small nuclear RNA), which is used, for example, in the splicing of pre-mRNA; snoRNA (small nucleolar RNA), which is used to process and chemically modify rRNA. In addition, viruses use RNA as a genetic material, which has motivated researchers to postulate that RNA-based life existed before DNA-based life. RNA makes up a few percent of a cell's dry weight. The prevalent species is rRNA, whereas mRNA makes up 3% to 6% of the total RNA.

The most common and simple secondary structure of RNA is the hairpin, which are formed by paring complementary bases separated 5 to 10 nucleotides from each other. Another common structure is the stem loop, where the bases that are paired are separated by more than 10 nucleotides and up to several hundreds. Figure 4.14 includes these and other commonly secondary structures found. Tertiary structure is found in the three most common RNAs. For example, tRNA has a characteristic three-dimensional structure that is essential for protein synthesis and rRNA shows well-defined, but local, tertiary structure.

Some RNAs have catalytic activity (they function as enzymes) and are called ribozymes. Some ribozymes catalyze self-splicing in which internal nucleotides of the RNA are removed and the two chains are joined again. The fact that RNA may have catalytic functions has reinforced the hypothesis of a RNA-based life.

Although most RNA molecules are single-stranded, double-stranded RNA is an intermediate in the replication of many viruses. This type of RNA is readily recognized by the infected cell, which degrades it into several small fragments (21-25 residues each). These fragments bind to complementary sequences of RNA present in the host cell and immediately degrade the host cell's RNA. This is called RNA interference and has been used by researchers to inhibit the expression of specific genes. Besides, double-stranded RNA induces the synthesis of α and β interferon. The binding of the interferons to the cell surface receptors leads to the activation of ribonuclease, which degrades ssRNA, and inhibits protein synthesis in the host. This sequence of events leads to the destruction of the virus (in the best case scenario).

Laboratory Experiments

This section describes a few experiments to determine critical properties of proteins. There is more than one technique that can be used to determine these properties and the choices presented here are not the only options available. For example, although molecular weights can be determined by sedimentation studies, gel electrophoresis, or static light scattering, only the last one is described.

Determination of Molecular Weights

The determination of molecular weights using static light scattering is quite straightforward and even an inexperienced operator can do routine analysis. The most important concern is in the preparation of the sample that must be "dust" free. The presence of minuscule quantities of dust will ruin a measurement because large particles, like dust, will dominate the scattering pattern.

Equipment Needed

There are two options for this. One possibility is to have a multiple angle laser detector (Wyatt Instruments, Santa Barbara) connected to an HPLC system, or to do batch measurements using a laser equipped with a goniometer or a multiple angle detector self standing. The protocol described below applies only to the first option. It is important to place an on-line filter before feeding the buffer to the column and carefully degas all buffers. A typical setup consists of an HPLC pump (isocratic) that feeds a GPC column. The eluent from the column is sent to the laser and from the laser to a UV or a refractive index detector. Although some recommend the use of an RI detector, our experience indicates that a UV detector works equally well.

Protocol

1. Choose a buffer compatible with your protein (the protein must be soluble and stable).
2. Filter the buffer first through a 0.45-μm filter and then through a 0.02-μm filter. For small amounts, a syringe filter will be sufficient. For large quantities use dedicated devices sold by chemical products suppliers.
3. Stabilize the column slowly. Sudden changes in pressure will shred the gel and debris will ruin the laser measurement. Flow rates of 1 mL/min are good enough for most samples.
4. Dissolve the protein at the highest desired concentration. It is recommended to do a couple of preliminary measurements to determine the appropriate concentration range.
5. Inject the sample into the port of the sample pump of your HPLC system.
6. Monitor the intensity scattered at various angles, θ (8 to 10 angles is more than sufficient).
7. Make a plot of the excess Raleigh ratio versus the $\sin^2(\theta/2)+kc$ for each concentration c (Zimm plots). The double extrapolation at zero angle and concentration yields 1/molecular weight. The slope of the line at zero angle yields the radius of gyration and the slope of the line at zero concentration yields the second virial coefficient.

Determination of Isoelectric Point

The easiest way to determine isoelectric points of a protein is by using isoelectrofocusing. This can be done in gels or in capillaries.

Isoelectric focusing separates proteins according to their isoelectric points. The separation is done in a non-sieving medium in the presence of carrier ampholytes, which establishes a pH gradient that increases from the anode to the cathode. Because a protein contains both positive and negative charge-bearing groups, the net charge of the protein will vary as a function of pH. The protein will migrate to a point within the gradient where its net charge is zero.

Equipment

A power supply and a electrophoresis cell. The protocol described below applies to the Mini-Cell (BioRad).

Materials

Acetic acid, Coomassie Blue G-250, Perchloric acid, isoelectric point markers, protein sample.

Protocol

1. Place the sample template on top of the polymerized gel. Apply ~ 1 μL of sample and of the standards. Let it rest for 5 minutes. Remove the template. Repeat the sample application at the other end of the gel.
2. Slide the lid of the cell toward the electrode plugs to remove it. Remove the graphite electrodes from the cell and rinse them with distilled water. Place them back. Lightly moisten the electrodes with water. Turn the gel with the adsorbed samples upside down and place it directly on top of the electrodes. You do not have a second chance to sit the plate correctly! Slide back the lid. Plug the power cables to the power supply. Use the following voltage program:
   a. Focus at 100 V for 15 minutes
   b. Increase voltage to 200 V for 15 minutes
   c. Increase the voltage to 450 V for an additional 60 minutes
   After focusing, turn off the power supply and unplug the power cables. Slide the lid and remove the gel. Separate the gel/gel support film from the plate. Remove the electrodes and clean them with water and a tissue. Rinse the chamber with distilled water.
3. Prepare a 0.025% Coomassie Blue G-250 solution in 3.5% Perchloric acid to stain the gel. A 7% acetic acid solution is needed to intensify and fix the stains. Immerse the gel in the Coomassie Blue G-250 "Quick Stain" and soak for one hour. Place the gels in the intensifier solution for approximately 15 minutes. Compare the position of the bands of the samples with those of the marker to determine the pI of the proteins contained in your sample

Determination of Diffusivity and Protein Aggregates

There are several instruments available in the market. One that is quite convenient for routine measurements is a backscattered optical instrument developed by Brookhaven Instruments (Brookhaven, MA). Several solid-state laser options are available and the temperature is efficiently controlled by a peltier heating element. The experiment below refers to this instrument.

Equipment

A laser power instrument and an autocorrelator. Buffers, polyacrylic acid cuvettes, and a protein sample.

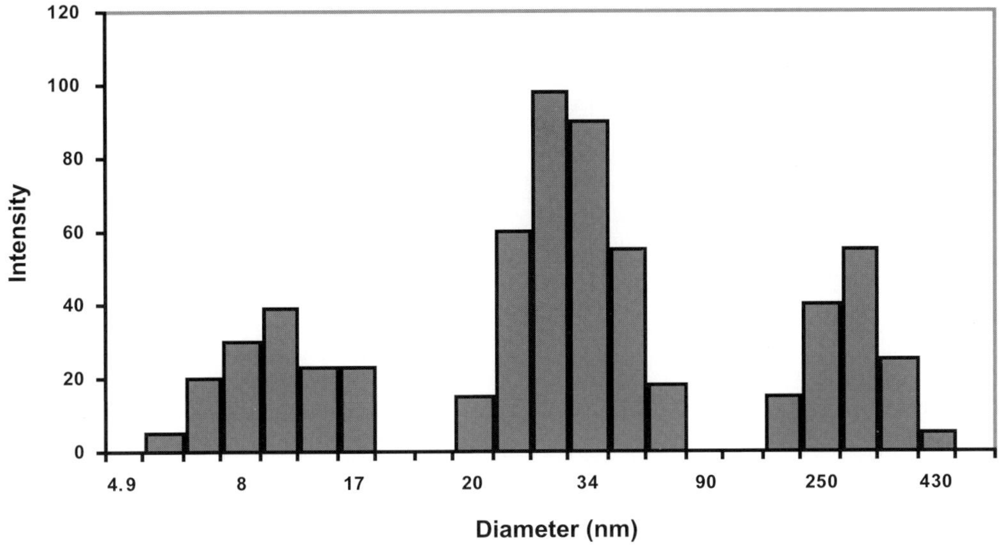

Figure 4.15. Size distribution determined by dynamic light scattering of a protein preparation showing the formation of aggregates.

Protocol

1. Follow steps one to three described in the static light scattering instrument experiment.
2. Do not filter the sample after the protein has been added because you will filter out the aggregates.
3. Place the cuvette in the cuvette holder.
4. Turn on the laser and allow the temperature to stabilize.
5. Start the collection of data (photon intensity vs. time) for one minute.
6. Monitor the correlation function. If the base line is within 0.05% from zero you have successfully done a dynamic light scattering experiment.
7. The software will fit the correlation function following various protocols. For monodisperse samples, the cumulant method is enough. For polydisperse samples, the Non-Linear Square approach is appropriate. You are expected to obtain a histogram as that shown in Figure 4.15 in the presence of aggregates.

Reconstitution of Membrane Proteins

Materials

Cytochrome c reductase, egg phosphatidylcholine, bovine brain phosphatidylserine, and Triton X-10. Sephadex G-200 is used in the gel-permeation experiments (Pharmacia). The BioBeads, SM-2 are from BioRad.

Method 1

Solubilize cytochrome c reductase (1.0 mg/mL) in 50 mM Tris-Acetate/0.5 mM EDTA/0.05% (w/v) Triton X-100 at a pH of 6.0. The enzyme-detergent complexes are mechanically shaken for five hours at 30°C and then stored at 4°C for no more than 4 days before use.

Dissolve phosphatidylcholine and phosphatidylserine (4:1 w/w) in chloroform and then twice dissolve in diethyl ether by rotary evaporation. Vesicles are formed spontaneously when an excess volume of 50 mM Tris-Acetate (pH 6.0)/0.5 mM EDTA/0.25% Triton X-100 aqueous buffer is added to the dry lipid and mechanically shaken for four hours at 30°C. The vesicles are used in the reconstitution experiments immediately afterward.

Two-dimensional crystals of cytochrome c reductase are grown by mixing the protein-Triton complex with the phospholipid-Triton solution (with a vesicle size of ~ 874.9 nm) for five hours. Triton X-100 is removed by dialysis in the presence of SM-2 BioBeads outside of the dialysis bag at 25 to 30°C for 48 hours. The buffer is changed once after 25 hours. The dialysis buffer is 50 mM Tris-Acetate (pH 7.0).

Method 2

Dissolve cytochrome c reductase at a concentration of 2.8 mg/mL in pH 7 buffer.

Multilamellar vesicles are prepared by the same method as in method 1, but an excess volume of 50 mM Tris-Acetate (pH 7.0)/0.5 mM EDTA/0.25% Triton X-100 aqueous buffer is added to the dry lipid and mechanically shaken for four hours at 30°C. After shaking, the vesicles are fractionated using gel-permeation chromatography. Pack a column (0.7 X 50 cm) with Sephedex-200 and equilibrate for two days at 2.2 mL/min. Load samples that do not exceed 1% to 2% of the total column volume. The average diameter of the lipid vesicles can be determined by dynamic light scattering and checked by transmission electron microscopy. Before fractionation, the average diameter of the vesicles as determined by dynamic light scattering should be approximately 310 to 850 nm. Vesicles with an average diameter of approximately 874.9 nm are used for reconstituting cytochrome c reductase. The polydispersity of the vesicles preparation should be less than 0.01, which is very close to having monodispersed vesicles in solution.

Two-dimensional crystals of cytochrome c reductase are grown by mixing the protein-Triton complex with the vesicle-Triton solution for five hours. The Triton X-100 is dialyzed in the presence of SM-2 (2 grams wet) BioBeads at 25 30°C for 48 hours, changing the buffer once after 24 hours. The dialysis buffer is 50 mM Tris-Acetate (pH 6.0). The protein-lipid complex is mechanically shaken gently for one hour at 25°C and then stored at 4°C.

Nomenclature

c_i concentration of species i (mol/L)
D^0 diffusion coefficient at infinite dilution (cm^2/sec)
f frictional coefficient
F Faraday's constant (96,485 C/mol)
I Ionic strength (mol/L)
M molecular weight (kg/kmol)
N_A Avogadro's number (6.023×10^{23} molecules/mol)
r distance (m)
R ideal gas constant (0.082 L-atm/K mol, 8.31 J/mol K, 83.14 10^6 cm^2 g/K mol sec^2)
s sedimentation coefficient (sec)
T absolute temperature (K)
v sedimentation velocity (m/sec)
$\bar{v}$ specific volume (m^3/kg)
z_i charge of species i

Greek Letters

ψ_E electrostatic potential (v)
κ^{-1} Debye length (m)
ε permittivity of the solvent (C^2/J m)
ρ density (kg/m^3)
ω angular velocity (rad/sec)

References

Balestrieri, C., G. Giovane, A. Irace, and L. Servillo. 1978. Second-derivative spectroscopy of proteins. *European Journal of Biochemistry* 90:433-40.

Bermudez, O., and D. Forciniti. 2004. Aggregation and denaturation of antibodies: a capillary electrophoresis, dynamic light scattering, and aqueous two-phase partitioning study. *Journal of Chromatography B* 807(1):17-24.

Bloemendal, M., and W. C. Johnson Jr. 1995. Structural information on proteins from circular dichroism spectroscopy possibilities and limitations. *Pharmaceutical Biotechnology* 7:65-100.

Boekema, E. J. 1990. The present state of two-dimensional crystallization of membrane proteins. *Electron Microscopy Reviews* 3:87-96.

Brange, J., L. Langkjaer, S. Havelund, and A. Volund. 1992. Chemical stability of insulin. 1. Hydrolytic degradation during storage of pharmaceutical preparations. *Pharmaceutical Research* 9:715-26.

Brunori M., B. Vallone, F. Cutruzzola, C. Travaglini-Allocatgelli, J. Berendzen, K. Chu, R. M. Sweet, and I. Schlichting. 2000. The role of cavities in protein dynamics: crystal structure of a novel photolytic intermediate of myoglobin. *Proceedings of the National Academy of Science USA* 97:2058-63.

Cleland, J. L., and D. I. C. Wang. 1993. Cosolvent effects on refolding and aggregation. *ACS Symposium Series* 516:151-66.

Cromwell, M. E. M., E. Hilario, and F. Jacobson. 2006. Protein aggregation and bioprocessing. *The AAPS Journal* 8:E572-9.

Dousseau, F., and M. Pezolet. 1990. Determination of the secondary structure content of proteins in aqueous solutions from their amide I and amide II infrared bands. Comparison between classical and partial least-squares methods. *Biochemistry* 29:8771-9.

Du, F., Z. Zhou, Z. Y. Mo, J. Z. Shi, J. Chen, and Y. Liang. 2006. Mixed macromolecular crowding accelerates the refolding of rabbit muscle creatine kinase: implications for protein folding in physiological environments. *Journal of Molecular Biology* 364:469-82.

Engelhardt, H., T. Scheybani, W. Von Gustedt, and W. Baumeitster. 1994. Formation of Two-dimensional Crystals of Membrane-Anchored and Water-Soluble Proteins. *Materials Research Society Symposium Proceedings* 330:201-8.

Forciniti, D. 1994. Protein refolding using aqueous two-phase systems. *Journal of Chromatography A* 668:95-100.

Gigant, B., P. A. Curmi, C. Martin-Barbey, E. Charbaut, S. Lachkar, L. Lebeau, S. Siavoshian, A. Sobel, and M. Knossow. 2000. The 4 angstrom x-ray structure of a tubulin: stathmin-like domain complex. *Cell* 102:809-16.

Greenfield, N. J. 1996. Methods to estimate the conformation of proteins and polypeptides from circular dichroism data. *Analytical Biochemistry* 235:1-10.

Havel, H. A. 1996. Derivative near-ultraviolet absorption techniques for investigating protein structure. In *Spectroscopic Methods for Proteins in Solution*, edited by H. A. Havel, pp. 62-8. New York: VCH.

Hovmoller, S., M. Slaughter, J. Berriman, B. Karlsson, H. Weiss, and K. Leonard. 1983. Structural studies of cytochrome c reductase: improved membrane crystals of the enzyme complex and crystallization of a subcomplex. *Journal of Molecular Biology* 165:401-6.

Jap, B. K., M. Zulauf, T. Scheybani, A. Hefti, W. Baumeister, and U. Aebi. 1992. 2D crystallization: from art to science. *Ultramicroscopy* 46:45-84.

Jiang, Y., Y. B. Yan, and H. M. Zhou. 2006. Polyvinylpyrrolidone 40 assists the refolding of bovine carbonic anhydrase B by accelerating the refolding of the first molten globule intermediate. *Journal of Biological Chemistry* 281:9058-65.

Joessang, T., J. Feder, and E. Rosenqvist. 1985. Heat aggregation kinetics of human IgG. *Journal of Chemical Physics 82*(1):574-89.

Jungbauer, A., W. Kaar, and R. Schlegl. 2004. Folding and refolding of proteins in chromatographic beds. *Current Opinion in Biotechnology* 15:487-94.

Kendrick, B. S., J. F. Carpenter, J. L. Cleland, and T. W. Randolph. 1998. A transient expansion of the native state precedes aggregation of recombinant human interferon-γ. *Proceedings of the National Academy of Science* 95:14142-6.

Lai, W. B., and A. P. J. Middelberg. 2002. The production of human papillomavirus type 16 L1 vaccine product from *Escherichia coli* inclusion bodies. *Bioprocess and Biosystems Engineering* 25:121-8.

Lee, R. C., F. Despa, L. Guo, P. Betala, A. Kuo, and P. Thiyagarajan. 2006. Surfactant copolymers prevent aggregation of heat denatured lysozyme. *Annals of Biomedical Engineering* 34:1190-1200.

Leonard, K., P. Wingfield, T. Arad, and W. Weiss, 1981. Three-dimensional structure of ubiquinol: cytochrome c reductase form neurospora mitochondria determined by electron microscopy of membrane crystals. *Journal of Molecular Bioliogy* 149:259-74.

Lewis, U. J., E. V. Cheever, and W. C. Hopkins. 1970. Kinetic study of the deamidation of growth hormone and prolactin. *Biochimica et Biophysica Acta* 214:498-508.

Lin, K., V. L. Rath, S. C. Dai, R. J. Fletterick, and P. K. Hwang. 1996. A protein phosphorylation switch at the conserved allosteric site in GP. *Science* 273:1539-41.

MacRaild, C. A., C. R. Stewart, Y.-F. Mok, M. J. Gunzburg, M. A. Perugini, and L. J. Lawrence. 2004. Non-fibrillar components of amyloid deposits mediate the self-association and tangling of amyloid fibrils. *Journal of Biological Chemistry* 279(20):21038-45.

Maina, C. V., P. D. Riggs, A. G. Grandea III, B. E. Slatko, L. S. Moran, J. A. Tagliamonte, L. A. Mcreynolds, and C. D. Guan. 1988. An *Escherichia coli* vector to express and purify foreign proteins by fusion to and separation from maltose binding protein. *Gene* 74:365-73.

Manning, M. C., K. Patel, and R. T. Borchardt. 1989. Stability of protein pharmaceuticals. *Pharmaceutical Research* 6:903-18.

Ong, E., J. M. Greenwood, N. R. Gilkes, D. G. Kilburn, R. C. Miller, and R. A. Warren. 1989. The cellulose-binding domains of cellulases: tools for biotechnology. *Trends Biotechnology* 7:239-43.

Peterson, S. A., T. Klabunde, H. A. Lashuel, H. Purkey, J. C. Sacchettini, and J. W. Kelly. 1998. Inhibiting transthyretin conformational changes that lead to amyloid fibril formation. *Proceedings of the National Academy of Science USA* 95:12956-60.

Ramachandran, G. N., and V. Sasisekharan. 1968. Conformation of polypeptides and proteins. *Advances in Protein Chemistry* 23:283-438.

Rigaud, J. L., G. Mosser, J. J. Lacapere, O. Olofsson, D. Levy, and J. L. Ranck. 1997. Bio-beads: an efficient strategy for two-dimensional crystallization of membrane proteins. *Journal of Structural Biology* 118:226-35.

Rigaud, J. L., D. Levy, G. Mosser, and O. Lambert. 1998. Detergent removal by non-polar polystyrene beads: applications to membrane protein reconstitution and two-dimensional crystallization. *European Biophysics Journal* 27:305-19.

Rosenqvist, E., T. Joessang, and J. Feder. 1987. Thermal Properties of Human IgG. *Molecular Immunology* 24(5):495-501.

Skerra, A., and T. M. G. Schmidt. 1999. Applications of a peptide ligand for streptavidin: the Strep-tag. *Biomolecular Engineering* 16:79-86.

Smith, B. C. 1996. *Fundamentals of Fourier Transform Infrared Spectroscopy.* Boca Raton, FL: CRC Press.

Smith, P. A., B. C. Tripp, E. A. DiBlasio-Smith, Z. Lu, E. R. La Vallie, and J. M. McCoy. 1998. A plasmid expression system for quantitative in vivo biotinylation of thioredoxin fusion proteins in *Escherichia coli*. *Nucleic Acids Research* 26:1414-20.

Stiborova, H., J. Kostal, A. Mulchandani, and W. Chen. 2002. One-step metal-affinity purification of histidine-tagged proteins by temperature-triggered precipitation. *Biotechnology and Bioengineering* 82(5):605-11.

Further Reading

Creighton, T. E. 1993. *Proteins: Structures and Molecular Properties,* 2nd ed. New York: W. H. Freeman and Company.

Chapter 5
Cell Disruption

Introduction

Some of the chemicals of microbial origin produced industrially today are of the extracellular type (i.e., the microbial cell produces and secretes them into the surrounding environment). A much larger proportion of the potentially useful microbial products are retained within the cells. For example, a vast majority of the enzymes known are intracellular.

Although intracellular products can be made extracellular by genetically modifying the host organism, cells are commonly disintegrated by physical, chemical, or biological means to release their contents into the surrounding medium. Besides, making the cell fully permeable to any significant fraction of the intracellular products and enzymes by genetic manipulation would not only be difficult, but also will imply cell death. It is in this context that the unit operation of microbial cell disruption for intracellular product isolation is of increasing importance.

The objective of any disruption procedure is to solubilize the maximum amount of product while maintaining its maximum biological activity. The main constraint to achieve that goal is to limit the detrimental effects of disruption on subsequent separation steps. An inevitable consequence of cell/tissue disruption is the release of contaminants.

There are several physical and chemical methods that can be used for cell disruption (Sadana 1994; Middelberg 1995; Hatti-Kaul and Mattiasson 2003): bead mills, high pressure homogenizers, solvent extraction, osmotic shock, ultrasonic disintegrator, and so on. They cover a range of scales and although a few can be used in laboratory and industrial scale, most are either laboratory or industrial scale. The availability of the same instrument in both laboratory and process scales is important because it facilitates scale-up.

Mechanical disruption is the simplest approach to make the cellular membrane permeable. Common devices found in laboratories are the Waring Blender and the Polytron. The Waring Blender is similar to a standard household blender, whereas the Polytron consists of a series of rotating blades that grind and disperse cells and tissues.

The simplest liquid homogenizer is the Dounce homogenizer. It consists of a round glass pestle that is manually driven into a glass tube that ends in a "bubble shape." A variation of it is the Potter-Elvhjem homogenizer in which the glass pestle is replaced by a Teflon one. A slightly more "primitive" approach is the use of a mortar and pestle, which are normally used to disrupt tissues (particularly plants). A common pretreatment is to freeze the tissue in liquid nitrogen, which facilitates the disruption.

The most common high pressure homogenizer found in laboratories is the French press, in which the cell suspension is forced by very high pressures (up to 300 MPa) through a hole. The cytoplasmatic fluid inside the cells is pressurized too. The sudden decompression as the suspension leaves the valve burst the cells open. Two of the advantages of the French press are the negligible generation of heat and the fact that it leaves the cell nucleus undisturbed. Unfortunately, it is limited to laboratory scale because the largest sample size that commercial French presses are able to treat is about 35 mL. The pressure drop needed to disrupt the cells depends on the kind of microorganism, for yeast and bacteria the use of 150 MPa is recommended, whereas 300 MPa is recommended for plant cells and nucleic materials.

The Mautin Gaugin high-pressure homogenizer is a device related to the French press, but is more of a common sight in industry and it is described in detail later in this chapter. There are some small instruments available for laboratory use. For example, BEE International Inc (South Easton, MA) commercializes a high pressure homogenizer that can process samples as small as 20 mL.

Bead mills fall somewhere in between mechanical and liquid homogenizers. In this kind of devices, the cell suspension is mixed with small plastic, glass, or metal beads and stirred either by vibration or by circulating the mixture through a screw type conveyor. The cells are broken by direct mechanical action (attrition in between the beads) and by shear forces. It is one of the most common mills found in industry and available also at laboratory scale.

Sonicators consist of a probe that vibrates at very high frequencies (18 kHz to 1 MHz). The vibration of the probe produces gas bubbles (a phenomenon known as cavitation) that grow during the rarefying phase of the sound wave and collapse during the compression phase. The implosion of those bubbles generates a violent shock wave equivalent to thousands of atmospheres. This high energy wave disrupts the cells. Because of the violent nature of the implosions, local temperatures as high as 1000°C have been reported. Obviously, such a temperature increase is very damaging to proteins and other delicate biological products. An additional problem is the generation of free radicals like singlet oxygen and hydrogen peroxide. N_2O has been used as a free radical scavenger to eliminate the deleterious effect of free radicals. Because sonication is such a harsh disruption method, it creates very small particulates that make further downstream processing difficult. They are very popular in laboratory scale but they have not found a place in industry because of the problems described above. Another method limited to laboratories is freeze and thaw. When cells are frozen, little ice crystals form in the cell cytoplasm. As the cells are put through multiple thawing and re-freeze cycles, the ice crystals grow and finally disrupt the cell membrane. Although efficient and relatively gentle, the method is very tedious because multiple cycles are needed.

Osmotic shock consists of the manipulation of the osmotic pressure across the cell membrane to disrupt cells. The cells are equilibrated against a high sucrose medium and then diluted quickly. The increase in pressure in the cytosol caused by the sudden dilution of the media damages the cell membrane. As such it is only good for periplasmic proteins. As any other disruption method, the osmotic shock efficiency depends on the kind of microorganism used. For example, some Gram-positive bacteria are able to resist up to 2 MPa (20 atm) of osmotic pressure.

Independently of the mechanical device used, additives are frequently added either to facilitate disruption or to prevent damage of valuable compounds. Typical additives are hypotonic buffers, lysozyme, DNA- or RNA-degrading enzymes, and protease inhibitors. Protease inhibitors are added to protect the product from photolytic degradation. Lysozyme at a concentration of 200 μg/mL is frequently added to bead mills particularly when disrupting yeast. The purpose of the addition of this glycolytic enzyme is to accelerate the disruption of the cell membrane. The viscosity of the cell suspension increases because of the release of nucleic acids, this viscosity increase may be lessened by adding DNAse (25-50 μg/mL) or RNAse (50 μg/mL).

The use of only chemicals or biological agents for cell disruption is not common in industry but

Table 5.1. Comparison Between Several Disruption Methods.

Method	Scaling	Use	References
Bead Mill	Good	Industrial/Lab	Limon-Lason et al. 1979; Shutte et al. 1983; Woodrow and Quirk 1982; van Gaver and Huyghebaert 1990; Bierau, Hinton, and Lyddiatt 2001
High Pressure Homogeneizer	Fair	Industrial/Lab	Engler and Robinson 1981
Ultrasonic Disintegrator	Poor	Laboratory	Doulah 1978; Feliu and Villaverde 1994.
Osmotic Shock	Poor	Laboratory	Coakley, Brown, and James 1974
French Press	Fair	Laboratory	Emanuel and Chaikoff 1957
Detergents	Good	Laboratory/Industrial	
Nitrogen Cavitation	Poor	Laboratory	Hunter and Commerford 1961
Enzymatic Lysis	Fair	Laboratory	Asenjo and Dunnill 1981; Bermudez and Forciniti 2004; Van Hee et al. 2004
Freeze and Press	Poor	Laboratory	Magnusson and Edebo 1974
Solvent Extraction	Good	Pilot Plant	

they are usually a common sight in laboratory scale. Some of the chemicals used are toluene, ether, DMSO, benzene, methanol, chloroform, detergents, antibiotics, thionins, and ethylenediaminetetraacetic acid (EDTA). All of them increase the permeability of the cell membrane, but by different mechanisms. EDTA is used to permeate gram-negative microorganisms because it binds to Ca^{2+} and Mg^{2+}, removing lipopolysaccharides. On the other hand, organic solvents like benzene or chloroform just dissolve the lipid bilayer. Biological means are even less common in industry because of the high cost associated with them. Examples of the use of biological means are the use of streptolysin or viruses at a laboratory scale to permeate mammalian cells. A more common biological mean is the use of enzymes. The enzymes most frequently used are lysozyme, β(1-6) and β(1-3) glycanases, proteases, and mannose.

Another method to disrupt cells is thermolysis. Gram-negative bacteria (Gram-positive bacteria seem to be more tolerant) can be heated up to 50 to 55°C (Katsui et al. 1982; Pisabarro, de Pedro, and Vazquez 1985) to release periplasmic proteins. Higher temperatures can be used to release cytoplasmic proteins. One of the advantages of this method is that the host organism is killed and preferential protease inactivation may be achieved. One of its disadvantages is the potential thermal denaturation of the product and an increase in viscosity of the suspension.

If the cells are disrupted to isolate a particular protein for research purposes (i.e., the likelihood of moving the process to production scale is null) any disruption method can be used. Moreover, because they do not require specialized equipment, chemical or biological methods are very suitable for laboratory scale. If the target protein has the potential of commercial use, a different perspective is needed. Table 5.1 compares several disruption procedures from a scale-up point of view. Bead mills together with high pressure homogenizers have the highest potential and are the preferred choices in industry today. Table 5.2 summarizes the most important characteristics of these unit operations. The final choice in between the two must be made based upon the production scale, the kind of microorganism to be used, and the location of the target product within the cell.

Independently of the type of disruption method chosen, other factors, such as the type of microorganism, the growth conditions and the location of products within the complex cell structure, will affect the efficiency of the disruption process. For example, Figure 5.1 shows the logarithm of the amount of β-galactoside released versus the number of passes through a high pressure homogenizer for *Escherichia coli* grown both in a well-defined and in a complex medium (Chisti and Moo-Young 1986). These differences are significant and suggest that bacteria have different cell morphologies when grown under different conditions. Although such a dramatic change is caused by a drastic

Table 5.2. Comparison Between Bead Mills and High Pressure Homogenizers.

Bead Mills	High Pressure Homogenizers
Available in all sizes.	Available in all sizes.
Easy sterilization.	Difficult to sterilize.
The disruption rate is a function of cells' concentration.	The disruption rate is not a function of cells' concentration.
There is a possibility of glass contamination.	Filamentous microorganisms can block this apparatus.
Large number of variables.	Small number of variables.

change in the growth conditions, less dramatic, but still significant changes are expected because of culture medium modifications.

Differences in the efficiency of the release of proteins are expected according to the location of the protein in the cell. The obvious case is that of membrane proteins. Less obvious is the fact that many enzymes are compartmentalized in the cell cytoplasm. Figure 5.2 illustrates this for six different enzymes (Follows et al. 1971). It is possible, also, that the decrease recovery in some cases is the result of differences in the stability of the enzymes. Still, for the most part, the release rate correlates with the location of the enzymes. For example, invertase and acid phosphatase are mostly located outside the cell; as such, they show the highest release rate versus all other proteins in the cell. Alcohol dehydrogenase (ADH) and glucose-6-phosphate dehydrogenase (G6PDH) were released at nearly the same rate as total protein. Finally, fumarase is located in the mitochondria, and therefore, its release rate is the lowest.

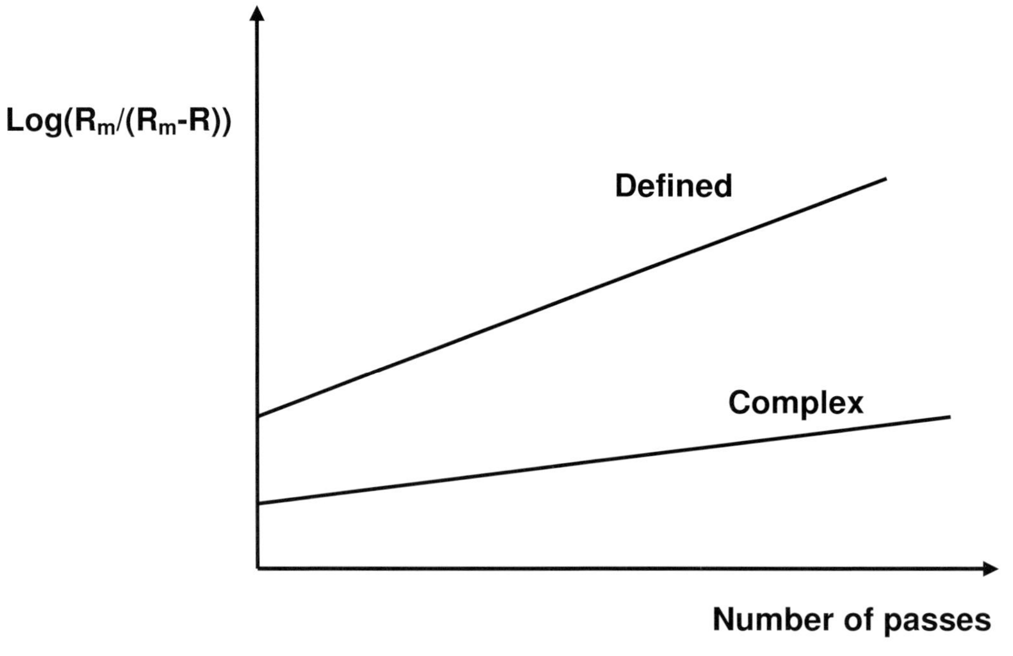

Figure 5.1. Disruption of *Escherichia coli* and monitoring of β-galactosidase for a well-defined medium and for a complex medium. Redrawn from Gray, P. P., P. Dunnill, and M. D. Lilly. 1972. The continuous-flow isolation of enzymes. In *Fermentation Technology Today*, edited by G. Terui, pp 347-51. Osaka, Japan: Society of Fermentation Technology. Copyright by the Society of Fermentation Technology; reproduced with permission.

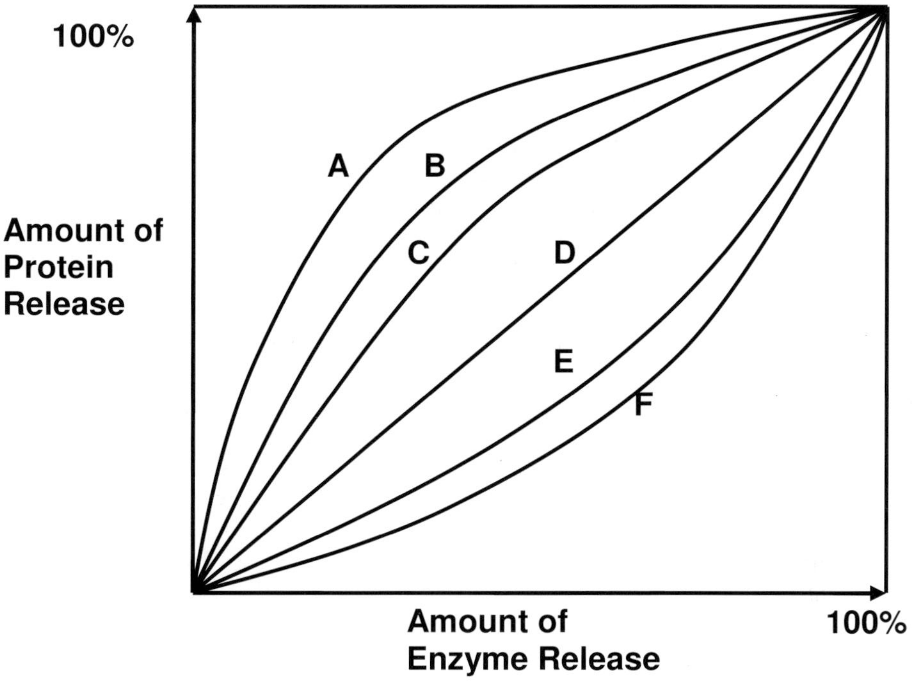

Figure 5.2. Amount of protein release vs. amount of enzyme release for A: fumarase, B: alakaline phosphatase, C: ADH, D: G6PDH, E: invertase, and F: acid phosphatase. Redrawn with permission from Follows, M., P. J. Hetherington, P. Dunnill, and M. D. Lilly. 1971. Release of enzymes from Bakers' yeast by disruption in an industrial homogenizer. *Biotechnology and Bioengineering* 13:549-60. Copyright John Wiley and Sons.

Before describing in detail the most common cell disruption methods, it is necessary to describe the main characteristics of the cell membrane and cell wall. Without at least a synopsis of this theme, it is impossible to choose rationally one cell disruption method over another.

Types of Cells

For many years, living organisms have been classified into two groups: eukaryotes and prokaryotes. Differences between both types affect their disruption. Eukaryotic DNA is enclosed by a membrane forming the nucleus. If the disruption method is gentle, the nuclear membrane is not damaged and, therefore, DNA is not released, simplifying further processing. Prokaryotes, on the contrary, do not have a nuclear membrane. Therefore, it is not possible to avoid DNA release during cell disruption. Prokaryotes have an outer coat called the cell wall. This wall is followed by the plasma membrane. Eukaryotes in general (except for example plants) do not have a cell wall but just the cell membrane; therefore, they are more fragile. Recently, living organisms have been re-classified into three groups: bacteria, archaea, and eukaryotes. The first two constituted before the prokaryotic group. Although this new classification is more proper, from the point of view of cell disruption, it is enough to consider the previous classification between prokaryotes and eukaryotes because of their differences in cell membrane.

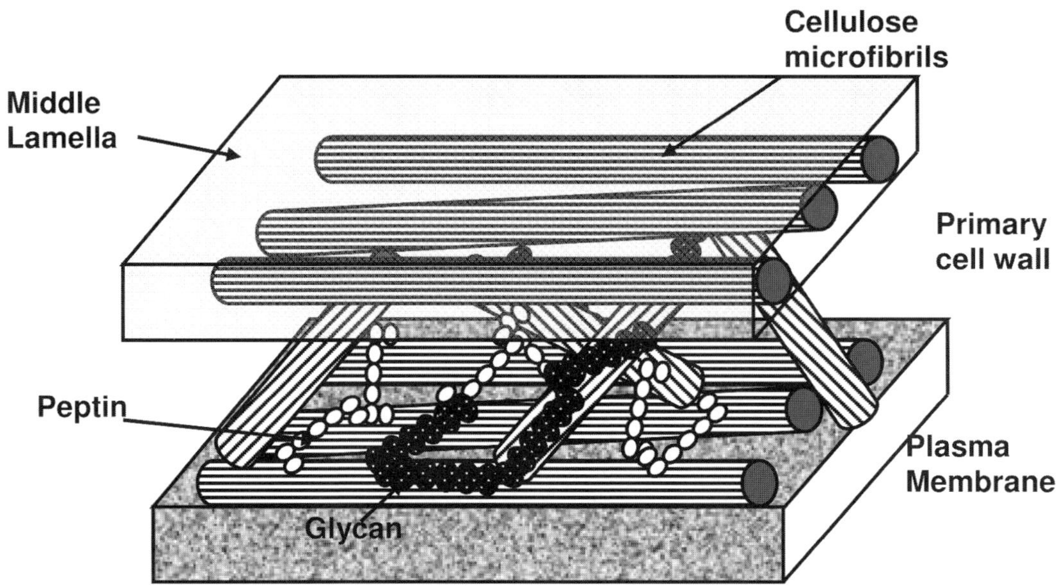

Figure 5.3. Plant cell wall

Physical Chemistry Properties of Cells (as They Relate to Disruption)

Cell walls are found in bacteria, plants, fungi, algae, and archaea. Their sole purpose is to protect the cells from the environment (including pathogens). Plant cells have the most complex cell wall. Their cytoplasm is separated from the environment by (inward to outward) a cell membrane, a primary cell wall, a middle lamella, and the secondary cell wall. Figure 5.3 shows a cartoon of a typical plant cell wall. The primary cell wall in plants is made mostly of cellulose. It is composed of cellulose microfibrils, kept in place by hydrogen bonds, hemicellulose, and pectin (a branched polysaccharide) and cross-linked glycans. The cross-linked glycans increase the tensile strengths of the microfibrils, whereas pectins allow the cell to resist compression. The most common proteins embedded into the cell wall are hydroxyproline-rich glycoproteins, arabinogalactan proteins, glycine-rich proteins, and proline-rich proteins. The primary cell wall is separated from the secondary cell wall by a pectin rich layer called middle lamella, a gelatinous membrane containing magnesium and calcium salts of pectic acid. The secondary plant cell wall (not shown in the figure) often has lignin, which adds considerable strength.

Algae cell walls contain also cellulose and glycoproteins. In addition, they have a variety of polysaccharides including manosyl fibrils, alginic acid, and sulfonated polysaccharides. Algae are one of the most resilient organisms to disrupt.

All bacteria cell walls contain peptidoglycan. In archaea, N-acetylalosaminuronic acid replaces N-acetyl muramic acid found in peptidoglycan. Bacteria are classified into two categories, Gram-positive and Gram-negative, according to their reaction to crystal violet dyes. Gram-positive bacteria's cell wall has a very thick layer of peptidoglycan. They also have a lipid bilayer cell membrane that is covered by a porous peptidoglycan layer, which does not exclude most antimicrobial agents. The peptidoglycans are linked to the cytoplasmic membrane by teichoic acids. In addition to the cell wall made of peptidoglycan and a phospholipid bilayer with membrane-spanning proteins, Gram-negative bacteria have an additional outer membrane. The additional outer membrane creates a periplasmic space between the cytoplasmic membrane and the outer membrane. Within this periplasmic space there is a thin layer of peptidoglycan. This outer membrane blocks antibiotics, dyes, detergents, and charge molecules, protecting the inner membrane and cell wall. Gram-negative bacteria are supposed to be resistant to lysozyme attack, whereas Gram-positive bacteria are susceptible to degradation by lysozyme. Still, lysis of Gram-negative bacteria by lysozyme is feasible at high salt concentrations and/or high shear.

E. coli is a typical Gram-negative bacteria; the inner cytoplasmatic membrane does not have any mechanical strength. After the outer membrane is broken, the contents of the cytoplasm are released unless the osmotic balance is maintained. The outer membrane is mostly composed of lipopolysacharides and divalent ions (like magnesium); therefore, solvents or chelating agents will disrupt it. It is worth mentioning, that other structures are present in the cell wall including: fimbrae (short protein tubes), pili (long protein tubes), S-layer (a two-dimensional array of crystalline proteins), capsules (impermeable structures made mostly by polysaccharides that are secreted by the cell), and slime (looser polysaccharide structure).

Some fungi species have cell walls composed by cellulose, glycosamine, and chitin. Most fungi associate into filamentous entities that make their rupture quite challenging because they produce clogging and impair the free circulation of the suspension through constrained spaces. *Saccharomyces cerevisiae* is a common fungi used in biotechnology. The walls of *S. cerevisiae* are more complex than those of *E. coli* and represent 15% to 25% of the cell dry weight. The wall is composed of glucan, mannoprotein, and chitin.

Cytoplasmic membranes of a thickness ranging from 7 to 10 nm are present in all cells. The cell membrane is the only "protective" barrier for mammalian cells, whereas in most bacteria and plant cells the cell membrane is encapsulated by the cell wall. Lipids, proteins, and carbohydrates comprise the cell membrane. Some or all of these components may be released to the medium upon disruption of the membrane. Therefore, it is worthwhile to define them better. The three main types of lipids are phospholipids, neutral lipids, and glycolipids. Phospholipids are the most abundant. The most common phospholipids are phosphatidylcholine, phosphatidylethanolamine, phosphatidylserine, and sphingomyelin. Phosphatidylglycerol and diphosphatidylglycerol are only found in bacteria and mitochondria. Cholesterol is an example of neutral lipids and it is found only in eukaryotic cells.

There is a myriad of proteins in the cell membrane but in general they fall into two categories: integral proteins and peripheral proteins. Integral proteins are intercalated into the lipid bilayer, whereas peripheral proteins are not embedded into the membrane but rather attach to the integral proteins. Integral proteins must be extracted by detergents that disrupt the lipid bilayer. After they are "released" from the membrane they are insoluble so they need to be complexed with detergents for further processing. Peripheral proteins can be released by disrupting their quaternary structures because they are associated with other proteins. Therefore, milder agents such as high ionic strength buffers or urea would be sufficient to release them. Furthermore, once released they are water soluble.

Most sugars in membranes are associated with proteins (glycoproteins), but some are associated with lipids (glycolipids). Therefore, the most common sugars found are those found in glycoproteins (see Chapter 4). Glycolipids mainly contain a single sugar (like glucose and galactose) or oligosaccharides of up to more than seven sugars.

The membrane is assembled in a typical double layer structure as that depicted in Figure 5.4. The phospholipids' charged head points outward (toward the exterior and interior of the cell), whereas the lipid tails of both layers face each other. This conformation is thermodynamically favored as the charged heads are hydrophilic and prefer to associate with water, whereas the lipid tails are hydrophobic and tend to associate with each other. The various lipids described above are not equally distributed in both layers. For example, in the influenza virus, the internal layer is rich in sphingomyelin, whereas phosphatidylethanolamine and phosphatidylserine predominate in the outside layer. The figure also shows an integral membrane protein whose ends protrude into the cytoplasm and into the extra cellular space.

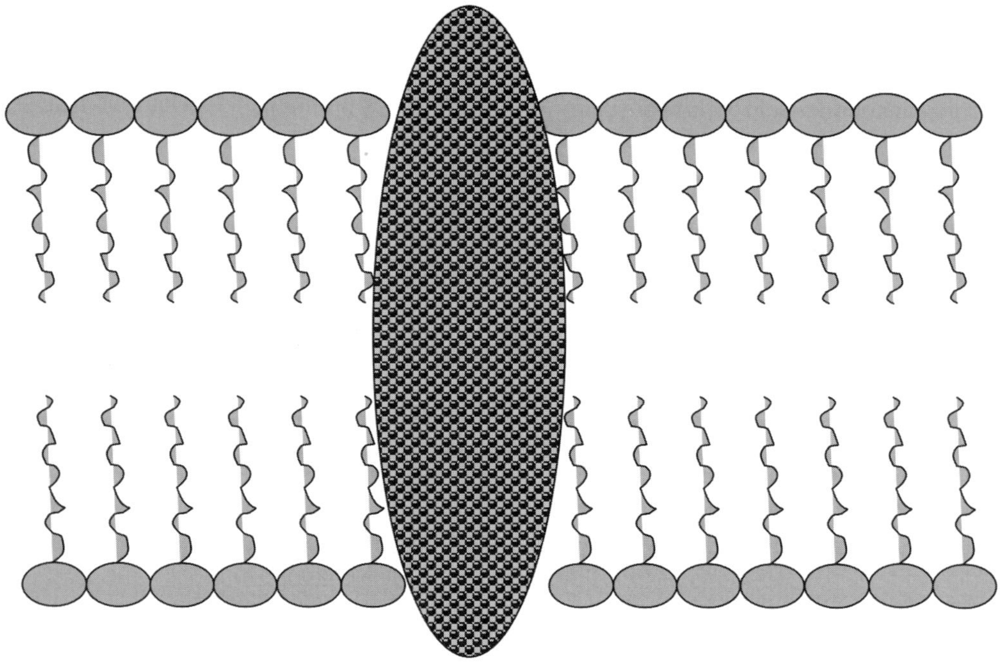

Figure 5.4. Schematic of a cell membrane showing a transmembrane protein

Monitoring Cell Disruption

Historically, cell disruption has been indirectly measured by monitoring the release of the target protein but true cell disruption can be quantified by optical methods, such as microscopy. Both methods may yield different results. Direct observation is better to understand the disruption process, whereas indirect methods are better for routine analysis. The typical direct method is to stain the cells with methylene blue dye or Gram staining to differentiate between intact and damaged cells. An alternative to microscope observation is to measure particle size distribution with particle analyzers or analytical centrifuges. In this case, the extent of disruption is reported as the fraction of cell debris to intact cells. Unfortunately, knowledge of particle size distribution has been only rarely incorporated into the mathematical description of the disruption method (Siddiqi, Tichener-Hooker, and Ayasi Shamlou 1996).

Indirect methods are the most popular. They consist of either measuring total protein content (using Lowry, Bradford, or BCA tests), or measuring enzymatic activity of the target protein using the appropriate assay, or both. Measuring total protein content and total enzyme activity is very convenient because it generates a base line (100% recovery) for the remaining purification processes.

Bead Mills

Cell disruption in bead mills is regarded as one of the most efficient techniques to free an intracellular protein. Various designs of bead mills have been used for microbial cell disruption, but the

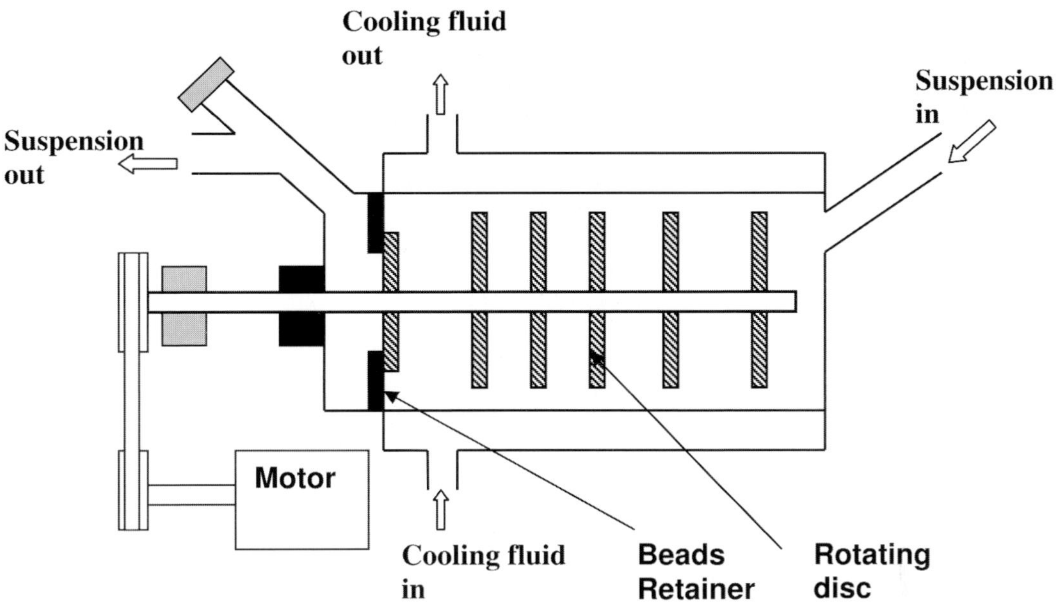

Figure 5.5. Schematic of a Dyno mill.

best-known are the Dyno mill and the Netsch mill. Figure 5.5 is a schematic of a Dyno mill. The equipment has a horizontal cylindrical chamber with a motor-driven central shaft supporting a collection of agitating elements. The chamber is filled to the desired level with steel or balloting glass beads, which provide the grinding action. The Dyno mill has efficient cooling systems and they are available in a variety of sizes. At the smallest end, one finds mills with a grinding container of 0.3 L (batch or continuous) followed in size by mills of 0.6 and 1.4 L (continuous). The inner cylinders are available in glass and in other materials. They are available at fixed and variable speeds and they can process up to 40 kg/h in continuous mode. At the other end, the model KD 600 has a product capacity of 2,200 to 11,000 lb/h and a volume of 600 liters.

There have been some theoretical models for this system (Linson et al. 1979; Mogren, Lindblom, and Hedenskog 1974), but most of the research has been experimental. Seven parameters have been found to be important in the design and operation of a bead mill:

i. Kind of microorganism. This is fixed by production necessity. If more than one microorganism is available, their different fragilities have to be considered. For example, for 15% cell slurries of *S. cerevisiae*, about 90% disruption has been reported (Rehacek and Schacfer 1977) at a residence time of 2.5 minutes in a 20-L capacity mill. Under the same conditions, only 84% of *Candida utilis* could be broken (Mogren, Lindblom, and Hedenskog 1974).
ii. Culture age. The fragility of a microorganism is related to its age. Changes in the microorganism age during the processing affect the microorganism resistance to disruption.
iii. Microorganism concentration. Opposite effects have been found (Mogren, Lindblom, and Hedenskog 1974; Currie, Dunnill, and Lilly 1972; Woodrow and Quirk 1982) of the suspension viscosity or microorganism concentration on the rate of disruption. However, the data are not directly comparable because different ranges of yeast concentration were used in each work.

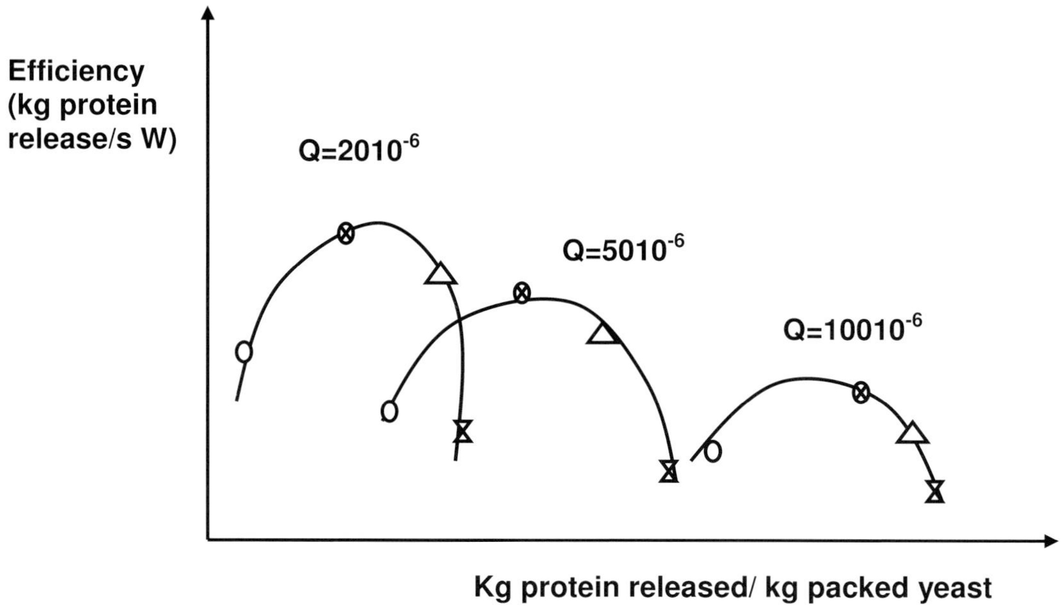

Figure 5.6. Efficiency of a bead mill at various flow rates and impeller tip speeds. Impeller speed: circle: 8 m/s; circle cross: 10 m/s; triangle: 15 m/s; sand clock: 20 m/s. Redrawn with permission from Limon-Lason, J., M. Hoare, C. B. Orsborn, D. J. Doyle, and P. Dunnill. 1979. Reactor properties of a high-speed mill for microbial cell rupture. *Biotechnology and Bioengineering* 21:745-74. Copyright John Wiley and Sons.

iv. Suspension flux. It is generally accepted that cell disruption rate, at constant impeller tip speed, is proportional to the inverse of suspension flux (Linson et al. 1979; Mogren, Lindblom, and Hedenskog 1974).

v. Temperature. One of the problems of operating a bead mill is temperature control. A temperature increase of 10 to 40°C has been found (Shutte et al. 1983; Currie, Dunnill, and Lilly 1972) when the cooling system is not efficient. This increase in temperature is not important if total protein released is a measure of performance. If total enzymatic activity is used as a measure of efficiency, temperature increase is critical for labile enzymes. Because intracellular enzymes are quite susceptible to heat or shear denaturation, we expect temperature control to be relevant in the performance of a bead mill.

vi. Rotor speed. The disruption rate is a strong function of rotor speed (Currie, Dunnill, and Lilly 1972; Woodrow and Quirk 1982). Speeds higher than 1,400 rpm do not increase the amount of protein in solution. Figure 5.6 illustrates the effect of tip speed on protein release at various flow rates. The figure shows that an optimum in protein release exists for each flow rate. A maximum is observed because an increase in rotation speed increases the disruption rate constant, but also increases flow rate, which in turn increases back mixing (decline in efficiency).

vii. Bed weight and bead size. Bed weight or the relation between bed weight and suspension volume (Hedenskog, Mogren, and Edebo 1970), and bead size are fixed based on the kind of microorganism and the suspension's average concentration (Currie, Dunnill, and Lilly 1972; Rehacek and Shacfer 1977). The disruption is faster with smaller beads (Dunnill and Lilly 1975). However, the smallest bead size that can be used is limited by beads' fluidization and sieve binding problems found with very small beads (Dunnill and Lilly 1975).

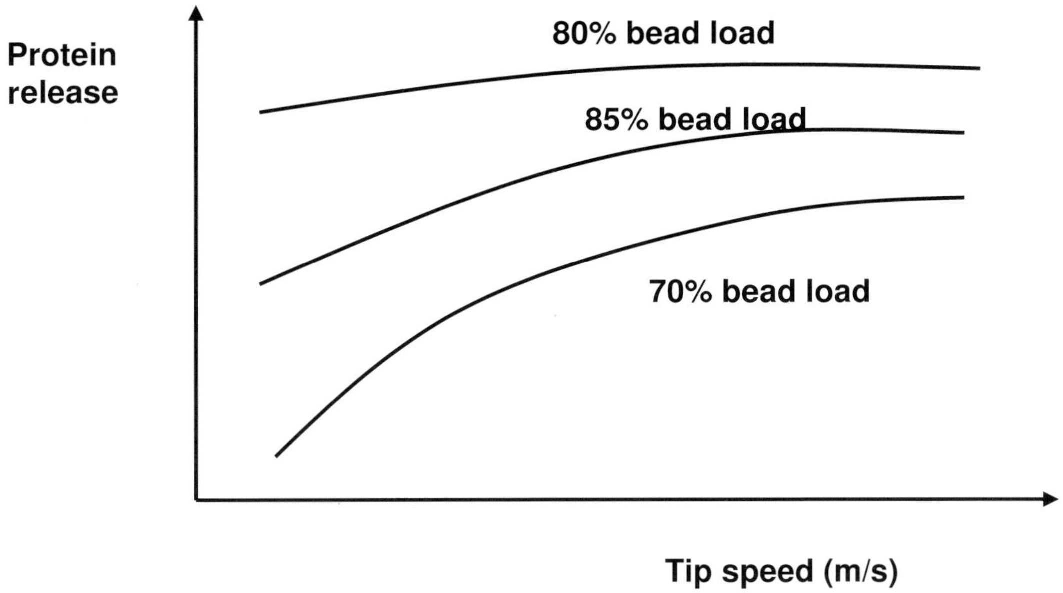

Figure 5.7. Protein release vs. impeller tip at various bead loads. Redrawn with permission from Shutte, H., K. H. Kroner, H. Hustedt, and M. R. Kula. 1983. Experiences with a 20 litre industrial bead mill for the disruption of microorganisms. *Enzyme and Microbial Technology* 5:143-8. Copyright by Elsevier B. V.

> Figure 5.7 illustrates the effect of bead load on the efficiency of a mill as a function of the tip speed. The figure shows that at high tip speeds the differences are less important than at low tip speeds where heavy loading of the mill increases the disruption efficiency dramatically. The trend shown in the figure may be an indication of a change in the disruption mechanism from mechanical abrasion caused by the beads at low tip speeds to shear forces at very high tip speeds. The bead size also affects the release differently depending on the location of the protein. For example, Shutte et al. (1983) found that the release of D-glucose-6-phosphate dehydrogenase (cytoplasmatic) was maximized with beads of 0.55 to 0.85 mm but α-D-Glucosidase (periplasmatic) was maximized with larger beads. The latter observation is expected because complete disruption is not needed for periplasmatic proteins.

A qualitative operational procedure for a bead mill is as follows. If the microorganism age increases during the operation, we can diminish the suspension flux or increase the rotor speed. If there is not any production constraint, the former option is more convenient. The temperature increase caused by an increase in the rotor speed decreases bioproduct recovery because of the intrinsic dependence of disruption with temperature and because of protein denaturation. The production level is better controlled by varying the rotor speed. The equipment has to have an efficient cooling system in order to avoid a decrease in the disruption kinetic constant and an increase in the denaturation rate. Shear induced denaturation of enzymes has been widely investigated. The most consistent conclusion is that shear denaturation takes place only in gas-liquid interfaces. Thus, if the enzyme is shear sensitive, we have to avoid the presence of air in the mill in order to minimize shear denaturation. Middelberg (1995) recommends to start exploring the disruption of a microorganism in a bead mill with high bead loading (80-90%), relatively high agitator speeds (tip velocity of 5 to 10 ms^{-1}), small bead diameters (0.5 mm) and moderate to high cell concentrations (40-50% wet weight).

Modeling

One of the most important uses of mathematical modeling of any device is the rationalization of experimental data and its extrapolation to conditions that are not covered by the experiments. On occasions, parametric studies using a robust mathematical model show the importance of parameters overlooked by experimentalists or prove that a combination of parameters rather than the individual ones affect the performance of the device. A known experimental fact for bead mills is that the kinetics of disruption is first order in the amount of intact cells. Still, some systems may not follow first order kinetics because of the location of the protein in the cell. First-order disruption kinetics have been observed in machines with predominant plug flow, whereas in machines in which the rotor design permits significant back mixing the disruption deviates from the first-order behavior (Dunnill and Lilly, 1975). A number of other authors have also reported first-order disruption in bead mills (Rehacek and Schaefer, 1977; Marffy and Kula, 1974).Let us assume that the maximum amount of protein C present in the cell is C_m and that the amount released is C; then, for a batch reactor the design equation is,

$$\frac{dC}{dt} = k(C_m - C) \qquad 5.1$$

in which k is the reaction rate constant. Integrating equation 5.1 yields,

$$\ln\left(\frac{C_m}{C_m - C}\right) = \ln D = kt \qquad 5.2$$

in which k is a function of temperature, impeller rotational speed, bead loading, bead size, and cell concentration. A more useful quantity is the disruption efficiency, defined by,

$$E = \frac{CmQ}{P'} \qquad 5.3$$

in which C is the amount of protein released per kg of packed yeast, Q is the flow rate, m is the microorganism concentration, and P′ is power consumption.

Limon-Lason et al. (1979) developed one of the most complete models of bead mills. They considered that each compartment (the space between disks in Figure 5.5) in a bead mill is equivalent to a continuously stirred tank reactor (CSTR). On the basis of tracer studies, they concluded that a model of CSTR in series with backflow is a good representation of the mill. This yields,

$$\frac{C_m}{C_m - C} = \left(1 + \frac{k\tau}{j}\right)^j \qquad 5.4$$

in which τ is the residence time (volume of the mill/flow rate) and j is the number of stirred tanks. In the limit of an infinite number of stirred tanks the mill would work as a plug flow reactor, which implies that all the cells will have the same residence time, which is highly desirable to control the degree of disruption and size of the debris.

An axial dispersion model to simulate a bead mill is a natural extension of the work by Limon-Lason. This model characterizes mass transport in the axial direction in terms of an effective or apparent longitudinal diffusivity D_L that is superimposed on the plug flow velocity. This model also assumes that the fluid velocity and reactant concentration are constant across the reactor diameter.

The magnitude of the dispersion is assumed to be independent of position within the vessel. Therefore, there will not be stagnant regions and no bypassing or short-circuiting of fluid flow in the model reactor.

We focus the following discussion on the disruption of baker yeast, whose kinetics of protein release by disruption in a high-speed bead mill has been fully investigated (Dunnill and Lilly 1975).

The non-steady state differential equation that represents the dispersion model in dimensionless form is:

$$-\frac{1}{Pe}\frac{\partial^2 C}{\partial Z^2}+\frac{\partial C}{\partial Z}+RC=-\frac{\partial C}{\partial t^*} \qquad 5.5$$

in which Pe is the Peclet group, Pe = Lv/D_L , L is a length term that characterizes the system, here taken for convenience as the mill's length; v is a characteristic velocity; R is the rate group for a first order reaction, R = kL/v; C is the fraction of protein remaining inside the cells; Z is a dimensionless distance in the axial direction varying between Z=0 and Z=1 and t is a characteristic time, t = t′v/L. Wehner and Wilhelm (1956) have shown that the "best" set of boundary conditions for Equation 5.5 is:

$$C(\infty) = 1 \qquad 5.6$$

$$C(0-)-\frac{1}{Pe_a}\frac{\partial C(0-)}{\partial Z}=C(0+)-\frac{1}{Pe_b}\frac{\partial C(0+)}{\partial Z} \qquad 5.7$$

$$C(0-) = C(0+) \qquad 5.8$$

$$C(1-)-\frac{1}{Pe_b}\frac{\partial C(1-)}{\partial Z}=C(1+)-\frac{1}{Pe_c}\frac{\partial C(1+)}{\partial Z} \qquad 5.9$$

$$C(1-) = C(1+) \qquad 5.10$$

$$C(\infty) \text{ is } \textit{finite} \qquad 5.11$$

Boundary conditions 5.7 and 5.9 are a result from the conservation of reactants at exit and entrance of the bed, taking into account flow and diffusion. Boundary conditions 5.8 and 5.10 arise from the intuitive argument that at steady state the concentration should be continuous between sections at the bed entrance and exit. Boundary conditions 5.7 and 5.9 are valid in steady and non-steady state. However, Danckwerts' (1953) approximation

$$\frac{\partial C(1-)}{\partial Z}=0 \qquad 5.12$$

is only valid at steady state for Pe finite and valid for steady and non-steady state for Pe infinite. The initial conditions are

$$C = 0 \qquad t^* = 0 \qquad \forall Z \qquad 5.13$$

$$C = 1 \qquad t^* = 0+ \qquad Z = 1 \qquad 5.14$$

To solve equation 5.5 we need to specify the Peclet number and we need an expression for the disruption rate constant (k). It is known that k is a function of several parameters: temperature, impeller rotational speed (Dunnill and Lilly 1975, Marffy and Kula 1974), bead loading (Dunnill and Lilly 1975, Shutte et al. 1983), bead size (Shutte et al. 1983, Marffy and Kula 1974), and cell concentration (Shutte et al. 1983). Limon-Lason et al. (1979) suggested,

$$k = Ku_t \tag{5.15}$$

in which K is a constant, which depends on the type of mill, and μ_t is the impeller rotational speed. Correlations for k with temperature and yeast concentration could be obtained by fitting Limon-Lason's results (1979). The following correlation of k with T was found,

$$k= a + b\,T + c\,T^2 \tag{5.16}$$

For the 5-L mill with stainless steel impellers, a flow rate of 66.7 10^{-6} m $^3/s$, u_t =8m/s and $Y=0.45$, the values of a, b, and c are a=0.02984, b=7.14286 10^{-5}, and c=-7.9365 10^{-5}. The correlation of k with yeast concentration is,

$$k = 0.065 - 0.03375[Y] \tag{5.17}$$

Equation 5.17 is valid for a 5-L mill, polypropylene impellers, and a flow rate of 20 10^{-6}m^3/sec. Equation 5.5 can be solved at steady state subjected to boundary conditions 5.7 and 5.12 analytically. The solution is:

$$C = \frac{4\beta \exp(Pe/2)}{\exp(\beta Pe/2)[1+\beta]^2 - \exp(-\beta Pe/2)[1-\beta]^2} \tag{5.18}$$

in which β is defined by

$$\beta = \left[1 + \frac{4k\theta}{Pe}\right]^{1/2} \tag{5.19}$$

and

$$\theta = \frac{L}{u} \tag{5.20}$$

Dispersion numbers (Pe) available from published data (Limon-Lason et al. 1979) can be used to calculate C from Equation 5.18. The predicted values of C versus fluxes agree well with experimental data and with the predicted values using a model of CSTR in series with backflow. The trend of C versus yeast concentration plots follows the experimental values for Y > 0.3. But, for Y < 0.3, the model predicts the opposite behavior. We used published data for C_M (maximum amount of available protein). In that work the authors found a value of C_M=0.076 for C < 0.3, and a value of 0.09 for C > 0.3. If C_M were not a step function of the packed yeast concentration, we would be able to explain using the previous model the experimental data. Figure 5.8 shows the concentration profile for several yeast concentrations. For any yeast concentration at z = 0.5, 90% of the final conversion is reached. Further inclusion of denaturation kinetics in Equation 5.5 will demonstrate if it is better to stop the disruption at lower levels of conversion.

Equation 5.5 at non-steady state can be solved using a finite difference algorithm. Wehner and Wilhelm (1956) point out that we cannot use boundary conditions 5.7 and 5.12 but in this case bound-

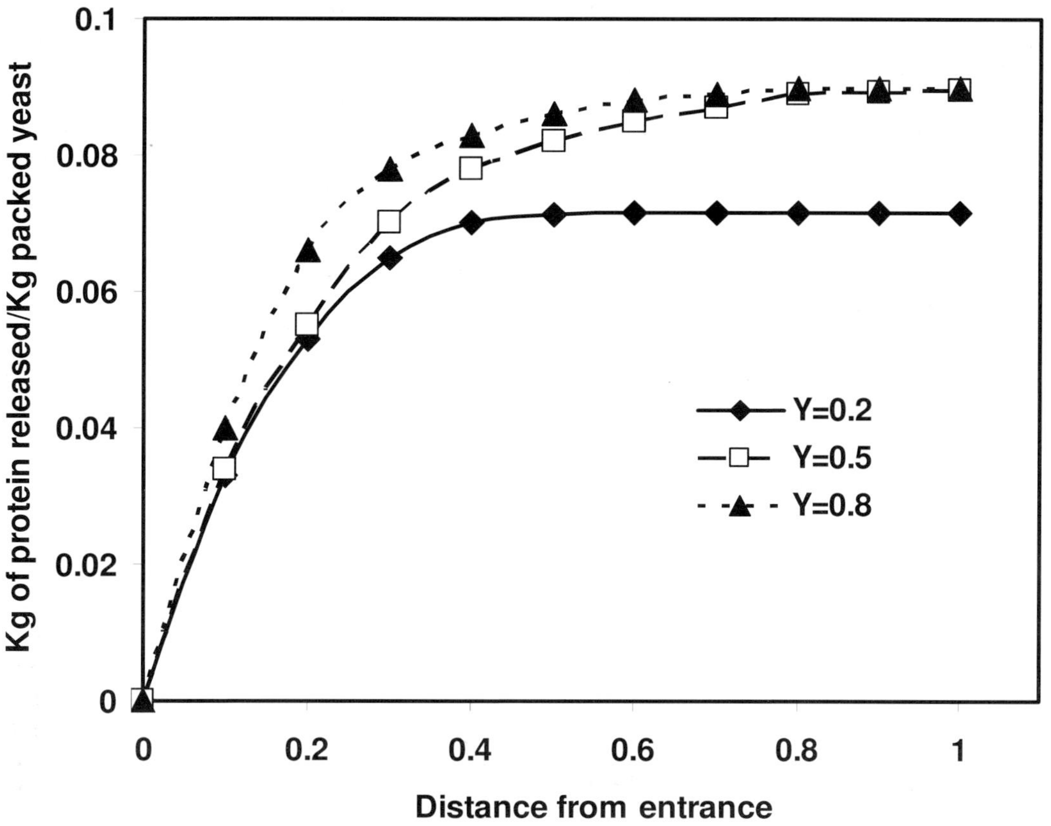

Figure 5.8. Amount of protein release vs. the length along the mill calculated using a dispersion model.

ary conditions 45.7 and 5.9 are still applicable. The total time elapsed to reach steady state is of 20 hours. The system was stable under different sets of parameters.

The dispersion model described above is able to simulate the dynamic and steady state behavior of bead mills. To obtain a more complete description of the system the model has to include enzyme denaturation kinetics and an energy balance. The theory predicts that 90% of the total cell disruption takes place in the first half of the mill. In the remaining half, the enzymes in solution are exposed to heat and shear effects, increasing the denaturation risks.

In general, enzymes have a marginal stability; that is, they are sensitive to shear and heat denaturation (Lencki, Tecante, and Choplin 1993). The simplest way to incorporate denaturation into Equation 5.5 is to write

$$R = \frac{kL}{v}C - \frac{k_d L}{v}\left(\frac{C}{R_M} - 1\right) \qquad 5.21$$

in which k_d is the denaturation constant. Temperature affects conversion, because the disruption kinetic constant and the suspension viscosity (reduced viscosity at higher temperatures would lead to greater back mixing) are a function of temperature. Incorporating an energy balance, we can account for temperature effects and can calculate the power needed to operate the system. The incorporation of the energy balance to the system brings an additional complexity to the model, because we need to solve the momentum balance in order to calculate the viscous dissipation term.

Detailed models help in the understanding of any process. However, they are not often useful from a practical viewpoint; that is, how can we use the model to optimize a cell disruption process. Empirical correlations of experimental data apply only to the process from where the correlation has been obtained and they may or may not be of use in scaling up. Still, they are convenient during the development stages of the disruption process. Because of the number of parameters that may affect the disruption efficiency, factorial design (discussed in detail in the appendix) is an attractive alternative. For example, Belo et al. (1996) used a 2^{6-2} fractional factorial design to optimize the release of cytochrome b5 by *E. coli*. The six factors considered were disruption time, cellular concentration, beads load, bead size, addition of toluene, and addition of lysozyme. After this incomplete design they performed a full factorial design 2^3 where the three factors were disruption time, beads load, and addition of lysozyme (the three factors showing positive response in the preliminary design). This reduced set of experiments demonstrated that the addition of lysozyme was statistically insignificant. The two-parameters "model" was fitted with a three-parameters polynomial. The optimum is obtained by maximizing the polynomial with respect to two variables: beads load and operating time. They confirmed the results of optimization by performing one additional experiment at the optimum conditions. Their work is a good example of the rational use of experimentation for the design and analysis of bead mills.

Homogenizers

High-pressure homogenizers are a common sight in the food processing industry. Actually, most of their development was motivated by the need of the milk processing industry. High-pressure homogenizers use a positive displacement pump to circulate the cell suspension through a specially designed valve. There are several devices available that differ, mostly, in the design of the valve. One of the most popular designs is the APV Gaulin homogenizer in which the suspension flows through a gap between the valve and the valve seat and leaves in the radial direction. After a short distance called the impact distance, the jet impacts on the "impact ring" and leaves the valve. The gap space can be adjusted by a flywheel attached to the valve rod, so if the flow rate is maintained constant, a decrease in the valve gap increases the velocity of the jet and increases the inlet pressure. Another popular device is the Niro Soavi VHP homogenizer that operates under the same principles; however, its valve design is different. Niro Soavi produces equipment able to process from 80 to 750 L/h. A high pressure homogenizer whose "valve" design is completely different is produced by Microfluidics (Newton, MA). The valve is replaced by a series of microchannels where the fluid is accelerated by a high-pressure reciprocal pump. The advantage of this design is that it has no moving parts and therefore its cleaning is simple. Besides, in situ sterilization can be easily implemented.

The cell disruption mechanism in a high-pressure homogenizer seems to be a combination of cavitation, turbulence, shear impingement, and pressure drop. In spite of the apparent violence of the process, it has been argued that shear induced protein denaturation is not important. Several models have been proposed using for example grinding theory (Wong, O'Neill, and Middelberg 1997) or computational fluid dynamics (Miller, Rogowski, and Kelly 2002). However, because of its complexity, it is common practice to correlate the degree of disruption with operating parameters empirically. For example (Hetherington et al. 1971),

$$\log\left(\frac{C_m}{C_m - C}\right) = k' N P^{\alpha} \qquad 5.22$$

in which C_m is the maximum obtainable protein release after N passes, k' is a dimensional $[Pa^{-\alpha}]$ disruption constant, P is the operating pressure, α is an organism dependent parameter that meas-

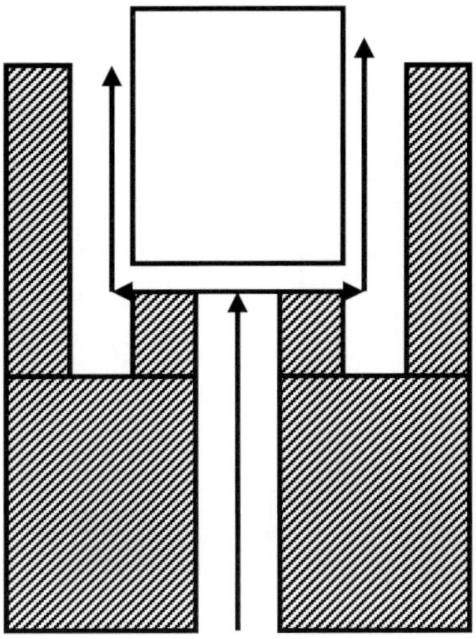

Figure 5.9. Schematic of a high pressure homogenizer valve. The arrows indicate the direction of the flow.

ures the resistance to disruption. For example $\alpha = 2.9$ for *S. cerevisiae* and $\alpha = 2.2$ for *E. coli*. The use of high-pressure homogenizers is limited to bacteria and yeast because mycelial-forming organisms block the nozzles. The disruption depends on the history of the culture but it is independent of cell mass concentration and increases with increasing temperature. Heat damage to proteins is less than with a bead mill because the temperature increase is less significant (about 2°C per 10 MPa).

The common operating strategy is to increase the flow rate by increasing the pressure drop. It is customary to pass the cell suspension multiple times through the valve but one must be careful not to produce materials so fine that would make impossible the separation of the solids from the supernatant via, for example, filtration or centrifugation.

Some disruption parameters have been correlated with geometric and operating variables. For example, Keshavarez (1991) showed that,

$$P_s \approx \frac{\kappa}{X^2 h^2} \qquad 5.23$$

in which P_S is the wall impact pressure, κ is a constant, X is the impact distance, and h is the gap spacing. Therefore, the impact pressure decreases with the square of the impact distance and gap spacing. However, the correlation with the square of the impact distance has been questioned by recent computational fluid dynamics calculations, which indicate that the pressure gradient decreases with the gap spacing to the power three according to,

$$G = \frac{\rho Q^2}{\beta h^3} \qquad 5.24$$

in which G is the pressure gradient, Q is the volumetric flow rate, ρ is the fluid density, and β is a constant.

An interesting approach to better understand how a high pressure homogenizer breaks cells is to use computational fluid dynamics, CFD. In this approach, the Navier-Stokes equations are numer-

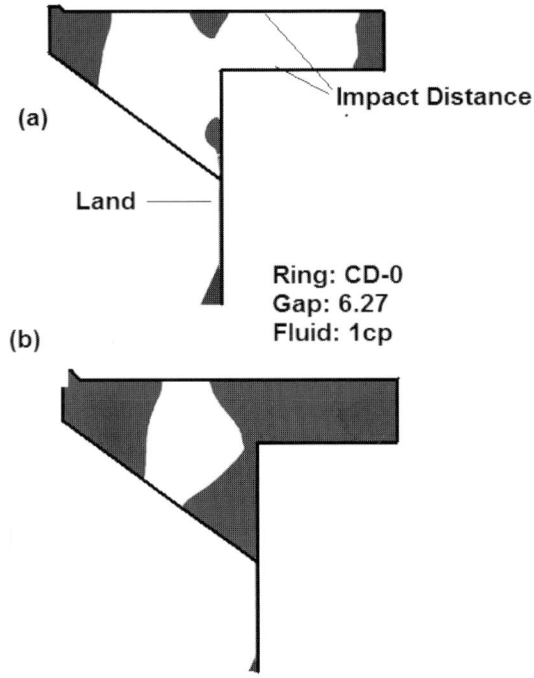

Figure 5.10. Location and size of the negative pressure (in white) zone. Gap width in μm. (a) Exit pressure: 5 psi bp, P_{max}: 7,006 psi. (b) Exit pressure 60 psi bp, P_{max}: 7,120 psi. Adapted with permission from Miller, J., M. Rogowski, and W. Kelly. 2002. Using a CDF model to understand the fluid dynamics promoting *E. coli* breakage in a high pressure homogenizer. *Biotechnology Progress* 18:1060-7. Copyright by the American Chemical Society.

ically solved for a particular geometry, for example, the valve of a high-pressure homogenizer. Recently, CFD (using Fluent, a standard software to perform CFD calculations) has been used to analyze the fluid dynamics of a APV Gaulin model 30 CD valve (Miller, Rogowski, and Kelly 2002). The study was focused on the effect of gap space, impact distance, and fluid viscosity on the parameters believed to be responsible for cell breakage. Figure 5.10 shows a calculated map of fluid velocities in the homogenizer valve. The white zones correspond to negative values of the pressure. Zones of negative pressures correspond to regions in the valve where cavitation occurs, which has been identified as a disruption mechanism in high pressure homogenizers (Lander et al. 2000). The main conclusions of this work were that impact distance only affected wall impact mechanism, but fluid viscosity affected all mechanisms. Unfortunately, all work was done with constant viscosity. This is important because the viscosity of the suspension increases, as a result of the freeing of nucleic acids, as the cells are disrupted. They found a potentially valuable correlation between cell breakage and impact pressure as,

$$\% \text{ cell breakage} = \text{maximum impact pressure (psig)} - 157.8 \qquad 5.25$$

The above models do not account for the properties of the cells in an explicit manner. It is known, for example, that cells in the stationary phase are more difficult to disrupt because of the cross linking nature of the peptidglycan layer (Middelberg, O'Neill, and Thomas 1994). The disruption rate

Table. 5.3. Parameters in Equation 5.28 (adapted with permission from Middleberg, A. P. J. 1995. Process-scale disruption of microorganisms. *Biotechnology Advances* 13:491-551). Copyright by Elsevier B. V.

Parameter	Value		
	Escherichia coli	Baker's Yeast	Brewer's Yeast
m	12.6	$12.2Y^{-0.18}$	$12.2Y^{-0.18}$
n	0.393	0.38	0.38
d	7.85	10.6	10.6
$\bar{S}$	33X-8.0L+48.82	61-64	51
σ	3.82	10-14	4

Y is the impact distance, X is the fractional peptidoglycan cross-linkage, and L is the average cell length.

can be estimated by the correlation between the strength distribution of the cells, $f_S(S)$ and the stress imposed by the homogenizer $f_D(S)$. For N passes the extent of the disruption is,

$$D = 1 - \int_0^\infty \left(1 - f_D(S)\right)^N f_s(S)dS \tag{5.26}$$

Of course, the problem is to know the distribution of stresses. For *E. coli* it has been proposed (Middelberg, 1992; Hull and Middelberg 1993),

$$f_S(S) = \frac{1}{\sigma\sqrt{2\pi}}\exp\left[\frac{-(S-\bar{S})^2}{2\sigma^2}\right] \tag{5.27}$$

in which $\bar{S}$ is the mean effective strength of the population and σ is a measure of the distribution of stresses. Considering the impact of small cylinders against a plane surface the following expression has been proposed for $f_D(S)$ (Middelberg 1992),

$$f_D(S) = \frac{\left(mP^n\right)^d}{S^d + \left(mP^n\right)^d} \tag{5.28}$$

Values of the parameters m, n, and d for several microorganisms are shown in Table 5.3 (Middelberg 1995).

Chemical and Biochemical Methods

Chemical methods for cell disruption include the use of antibiotics, organic solvents, phage lysis, autolysis, chelating agents, hydroxide and hypochlorite, detergents, and enzymes.

The mechanism of cell disruption by antibiotics depends on their kind. For example, β-lactam antibiotics inhibit peptidoglycan synthesis and polymixin disorganizes both the outer and inner membranes. Antibiotics are limited to laboratory processes because of their high cost. The use of organic solvents, although relatively common in the lab, is unlikely to gain a place in industry. Most of the time, the release is low and most solvents (such as toluene) are toxic. Another method of low industrial interest is the use of phages to lyse cells. Phage contamination in the cell culture spells disaster, as the cells will be killed before the process is completed. One alternative is to clone the phage gene directly in *E. coli* (Sanchez-Ruiz 1989), but requires further exploration to determine its validity in an industrial setting.

Chelating agents like EDTA are able to disrupt cell walls containing divalent ions. The complexation of the cations by EDTA releases patches of polyliposacharides. EDTA alone is unable to disrupt the inner membrane. Therefore, it is often used in conjunction with other methods.

Alkaline cell disruption is very harsh and viable only when the product can tolerate extreme pHs. Alkaline lysation works by saponification of the lipids of the cell wall. Hypochlorite (3%) has been used for the release of commercially important polymers such as polyhydroxyl butyrate (Hahn et al. 1995).

Detergents are a mild or supplemental alternative to mechanical disruption. Detergents increase the permeability of the cell membrane by disrupting lipid-lipid, protein-protein, and protein-lipid interactions. They are used also to solubilize membrane proteins and lipids, control protein crystallization, prevent nonspecific binding, and are used as additives in electrophoresis.

Detergents are amphiphatic compounds that are used to solubilize fats in household products. Amphiphatic compounds are special types of molecules having a hydrophilic (or ionic) head and a hydrophobic (fat-loving) tail. When a detergent is in contact with fats its hydrophobic tail associates with the fat as the hydrophilic tail remains in contact with water. When several detergent molecules surround a molecule of fat, they increase its solubility in water, because the fat is surrounded by a polar crown (a micelle). Because lipids mostly make up cell membranes, it seems logical that detergents are used for the disruption of the membranes.

Several companies offer premade detergent-base cocktails. For example, Pierce offers Poppers Cell Lysis Reagents. Cocktails are available for Gram-negative bacteria, *S. aureus*, *E. coli*, Archaebacteria, nematodes, insect cells, yeast, mammalian cells, tissue (such as heart, liver, kidney and brain), and nuclear proteins.

To understand the effect of detergents on membranes, it is useful to consider the behavior of a detergent in an aqueous solution as its concentration is varied. Detergents at very low concentrations form a monolayer at the liquid/air interface; their tails are in contact with air while their heads are pointing toward the bulk of the water. As the concentration of the detergent is increased, they form aggregates called micelles. The monomers arrange in a spherical pattern with their tails pointing toward the center of the sphere and their heads in contact with the water. The critical micelle concentration, CMC, is the minimum concentration of detergent needed for the formation of micelles. The number of monomers per micelle is called the aggregation number. Finally, the critical micelle temperature is the lowest temperature at which micelles can form. A detergent solution is clear; as the CMT is reached the solution becomes turbid, the temperature at which turbidity starts is called the cloud point. Thermodynamics indicate that the cloud point is a lower temperature critical point. At temperatures below the CMT the solution split into two phases, one phase is rich in water and the other is rich in the detergent. The aggregation number, the CMC, and the CMT are used to characterize a detergent. CMC is an indicator of the strength at which detergent binds to protein, low values indicate strong binding and high values indicate weak binding. It can be understood as a numerical value of hydrophilicity. The CMC of nonionic detergents in general decreases with increasing temperature, whereas the CMC of ionic detergents decreases with an increase in ionic strength. Finally, the aggregation number increases with increasing ionic strength.

Detergents can be nonionic, cationic, anionic, and zwitterionic. Zwitterionic detergents are milder and less denaturing than ionic detergents and they are used to solubilize membrane proteins. Some detergents are "mild," such as sodium cholate and sodium deoxycholate, whereas others, such as SDS, are very strong and denaturant. Table 5.4 summarizes the most common detergents as well as their main characteristics. The feasibility of removing the detergent by dialysis is included in the last column. This feasibility strongly depends on the size of the detergent and the size of the micelles. CHAPS and octyl-β-glucoside form small micelles and they can be removed by dialysis or

Table 5.4. Common Detergents (adapted with permission from Pierce, part of Thermo Fisher Scientific, Rockford, IL). Copyright Thermo Fisher Scientific.

Detergent	Type	Aggregation Number	MW	CMC mM	CMC % w/v	Cloud Point °C	Dialyzable
Triton X-100	NI	140	647	0.24	0.0155	64	No
Triton X-114	NI	—	537	0.21	0.0113	23	No
NP-40	NI	149	617	0.29	0.0179	80	No
Brij-35	NI	40	1225	0.09	0.1103	>100	No
Brij-58	NI	70	1120	0.077	0.0086	>100	No
Tween-20	NI	—	1228	0.06	0.0074	95	No
Tween-80	NI	60	1310	0.012	0.0016	—	No
Octyl ß-glucoside	NI	27	292	23-25	0.6716-0.7300	>100	Yes
Octylthio ß-glucoside	NI	—	308	9	0.2772	>100	Yes
SDS	AI	62	288	6-8	0.1728-2304	>100	Yes
CHAPS	ZW	10	615	8-10	0.4920-0.6150	>100	Yes
CHAPSO	ZW	11	631	8-10	0.5048	90	Yes

AI = ionic; CMC = critical micelle concentration; NI = non-ionic; ZW = Zwitterionic.

gel filtration even when the CMC has been exceeded. Triton X-100, Triton-X-114, and NP-40 are used to solubilize membrane proteins under non-denaturing conditions. Because they are bulky, they are difficult to remove by dialysis. Ion exchange chromatography, gel permeation, or precipitation can be used instead. Brij-58, a cetyl ether (C16), and Brij-35, a lauryl ether (C12), are difficult to remove. Brij-35 is used in HPLC to prevent non-specific interactions to gel filtration and affinity gels. Octyl β-glucoside and Octyl β-thioglucopyranoside are also mild and are used in the solubilization of membrane proteins. Both are easily removed by dialysis. Tween-20 and 80 are non-denaturing polyoxyethylene sorbitan esters of fatty acids. CHAPS and CHAPSO are used to solubilize membrane proteins and receptors and are also easily removed by dialysis. If the integrity of the protein must be preserved, ionic detergents such as SDS must be avoided.

SDS is a good example of detergents that are a strong denaturant. It binds cooperatively to membrane and non-membrane proteins below the CMC and alters most proteins into rigid rods whose length is proportional to molecular weight. Because of this, it is commonly used in electrophoresis.

Detergents are particularly handy to isolate membrane proteins (Von Jagow, Link, and Schagger 2003). This has a potentially high economic impact because membrane proteins are approximately 30% of the eukaryotic proteome and are a key target in drug discovery. The most common detergent used for this purpose is Triton. Triton X-100 does not bind soluble proteins and it is noncooperative; that is, the binding of a detergent molecule to the protein is independent of the previous binding of other detergent molecules. Triton X-114 is another member of the Triton family whose solutions are homogeneous at 0°C but above 20°C separate into two clear, one aqueous rich and another detergent rich, phases.

The ratio between detergent and membrane lipids is key for the isolation of membrane proteins. Detergent monomers partition into the membrane bilayer. Thus, the initial stage is lysation or rupture of the membrane. At detergent membrane lipid ratios of 0.1:1 (up to 1:1) the lipid bilayer remains intact, but extraction of some membrane proteins occur. A ratio of 2:1 solubilizes the membrane forming mixed micelles, including phospholipids-detergent micelles, detergent-protein micelles, and lipid-detergent-protein micelles. At a 10:1 ratio, the detergent replaces all native membrane interactions. The optimum amount of detergent depends on the CMC, aggregation number, temperature, type of membrane, and detergent. The goal should be to isolate one membrane per micelle.

One of the most popular enzymes used alone or in cocktails to disrupt cells is lysozyme. Lysozyme (peptidoglycan N-acetylmuramoylhydrolase) has a molecular weight of 14,300 Da and a pI of 11.35. It is active from pH 6.0 to 9.0 and over an ionic strength range of 0.01 to 0.1M, depending on the pH. The enzyme works by hydrolyzing the β(1→4) glycosidic bond between

N-acetylglucosamine and N-acetylmuramic acid in the polysaccharide backbone of peptidoglycan. The way in which lysozyme lyses bacteria is by hydrolyzing the peptidoglycan present in bacterial cell walls. It is commonly used in the lysation of Gram-positive microbial cells because they contain a large amount of peptidoglycan. Gram-negative bacteria are less prone to disruption by the enzyme, because they contain a smaller amount of peptidoglycan and are further protected by their outer membrane. However, they can be made more susceptible to lysation by lysozyme with the addition of EDTA, which chelates metal ions present in the outer membrane, or by increasing the ionic strength.

Use of Additives During Cell Disruption

Proteins may be irreversibly damaged during cell disruption. Besides the possibility of denaturation by shear or foaming in mechanical devices and the potential of hot spots in bead mills that exceed the melting point of the protein, cell membrane disruption may alter the metabolic function of some cellular compounds. Moreover, cell disruption may cause a change in pH and the release of potentially harmful compounds. The major concern is the release of proteases and peptidases because they are proteolytic enzymes. Therefore, it is common practice to add additives to eliminate the possibility of harm. The type of additive depends on the desire effect and on the type of cells. The addition of buffers or polyols is recommended to eliminate pH variations. The disruption of plant cells releases polyphenolic compounds that are enzyme inhibitors. The addition of polyvinylpyrrolidone and ascorbic acid is needed to reduce polyphenol oxidase action. Mercaptoethanol, dithirothreitol, or ascorbic acid may be added to reduce the risks of oxidation. Protease activity is regulated in the intact cell by compartmentalization or by inhibitors to prevent damage of proteins. Cell lysation disturbs this and the proteolytic enzymes may destroy the product of interest. To eliminate, or lessen, this problem, protease inhibitors are added to the disruption mixture. Protease inhibitors are biological or chemical compounds that reversibly or irreversibly bind to the protease inhibiting its enzymatic function. Table 5.5 summarizes some commonly found ones.

Small-scale disruption devices are normally glass made. However, large scale devices are made by metal. These metals- like iron, copper, or nickel-may leach into the cell suspension during the disruption process. The addition of chelating reagents, typically EDTA, is highly recommended to eliminate this problem.

Table 5.5. Common Protease Inhibitors.

Family	Enzyme	Inhibitors
Serine protease trypsin	Subtilisin Thrombin Protease factor Xa	Phenylmethlysulfonyl fluoride Chloromethylketones Benzamidine Aprotinin
Thiol protease (Cysteine protease)	Papain Calpain I and II Cathepsins Caspase-1	Iodoacetate Leupeptin
Acid proteases Carboxyl or aspartyl proteases	Pepsin Rennin Cathepsin D	Pepstain A
Metalloproteases Zinc proteases	Carboxylpeptidase Collagenase Thermolysin	EDTA 1,10 phenathroline

EDTA = ethylenediaminetetraacetic acid

Laboratory Experiments

Two laboratory experiments are described in the following pages. In the first experiment, we present a protocol to disrupt a yeast, *Candida boidinii*, using a bead mill. In the second experiment, we present a protocol to homogenize a tissue, bovine eye lenses.

Disruption of Candida boidinii *in a Bead Mill*

The objective of this experiment is to disrupt *Candida boidinii* to release alcohol dehydrogenase.

Equipment

A refrigerated bead mill. Beads of various materials and sizes. The protocol described below applies to a bead mill refrigerated with CO_2.

Materials

Candida boidinii ATCC 18810; Medium: YM broth (Difco 0711) in 10% methanol.

Methods

Because the production and recovery of the enzyme is a function of the metabolic state of the cell culture, it is convenient to prepare a growth curve first.

a. Separate the cells from the supernatant by centrifugation at 5000 x g for 30 minutes.
b. Place 50 mL of the supernatant in the mill reservoir and add the beads.
c. Turn on the CO_2 flux to refrigerate the mixture before the milling start. A minute should be enough. Longer times will freeze the sample.
d. Turn the mill on for no more than one minute.
e. Remove the container, decant the beads and transfer the supernatant to a centrifuge tube.
f. Centrifuge at 10,000 x g for half an hour at 5°C.
g. Remove a sample of the supernatant and determine total protein content using Bradford test and alcohol dehydrogenase activity test using the enzymatic assay available from Sigma (St. Louis, MO). The enzymatic assay must be performed immediately because ADH is not a very stable protein.
h. There are several parameters that you can change: milling time; the type, concentration, and size of beads; temperature; cell suspension concentration; metabolic state of the yeast. The optimum milling conditions can be determined by performing a 2^8 full factorial experiment and analyzing the statistical model.

Disruption of Lenses in a Wheaton Homogenizer

Equipment

A Wheaton glass tissue grinder.

Materials

Calf lenses can be bought from Antech Animal Technologies, Inc. (Tyler, TX). Otherwise, a "neighborhood" slaughtering house should be able to provide cow eyes from which lenses can be removed from the inside out.

Methods

a. Place eight to nine lenses (note the weight) in the homogenizer and add four volumes (4 mL/g lens) of 100 mM sodium phosphate buffer containing 3 mM sodium azide. Adjust the pH to 7.0 with HCl.
b. Place the grinder into an ice bath and manually homogenize the sample with the pestle. Homogenization is finished when the lenses are completely dissolved into the buffer.
c. Centrifuge the homogenized lenses at 21,000 x g for one hour at 5°C. Centrifugation is performed to separate the soluble proteins from the insoluble membrane material.
d. After centrifugation is completed, the insolubles are discarded and the supernatant is placed in a small, plastic, capped bottle. The supernatant should be frozen if not used immediately.
e. The total amount of protein in the supernatant must be measured by the BCA test. The supernatant will be used in some experiments described in the following chapters.

Summary

There are several methods available for the disruption of cells. All of them are suitable for laboratory experiments but only a handful are being adapted in industry. They are the bead mill, the high pressure homogenizer and the microfluidic device. Reasonably good models exist to facilitate scaling up. Still, direct experimentation in laboratory scale is needed to find the optimal conditions for a particular organism and a particular protein or enzyme. Chemical and biological methods may be used in conjunction with mechanical methods if they are economically and environmentally reliable.

Nomenclature

C_m	maximum obtainable protein released (kg/kg of microorganism)
C	amount of protein released (kg/kg of a microorganism)
G	pressure gradient (Pa/m)
E	disruption efficiency
f_S (S)	strength distribution of a cell population
f_D (S)	distribution of stresses applied to a cell population
h	gap spacing (Length)
j	number of stirred tanks
k	first order reaction rate constant (1/sec)
k_d	denaturation constant (1/sec)
k′	dimensional [$Pa^{-\alpha}$] disruption constant
L	characteristic length
m	microorganism concentration (kg/L)
P	operating pressure (Pa)
P_s	wall impact pressure (Pa)
P′	power consumption (watt)
Pe	Peclet number, $Pe = Lu/D_L$
Q	volumetric flow rate (m^3/sec)
R	rate group for a first order reaction, $R = kL/u$
$\bar{S}$	mean effective strength of the population
t*	characteristic time, $t^* = t'u/L$
t	time (s)

T	temperature (C)
u_t	impeller rotational speed (rad/sec).
v	characteristic velocity (m/sec)
x	impact distance (m)
Y	yeast concentration
Z	dimensionless distance in the axial direction

Greek Letters

τ	residence time (volume of the mill/flow rate)
σ	distribution of stresses
ρ	fluid density (kg/m^3)

References

Asenjo, J.A., and Dunnill, P. 1981. The isolation of lytic enzymes from cytophaga and their application to the rupture of yeast cells. *Biotechnology and Bioengineering* 23:1045-56.

Belo, I., J. A. L. Santos, J. M. S. Cabral, and M. Mota. 1996. Optimization study of *Escherichia coli* TB1 cell disruption for cytochrome b_5 recovery in a small-scale bead mill. *Biotechnology Progress* 12:201-4.

Bermudez, O. and D. Forciniti. 2004. Decontamination of surfaces by lysozyme encapsulated in reverse micelles. *Journal of Chromatography* B, 807(1):95-103.

Bierau, H., R. J. Hinton, and A. Lyddiatt. 2001. Direct process integration of cell disruption and fluidized bed adsorption in the recovery of labile microbial enzymes. *Bioseparation* 10:73-85.

Chisti, Y., and M. Moo-Young. 1986. Disruption of microbial cells for intracellular products. *Enzyme Microbiology Technology* 8:194-201.

Coakley, W.T., R. C. Brown, and C. J. James. 1974. Optimization of the release of undegraded material extracted from cells by ultrasound. *Biotechnology and Bioengineering* 16:659-73.

Currie, J. A., P. Dunnill, and M. D. Lilly. 1972. Release of protein from bakers' yeast (*Saccharomyces cerevisiae*) by disruption in an industrial agitator mill. *Biotechnology and Bioengineering* 14:725-36.

Danckwerts, P. V. 1953. Continuous-flow systems. Distribution of residence times. *Chemical Engineering Science* 2:1-13.

Doulah, M. S. 1978. Application of the statistical theory of reliability to yeast cell disintegration in ultrasonic cavitation. *Biotechechnology and Bioengineering* 20:1287-9.

Dunnill, P., and M. D. Lilly. 1975. Protein extraction and recovery from microbial cells. In *Single Cell Protein* II, edited by S. Tannenbaum and D. I. C. Wang, pp. 179-207. Cambridge, MA: MIT Press.

Emanuel, C. F., and I. L. Chaikoff. 1957. An hydraulic homogenizer for the controlled release of cellular components from various tissues. *Biochimica et Biophysica Acta* 24:254-61.

Engler, C. R., and C. W. Robinson. 1981. Disruption of *Candida utilis* cells in high pressure flow devices. *Biotechnology and Bioengineering* 23:765-80.

Feliu, J. X., and A. Villaverde. 1994. An optimized ultrasonication protocol for bacterial cell disruption and recovery of β-galactosidase fusion proteins. *Biotechnology Techniques* 8:509-14.

Follows, M., P. J. Hetherington, P. Dunnill, and M. D. Lilly. 1971. Release of enzymes from Bakers' yeast by disruption in an industrial homogenizer. *Biotechnology and Bioengineering* 13:549-60.

Gray, P. P., P. Dunnill, and M. D. Lilly. 1972. The continuous-flow isolation of enzymes. In *Fermentation Technology Today*, edited by G. Terui, pp 347-51. Osaka, Japan: Society of Fermentation Technology.

Hahn, S. K, Y. K. Chang, B. S. Kim, and H. N. Chang. 1995. Recovery and characterization of poly(3hydroxybutyric acid) synthesized in *A. eutrophus* and recombinant *E. coli. Applied and Environment Microbiology* 61:34-9.

Hatti-Kaul, R., and B. Mattiasson. 2003. Release of protein from biological host. *Biotechnology and Bioprocessing* 27:1-27.

Hedenskog, G., H. Mogren, and L. Edebo. 1970. A method for obtaining protein concentrates from microorganism. *Biotechnology and Bioengineering* 12:947-59.

Hetherington, P. J., M. Follows, P. Dunnill, and M. D. Lilly. 1971. Release of protein from baker's yeast (*Saccharomyces cerevisiae*) by disruption in an industrial homogenizer. *Transactions of the Institution of Chemical Engineers and the Chemical Engineer* 49:142-8.

Hull, A. S., and A. P. J. Middelberg. 1993. Evidence that septated *E. coli* disrupt preferentially during high-pressure homogenization. *Transaction of the Institution of Chemical Engineers. Part C: Food and Bioproducts Processing* 71:264-6.

Hunter, M. J., and S. L. Commerford. 1961. Pressure homogenization of mammalian tissues. *Biochimca et Biophysica Acta* 47:580-6.

Katsui, N., T. Tsuchido, M. Takano, and I. Shibasaki. 1982. Heat-induced blebbing and vesiculation of the outer membrane of *Escherichia coli. Journal of Bacteriology* 151:1523-31.

Keshavarez Moore, E. 1991. New findings on fundamentals of cell disruption in high pressure homogenizers. *Biology of Recombinant Microorganisms and Animal Cells* 34:27-92.

Lander, R., W. Manger, M. Scouloudis, A. Ku, C. Davis, and A. Lee. 2000. Gaulin homogenization: A mechanistic study. *Biotechnology Progress* 16:80-5.

Lencki, R. W., A. Tecante, and L. Choplin. 1993. Effect of shear on the inactivation kinetics of the enzyme dextransucrase. *Biotechnology and Bioengineering* 42:1061-7.

Limon-Lason, J., M. Hoare, C. B. Orsborn, D. J. Doyle, and P. Dunnill. 1979. Reactor properties of a high-speed mill for microbial cell rupture. *Biotechnology and Bioengineering* 21:745-74.

Magnusson, K. E., and L. Edebo. 1974. Estimation of the disruption in freeze-pressed Accharomyces cerevisiae by an electronic particle counter. *Biotechnology and Bioengineering* 16:1273-82.

Marffy, F., and M. R. Kula. 1974. Enzyme yields from cells of brewer's yeast disrupted by treatment in a horizontal disintegrator. *Biotechnology and Bioengineering* 16:623-34.

Middelberg, A. P. J. 1992. A model for the disruption of *Escherichia coli* by high pressure homogenization. Ph.D. thesis, University of Adelaide, Adelaide, Australia.

Middelberg, A. P. J., B. K. O'Neill, and C. J. Thomas. 1994. A simplified model for the disruption of *Escherichia coli*: the effect of cell septation. *Biotechnology Progress* 10:109-13.

Middelberg, A. P. J. 1995. Process-scale disruption of microorganisms. *Biotechnology Advances* 13:491-551.

Miller, J., M. Rogowski, and W. Kelly. 2002. Using a CDF model to understand the fluid dynamics promoting *E. coli* breakage in a high pressure homogenizer. *Biotechnology Progress* 18:1060-7.

Mogren, H., M. Lindblom, and G. Hedenskog. 1974. Mechanical disintegration of microorganisms in an industrial homogenizer. *Biotechnology and Bioengineering* 16:261-74.

Pisabarro, A. G., M. A. de Pedro, and D. Vazquez. 1985. Structural modifications in the peptidoglycan of *Escherichia coli* associated with changes in the state of growth of the culture. *Journal of Bacteriology* 161:238-42.

Rehacek, J., and J. Schacfer. 1977. Disintegration of microorganisms in an industrial horizontal mill of novel design. *Biotechnology and Bioengineering* 19:1523-34.

Sadana, A. 1994. Bioseparation steps for proteins and other biological products. Part 1: Cell disruption. *BioPharm* 7(2):26, 27, 30, 32, 34-36.

Sanchez-Ruiz, S. A. 1989. Studies on cell disruption and cell debris removal in downstream bioprocessing. M.Sc. thesis, MIT, Cambridge, MA.

Shutte, H., K. H. Kroner, H. Hustedt, and M. R. Kula. 1983. Experiences with a 20 litre industrial bead mill for the disruption of microorganisms. *Enzyme and Microbial Technology* 5:143-8.

Siddiqi, S. F., N. J. Tichener-Hooker, and P. Ayasi Shamlou. 1996. Simulation of particle size distribution changes occurring during high-pressure disruption of bakers' yeast. *Biotechnology and Bioengineering* 50:145-50.

Van Gaver, D., and A. Huyghebaert. 1990. Optimization of yeast cell disruption with a newly designed bead mill. *Enzyme and Microbial Technology* 13:665-71.

Van Hee, P., A. P. J. Middelberg, R. G. J. M. Van der Lans, and L. A. M. Van der Wielen. 2004. Relation between cell disruption conditions, cell debris particle size, and inclusion body release. *Biotechnology and Bioengineering* 88:100-10.

Von Jagow, G., T. A. Link, and H. Schagger. 2003. *Purification Strategies for Membrane Proteins. Membrane Protein Purification and Crystallization,* 2nd edition. San Diego, CA: Elsevier Science, pp. 1-18.

Wehner, J. F., and R. H. Wilhelm. 1956. Boundary conditions of flow reactor. *Chemical Engineering Science* 6:89-93.

Wong, H. H., B. K. O'Neill, and A. P. J. Middelberg. 1997. A mathematical model for *Escherichia coli* debris size reduction during high pressure homonenisation based on grinding theory. *Chemical Engineering Science* 52:2883-90.

Woodrow, J. R., and A. V. Quirk. 1982. Evaluation of the potential of a bead mill for the release of intracellular bacterial enzymes. *Enzyme and Microbial Technology* 4:385-90.

Chapter 6
Separation of Solids

Introduction

The separation of solids from a mother liquor is a common bioprocessing operation. Some examples include the separation of an enzyme of interest from the cell culture broth, the separation of cell debris from its supernatant containing an intracellular enzyme, and the separation of a protein precipitate from its mother liquor. The simplest way to separate a solid from a liquid is by allowing the solid to sediment under gravity. Unfortunately, gravity-driven sedimentation is a slow process. Therefore, the sedimentation rate is increased by imposing a centrifugal force (centrifugation). An alternative to centrifugation is filtration, which is a well-established technique that is used in a variety of industries, including service industries such as public utilities. Although the filtration equipment may be quite conventional, new and very selective filters have been developed recently. These new developments have found new uses for filtration. Although the main operations for the separations of solids are filtration and centrifugation, liquid/liquid separations may be an alternative for very large processing volume.

In both filtration and centrifugation, the quality of the separation depends on the size and stability of the solid particles. That is the larger the particle the better (or faster) the separation is. The solids found in bioseparation have a variety of sizes. They range from whole cells that are of micron size to proteins that are in the nanometer range. The size of cell debris falls somewhere in between with sizes from a few nanometers to microns. One way of increasing the size of the solids is to promote their association in larger entities or flocks. Because this is common practice in industry, we dedicate the first portion of this chapter to the study of flocculation. This is followed by a relatively brief treatment of centrifugation. Then, the chapter presents the principles of filtration, covering conventional filtration, ultrafiltration, dialysis, reverse osmosis, and some new developments in ultrafiltration. Finally, the chapter discusses the basic principles and the application of liquid/liquid extraction for the separation of solids.

Flocculation

The objective of flocculation is to increase the size of particles to facilitate centrifugation or filtration. One of the most common uses of flocculation is in water purification facilities, where salts (mostly aluminum acetate) are used to coagulate particulates present in the water. The flocculation techniques used in water purification facilities and in bioprocessing are the same and are based in both cases in the efficiency of a material to destabilize colloids.

It is customary to consider proteins and other bioparticles colloids. It is important, however, not to fall into the common mistake of considering them true colloids. Bioparticles and true colloids have a net surface charge, however, colloids are rigid and bioparticles are deformable. Besides, the surface charge in true colloids is uniformly distributed, whereas in cell debris and proteins the surface charge is distributed in negative and positive patches that are not evenly distributed on the surface. Still, the presence of a relatively heavy charge surface and their size gives proteins and cell fragments similar characteristics to colloids.

The Derjaguin, Landau, Verwey, and Overbeek (DVLO) theory was explained in detail in Chapter 4 regarding protein solubility. It was pointed out that DVLO theory can be used to predict the "critical flocculant concentration"—that is, the minimum amount of neutral electrolyte needed to initiate coagulation of the colloid. DVLO theory predicts that the critical flocculant concentration depends on $1/Z^6$, where Z is the charge of the electrolyte. An empirical extension of this finding is called the Schulze-Hardy Rule, which establishes a general guideline to choose a flocculant. We can state it as:

> If one needs a concentration of 1 to coagulate a colloid for an electrolyte of charge 3, then as we decrease the charge to 2 and 1 we will need ten and a thousand times more electrolyte respectively.

Four mechanisms are used to explain the coagulation of colloids by the addition of flocculants. These four are double-layer compression, adsorption and charge neutralization, enmeshment in a precipitate, and interparticle bridging. Only one mechanism may be present, but often more than one mechanism may be acting synergistically. Chapter 4 explains double-layer compression in detail. Adsorption and charge neutralization are actually quite straightforward to understand. Suppose that the surface of a colloid is negatively charged. The addition of an electrolyte, whose cation has a strong affinity for the negatively charged groups on the surface of the colloid, will partially neutralize the colloid's charge and subsequently decrease the zeta potential. This decrease in the zeta potential allows the colloids to overcome the energy barrier that keeps them in solution. Potentially useful salts for charge neutralization are those whose ions have high affinity for the colloid charges. Hydroxymetallic compounds and polyelectrolytes, which have a strong affinity for the colloid surface, are used. Hydroxymetallic compounds are formed when salts of some metals are dissolved in water. Typical examples are aluminum or iron salts such as $Al_2(SO_4)_3$, $FeCl_3$, and $Fe_2(SO_4)_3$. For example, when $Al_2(SO_4)_3$ is dissolved in water the following hydroxymetallic compounds are formed: $Al_6(OH)_{15}^{3-}$, $Al_7(OH)_{17}^{4+}$, $Al_8(OH)_{20}^{4-}$, and $Al_{13}(OH)_{34}^{5+}$. The first and third species will neutralize positively charged colloids, whereas the second and last species will neutralize negatively charged ones. Because the pH of the media determines the dominant species, a judicious choice of the pH and of the concentration of the aluminum salt is needed to precipitate a particular colloid. Although the selection of the pH is straightforward, caution must be used when using this kind of salt, because adding an excess of the metal salt may produce charge reversal of the colloid, and therefore, the colloid will be re-stabilized. Swept precipitation is the dragging of a colloid by the precipitate of other species such as a salt or a polymer. Again, a metal salt when added at high enough concentrations that will follow this mechanism is aluminum sulfate. Aluminum at basic pH precipitates as $Al(OH)_3$, which is an amorphous and gelatinous solid that forms flocks and settles.

This mechanism will prevail under alkaline conditions. Finally, large hydroxymetallic ions such as $Al_7(OH)_{17}^{4+}$ may bridge several particles together. This will cause an increase in size pulling the colloidal particles out of solution.

Ionic polymers or poly(electrolytes) can be used instead of inorganic or organic salts to destabilize colloids. Ionic polymers may coagulate colloids by double-layer compression, specific ion adsorption, sweep flocculation enmeshment, polymer charge patch formation, or polymer bridging. Examples of poly(electrolytes) are alkyl polyamines, aminoacrylate-aceylamide copolymer, and tannins. Ionic polymers are not popular because of difficulties arising in their postcoagulation separation and because of their cost. Many of them will adsorb to protein molecules so strongly that they will disrupt their structure.

Flocculation of bacteria is particularly important in the capture of the microorganisms from fermentation broths. O'Mahony et al. (2006) published an exhaustive study of the effect of several flocculants to increase the capture of bacteria. The study was limited to *Escherichia coli* suspensions with an optical density lower than 15, but considering the importance of this bacterium in biotechnology, the results are quite valuable. They found that the best results were obtained when the pH of the cell suspension was adjusted to 8.0 and after pretreatment with sodium-activated bentonite and PEI. Diatomaceous earth (20g/L) was used as a filter aid.

Centrifugation

Conceptually, a centrifuge is a spinning settling tank. The rapid rotation of the entire unit replaces gravity by a controllable centrifugal or radial force. Because this force is several times larger than gravity, centrifugation increases the sedimentation rate of particles. Centrifuges are built in different sizes and configurations and are available in a range of maximum velocities. In laboratory scale, they consist of a spinning rotor where bottles of various sizes (from a few of milliliters to one liter) are placed. The speed ranges from a few hundred to 100,000 relative centrifugal force in ultracentrifuges, where the relative centrifugal force is the acceleration of the centrifuge express as a multiple of gravity. Table 6.1 summarizes the main characteristics of centrifuges used in bioprocessing. Industrial centrifuges are normally continuous and are dominated by two configurations: tubular and disc stack. Disc stack centrifuges are appropriate to sediment suspensions whose average particle size ranges from 0.1 μm up to half a millimeter at concentrations that do not exceed 25% solids. For larger particle sizes (ranging from 10 μm to several mm) and higher solid concentrations (up to 50%) decanter centrifuges are recommended. For particles below 1 μm, filtration is a better choice unless the density of the particles is very different from the one of the solvent, which is not the case in biotechnology. Because it is convenient to express sedimentation rates that occur in centrifuges in terms of gravity-driven sedimentation rates, we explore the latter first.

Gravity-Driven Sedimentation

Buoyancy and drag are the two forces acting on a particle suspended in a liquid phase. These forces determine the settling rate of the particle. Consider a particle suspended in a fluid subjected to a force field, which points downward. Because the volume occupied by the particle was occupied by the fluid before its insertion, the particle feels a force caused by the displaced fluid in the opposite direction to the force field. The difference between those two forces is called buoyancy and it is given by,

$$F_B = \left[\frac{\pi d^3}{6}(\rho_s - \rho)\right]\vec{a} \qquad 6.1$$

Table 6.1. Characteristics of Continuous Feed Centrifuges (from Axelsson, H., and B. Madsen. 2006. *Centrifuges, Sedimenting. Ullman's Encyclopedia*. New York: John Wiley & Sons, pp. 89-117). Copyright by John Wiley & Sons; reproduced with permission.

Type	Solids Discharge	Feed L/h	Solids vol%	Solids L/h	Σ value (m^2)	Max Z	Consistency of Solids
Disc solid bowl	Manual	20-100,000	<1	0	1000-300,000	10,000	Firm paste
Solids-ejecting radial	Intermittent	20-100,000	<25	<3000	1000-170,000	14500	Thick flow-ring slurry
Solids-ejecting, axial	Intermittent	1000-150000	<15	<1000	110,000-220,000	15,000	Thick flow-ring slurry
Nozzle discharge	Continuous	1000-180,000	4-30	>150 <40,000	69,000-180,000	15,000	Thick flow-ring slurry
Peripheral nozzle	Continuous	300-310,000	2-30	>3,000 <140,000	35,000-180,000	11,000	Thin slurry
Decanter centrifuge	Continous with scroll	300-200,000	5-50	<50,000	400-25,000	10,000	Thick slurry to semisolid
Tubular bowl	Manual	20-7000	<1	0	1,400-4,500	31,000	Firm paste
Multichamber bowl	manual	100-20000	<5	0		9000	Firm paste
Centritech machines	Intermittent	5-100	<1	<15	—	100	
Inverted solid bowl	Intermittent with scraper	10-6000	1-30	<1000	800-20,000	20,000	

in which d is the diameter of a particle, ρ_S is the density of the particle, ρ is the density of the solvent, and $\vec{a}$ is the acceleration as a result of the force field. As the particle moves downward under the effect of the force field, the surface of the particle feels a frictional force caused by the presence of the solvent molecules (equivalent to the drag force that a car feels as it accelerates in the highway). The drag force is,

$$F_D = 3\pi d\mu v_p \quad 6.2$$

in which μ is the viscosity of the solvent and v_p is the velocity of the particle. After a constant settling velocity is reached, $F_B=F_D$. Therefore,

$$3\pi d\mu v_p = \left[\frac{\pi d^3}{6}(\rho_s - \rho)\right]\vec{a} \quad 6.3$$

Under gravity $(\vec{a} = \vec{g})$ and the settling velocity is:

$$v_g = \left[\frac{d^2}{18\mu}(\rho_s - \rho)\right]\vec{g} \quad 6.4$$

Equation 6.4 shows that the sedimentation rate is a function of the density difference between the particles and the solvent, the viscosity of the solvent, and the diameter of the particle squared. The viscosity and the density difference are fixed for a given process (for particles of biological origin or proteins that density difference is a very small number). Therefore, the sedimentation rate under a gravitational field can only be increased by increasing the size of the particles, which is done in practice by adding flocculants as explained in the previous section.

Centrifugal Force-Driven Sedimentation

Under a centrifugal force Equation 6.3 becomes,

$$v_w = \left[\frac{d^2}{18\mu}(\rho_s - \rho)\right]\omega^2 r_i \quad 6.5$$

in which r_i is the distance from the center of the centrifuge to the particle and ω is the rotational speed (radian/time). In this case, the sedimentation velocity is also a function of the rotational speed and of the position of the particle with respect to the axis of rotation. Because the acceleration that the particle feels depends on its distance from the axis of rotation it is customary to characterize the centrifugal force by the g number or acceleration factor Z, which is defined by,

$$Z = \frac{r_i \omega^2}{g} \quad 6.6$$

A common device found in laboratories is the swing-out rotor. Neglecting the time needed for acceleration and de-acceleration (this can be easily incorporated) (Axelsson and Madsen 2006), the settling velocity for a particle in a bottle place in such a device is,

$$v_g = \frac{g}{t\omega^2}\ln\frac{r_2}{r_1} \quad 6.7$$

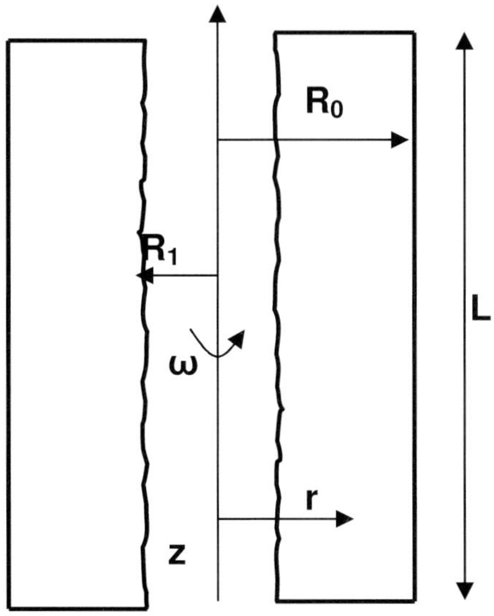

Figure 6.1. Tubular centrifuge. L is the length of the bowl, R_0 is its radius, and R_1 is the distance from the center of the centrifuge to the solid/liquid interface.

in which t is the centrifugation time needed for the particles to leave the volume element in the bottle defined by r_2 and r_1 and where r_i is radial distance from the center of rotation to the particle. The spinning bottle is a convenient lab tool to determine sedimentation velocities. It must be noticed that Equation 6.7 is valid only at low Re values (< 0.4) and dilute concentrations.

The choice of a centrifuge for a given task consists of choosing a machine with the desired centrifugal flow and in the case of continuous centrifuges the desired indicated processing capacity. The design equation for a particular centrifuge depends on the geometry of the instrument. The four main types of centrifuges, tubular, disc stack, multichamber bowl, and decanter, are described in the following paragraphs.

Figure 6.1 shows a tubular centrifuge. A laundry machine is the closest example for this kind of device. Just imagine that you are constantly pouring wet clothes (cells) into it. A variation of tubular bowl centrifuges is the Carr Powerfuge. This centrifuge is a high-speed centrifuge with baffles that facilitate solid retention. It comes available with constant solid removal via a vibrating knife, which scrapes solids away as the centrifuge rotates at low speeds. This competes with the disc stack centrifuges described later.

The mathematical description of a tubular centrifuge is quite straightforward. The flow is bidirectional because the suspension moves in the radial and axial directions. The starting working equation is,

$$v_w = \left[\frac{d^2}{18\mu}(\rho_s - \rho)\right]\omega^2 r_i \qquad 6.8$$

For a tubular bowl centrifuge, the fluid accelerates toward the walls of the centrifuge, in the radial direction; therefore,

$$\frac{dr_i}{dt} = \left[\frac{d^2}{18\mu}(\rho_s - \rho)\right]\omega^2 r_i \qquad 6.9$$

Using Equation 6.5,

$$\frac{dr_i}{dt} = v_g \frac{\omega^2 r_i}{g} \qquad 6.10$$

The flow of solids in the axial direction is given by:

$$\frac{dz}{dt} = \frac{Q}{\pi\left(r_0^2 - r_1^2\right)} \qquad 6.11$$

in which r_0 and r_1 are the radius of the cylinder and the radius of the liquid surface respectively and Q is the feed flow. Equation 6.10 can be divided by equation 6.11 to obtain a phase space (r-z) differential equation,

$$\frac{dr_i}{dz} = v_g \left(\frac{r_i \omega^2}{g}\right) \frac{\pi\left(r_0^2 - r_1^2\right)}{Q} \qquad 6.12$$

The trajectory of a particle along the centrifuge depends on its point of entry. The most difficult particle to sediment will be one entering at r_i=0 and z=0 and reaching the wall at the end of the tube of length L. Integration of Equation 6.12 for such a particle yields,

$$Q = v_g \left(\frac{\omega^2}{g \ln(r_0 / r_1)}\right) \frac{\pi L\left(r_0^2 - r_1^2\right)}{1} \qquad 6.13$$

or,

$$Q = v_g \left(\frac{2\pi L \bar{r}^2 \omega^2}{g}\right) = v_g \sum \qquad 6.14$$

in which $\bar{r}$ is the average log radius. Σ is called the "sigma value" of the centrifuge and it is used for scaling up purposes. It has units of surface and its meaning is that of the equivalent surface area of a static sedimentation tank with the same theoretical performance as the centrifuge. Notice that it is a function of the properties of the centrifuge only. The properties of the fluid to be centrifuged are included in v_g. The flow rate is given by process needs. Therefore, knowledge of the sedimentation rate under gravity allows the calculation of the "sigma value," which defines the characteristics of the centrifuge that is needed. Equation 6.14 is valid for any kind of centrifuge as long as the correct value of Σ is used. In practice, the efficiency of the centrifugal separation is less than the one given by Σ. Therefore, it is customary to calculate the flow-rate by,

$$Q_{actual} = \eta v_g \sum \qquad 6.15$$

in which η is the efficiency of the centrifuge. Depending on the centrifuge and the material the following efficiency values have been found: for disc-bowl centrifuges, η = 45 to 73%; for decanter centrifuges, η = 54 to 67%; and for tubular bowls, η = 90 to 98% (Axelsson and Madsen 2006). Most of the inefficiencies are caused by deviations from ideal flow conditions.

Disc stack centrifuges are one of the most efficient devices to increase the sedimentation rate of a solid suspension. Disc stack centrifuges can handle both solid suspensions and liquid/liquid dispersions. Figure 6.2 shows a disc centrifuge. Figure 6.3 shows this type of centrifuge. Discs are

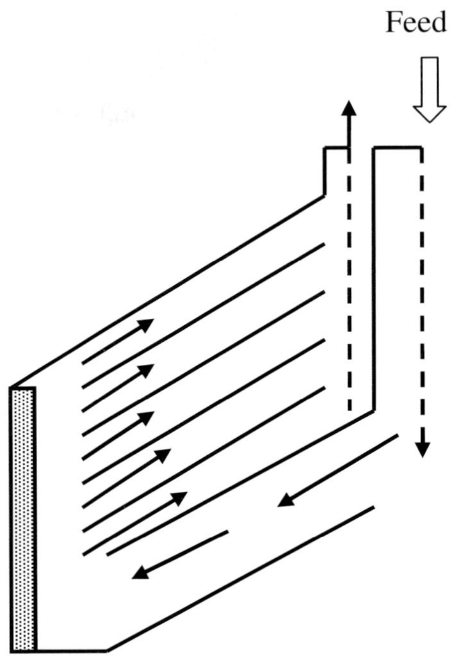

Figure 6.2. Disc stack centrifuge. A half-plane is shown. The arrows indicate the fluid flow directions.

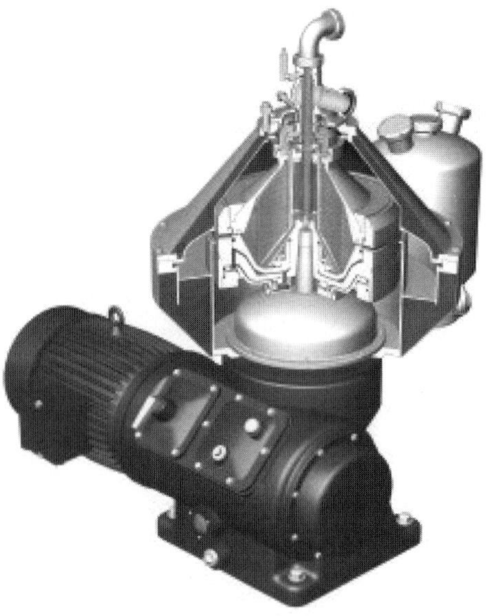

Figure 6.3. Disc stack centrifuge. Copyright Alpha Laval, with permission.

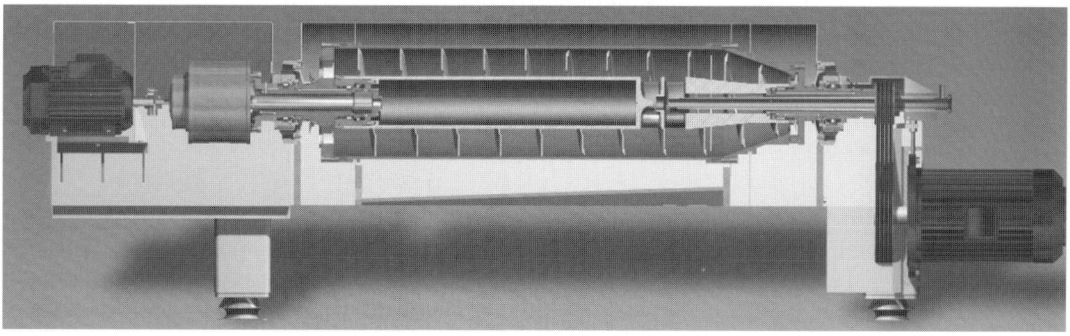

Figure 6.4. Decanter centrifuge Copyright Alpha Laval, with permission.

available in a variety of materials such as stainless steels, tungsten carbide, nickel based alloys, polymeric, and ceramic sealing compounds.

A disc centrifuge has four zones: 1) inlet, 2) disc stack, 3) liquid discharge, and 4) solid discharge. The suspension enters the centrifuge in the inlet zone, where the liquid is accelerated to the speed of the rotating bowl. It has been claimed that a well-designed inlet zone reduces foaming and minimizes temperature increase. In other words, the lower the energy dissipation, the better the centrifuge performs. The disc stack zone is where the separation between the clarified liquid and the sludge takes place. The interface between solids and liquid can be controlled. After the suspension is separated, the liquid is discharged in the liquid discharge section, whereas the solids go to the solid discharge zone. One of the main disc stack centrifuge makers, Alpha Laval, offers three designs: continuous solids discharge that is achieved by nozzles in the periphery of the bowl; intermittent solids discharge that is done by a series of ports that open and close at controlled intervals; and manual removal that is achieved by stopping the machine and removing the solids manually.

An analysis similar to that used for tubular centrifuges, but for a different geometry, yields the following design equation for a disc type centrifuge:

$$Q = v_g \left(\frac{2\pi n \omega^2}{3g} \left(r_0^3 - r_1^3 \right) \cot \theta_D \right) = v_g \Sigma \qquad 6.16$$

in which n is the number of discs, θ is the angle of the discs, r_0 is the distance from the axis of rotation to the beginning of the disc, and r_1 is the distance from the axis of rotation to the tip of the discs.

Other centrifuges are the decanter and the multichamber bowl. The decanter is a variation of the tubular centrifuge. It consists of a screw conveyor within a horizontal chamber that rotate at different speeds from each other. The screw conveyor removes the solids continuously. The liquid is removed from the center of rotation at one end of the bowl. Figure 6.4 shows a decanter produced by Alpha Laval. The Σ value for the decanter is (Axelsson and Madsen 2006),

$$\Sigma = \frac{\pi \varpi^2}{g} \left[L_1 \left(\frac{3}{2} r_2^2 \frac{1}{2} r_1^2 \right) + L_2 \left(\frac{r_2^2 + 3 r_2 r_1 + 4 r_1^2}{4} \right) \right] \qquad 6.17$$

in which L_1 and L_2 are the lengths of the cylindrical and conical parts respectively. r_1 is the inner radius of liquid and r_2 is the inner radius of the bowl, such as r_2-r_1 is the thickness of the liquid in the device. The multichamber bowl centrifuge is a solid-bowl centrifuge with two or more chambers. Because it has longer residence time than a bowl centrifuge, it has better clarification and dewatering performance. They are commonly used in the blood plasma fractionation industry. The Σ value for a multichamber bowl centrifuge is (Axelsson and Madsen 2006),

$$\Sigma = \frac{\pi\omega^2}{g}\frac{L}{3}\sum_{i=0}^{i=n}\frac{r_{2i+1}^3 - r_{2i+2}^3}{r_{2i+1} - r_{2i+2}} \qquad 6.18$$

in which indices with even numbers correspond to inner radius of the chamber, indices with odd numbers correspond to outer radius of the chamber, n+1 is the number of chambers, and L is the height of the chambers.

The use of the Σ value to scale up a centrifuge (including the efficiency of the centrifuge, a value normally available from the manufacturer) yields good results for precipitates that are not too sensitive to shear. Protein precipitates are normally fragile and moderate shear may disrupt them. The sedimentation velocity is determined by batch laboratory centrifuges in which the fluid is not moving; therefore, the precipitate is not subjected to shear. The use of this value in the scaling-up equation may lead to gross errors. Boychyn et al. (2001) have proposed to resolve this problem by mimicking the conditions of high shear present in a continuous centrifuge. To do this, they used computational fluid dynamics to determine the regions and the extent of energy dissipation in industrial centrifuges. Then, they used this information to pretreat the precipitate before it is placed in the centrifuge. In pretreatment the precipitate is subjected to the same energy dissipation as the one calculated for the continuous centrifuge. The inclusion of this correction into the design equation yields excellent results for some centrifuges and reasonable ones for other types. For example, in the sedimentation of yeast-protein precipitate, before correction for shear the predicted values for clarification for a bowl, a pilot multichamber bowl, and a disc centrifuge differ from the experimental values by ~+200%, ~+400%, and ~+60% respectively (the low shear of disc centrifuge explains the better agreement). Upon corrections, the calculated and experimental values are off by approximately -10% for all three.

Centrifugal Elutriation

Centrifugal elutriation is a technique that is used to separate cells according to their size. Because centrifugal elutriation is a specialized use of centrifuges and because it is a good example of the use of design equations in the selection of operating parameters of a centrifuge, a brief treatment of this technique follows. In centrifugal elutriation, a cell suspension is separated by size by "tuning" the centrifugal force and a fluid flow moving in the opposite direction. The key of the instrument is a specially designed centrifuge rotor that allows for the circulation of fluid through the cell and the optical monitoring of the various cell sizes as they are washed out by a buffer that circulates opposite to the centrifugal force. A careful balance between centrifugal force and buffer flow allows the fractionation of cells of a desired size. The sedimentation velocity is given by Equation 6.5. The velocity of the buffer is equal to the flow rate divided by the area of flow. If the centrifuge tube has a variable area, then both velocities are a function of the distance of separation from the axis of rotation. At any particular distance, if the flow velocity is larger than the sedimentation velocity, the cells will move against the centrifugal force (see Equation 6.8). The following example has been adapted from the Beckman (Fullerton, CA) web site. Suppose we want to separate 6.0- to 7.5-μm cells from a mixture with cell sizes ranging from 2.5 to 10.3 μm. We fix the rotation speed to 2500 rpm. Equation 6.8 can be used to calculate the sedimentation velocity for both types of cells. With that value we can calculate the elutriation velocity (the flow rate of the buffer). The experimental procedure consists of the establishment of an elutriation protocol as follows. First, a 150-mL sample is collected at a flow rate of 8 mL/min. This fraction contains cells that are smaller than 6 μm.

Then, a second fraction of 150 mL is collected at a flow rate of 13 mL/min, this fraction contains the cells of desired size. The remaining cells are discarded too because they are too big.

Filtration

Filtration is the oldest method used for the separation of solids and liquids. The developing of new and sophisticated filtering media (membranes), however, has revitalized the technique. Filtration can be classified into a) conventional and b) non-conventional filtration. Conventional filtration uses filtration media with coarse selectivity; that is, it cannot separate particles of similar size. In addition to a membrane or cloth, a cake is normally present in conventional filtration. The solids to be filtered may form this cake, but more often it consists of a mixture of the solids to be filtered and a filtration aid such as diatomaceous earths. Nonconventional filtration is quite different. The filtration media is a filter or membrane, whose lateral dimensions are much greater than its thickness (at a minimum 100x greater) and through which mass can be transferred under various driving forces. Membranes are able to fractionate particles of similar size. Furthermore, most of the time cakes are not formed, or they are undesirable.

Conventional Filtration

The most common filtration devices found in industry are: a) plate and frame and b) rotary vacuum filter. Precoating of the filter with, for example, diatomaceous earths or perlites is standard in bioseparation.

A plate and frame filtering device consists of a series of plates placed inside a shell. Each plate is made of a fine mesh usually covered by a filtration cloth. The liquid to be filtrated enters the device on the shell side, it moves through the walls of the plate into the plate center and it is extracted from the bottom part of the frames. The solids accumulate on the exterior wall of the plates. The main use of this type of device is the clarification of liquids when recovery of the solids is not important. The plates are commonly pretreated with diatomaceous earths or activated carbon. After the plates are covered by the filter aid, the suspension to be filtered is pumped into the unit. The filter can operate for prolonged periods of time without cleaning if the solids concentration in the liquid is not too high. Most of the units can be cleaned in place by reversing the direction of flow or by blowing air inside the frames. This action removes the filter aid and the accumulated solids, which are collected at the bottom of the device, from the surface of the plate. Eventually, the regeneration of the filter is not feasible anymore and the filter cloth needs to be changed.

Figure 6.5 shows a schematic of a rotary filter. The rotary filter consists of a hollow drum that rotates along its axis and is partially submerged during every revolution. Vacuum is applied to the inside of the drum. The drum is made of a mesh of stainless steel and it is covered initially by a filter aid. As the drum is immersed in the slurry to be filtered, the liquid passes through the filter aid cake and is removed from the inside of the drum. One cycle of the drum is divided into four stages: 1) cake formation, 2) cake washing, 3) cake drying, and 4) cake scraping.

In both, plate/frame and rotary vacuum filters the fluid flow is perpendicular to the filtration media. Therefore, the geometry is identical to standard dead end filters; that is, filtration through a Buchner funnel. The empirical relation between the filtrate velocity and the pressure drop across a filter is called Darcy's Law:

$$v = \frac{k_D \Delta P}{\mu l} \qquad 6.19$$

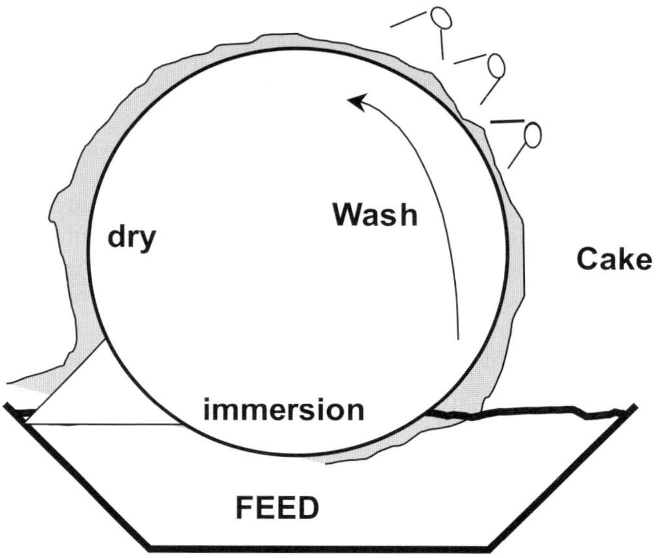

Figure 6.5. Side view of a vacuum rotary filter.

in which v is the velocity of liquid, ΔP is the pressure drop across a bed of thickness l, and k_D is the Darcy's Law permeability of the bed. For batch filtration, the velocity of the liquid is equal to,

$$v = \frac{1}{A}\frac{dV}{dt} \tag{6.20}$$

in which V is the volume to be filtered, and A is the area of flux. The flow finds obstacles, the membrane, and the cake deposited on top of it, which slow it down. It is convenient to define a combine resistance to the flow by,

$$\frac{1}{k_D} = R_M + R_C \tag{6.21}$$

in which R_M is the resistance of the filter medium and R_C is the resistance of the cake. Using Equations 6.19, 6.20, and 6.21, an equation for the rate of filtration is obtained,

$$v = \frac{1}{A}\frac{dV}{dt} = \frac{\Delta P}{\mu(R_M + R_C)} \tag{6.22}$$

The resistance of the cake can be written as,

$$R_c = \alpha\rho_c\frac{V}{A} = \alpha\frac{w}{A} \tag{6.23}$$

in which $\alpha = \alpha' \Delta p^s$ (α is the specific cake resistance, for an incompressible cake s = 0; otherwise, s is in between 0.1 and 0.8). ρ_c is equal to the mass of solids in the cake divided by the volume of filtrate. Filtration at various Δp allows the calculation of α' and s. For an incompressible cake, Equation 6.22 is,

$$\frac{1}{A}\frac{dV}{dt} = \frac{\Delta P}{\mu[\alpha \rho_c (V/A) + R_M]} \tag{6.24}$$

This equation can be integrated between t=0 where V = 0 and t=t in which the volume is V,

$$\frac{t}{V} = \frac{1}{K}(V + 2V_0) \tag{6.25}$$

with

$$K = \frac{2\Delta P A^2}{\alpha \rho_c \mu} \qquad V_0 = \frac{R_M A}{\alpha \rho_c} \tag{6.26}$$

Equation 6.25 can be used to calculate the filtration time. If $2V_0 << V$ (R_M is small) in Equation 6.25 we get:

$$t = \left(\frac{V}{A}\right)^2 \frac{\alpha \rho_c \mu}{2\Delta P} \tag{6.27}$$

Equation 6.27 allows the calculation of the filtration time versus the pressure drop. More importantly, the equation can be rearranged to produce a relationship between the specific resistance of the cake in terms of the experimental variables V, ΔP, and t. Therefore, a laboratory scale experiment can be used to determine the value of α and Equation 6.27 can be used to determine the filtration area given the process needs (flow rate).

For a vacuum drum filter, each of the phases of the cycle need to be treated separately. Assuming that R_M, the resistance of the filter, is small, the rate of filtration during the cake formation phase (Equation 6.22) can be written as,

$$\frac{1}{A}\frac{dV}{dt} = \frac{\Delta P}{\mu R_C} \tag{6.28}$$

Using Equation 6.23 for the cake resistance and integrating,

$$t_f = \frac{\mu \alpha' \rho_c}{2(\Delta P)^{1-s}} \left(\frac{V_f}{A}\right)^2 \tag{6.29}$$

it is common to express the time spent during cake formation as a fraction (β) of the time of the cycle, $t_f = \beta t_c$. Therefore,

$$\frac{V_f}{A t_c} = \left[\frac{2\beta(\Delta P)^{1-s}}{\mu \alpha' \rho_c t_c}\right]^{1/2} \tag{6.30}$$

Consider again Equation. 6.24,

$$\frac{1}{A}\frac{dV}{dt} = \frac{\Delta P}{\mu[\alpha\rho_c(V/A) + R_M]} \qquad 6.31$$

During the washing phase, the cake can be assumed to be incompressible and the resistance of the filtration media, R_M, small when compared with the resistance of the cake. Thus, the rate of washing during the washing phase is

$$\left(\frac{1}{A}\frac{dV}{dt}\right)_{V=Vf} = \frac{\Delta P}{\mu\alpha\rho_c(V_f/A)} \qquad 6.32$$

The volume of washing is,

$$V_w = t_w\left(\frac{dV}{dt}\right)_{V=Vf} = \frac{At_w\Delta P}{\mu\alpha\rho_c(V_f/A)} \qquad 6.33$$

Substituting Equation 6.30 into Equation 6.33 we get,

$$\frac{V_w}{A} = \left[\frac{\Delta P^{1-s}}{2\mu\alpha'\rho_c t_f}\right]^{1/2} t_w \qquad 6.34$$

Taking the ratio of Equations 6.29 and 6.34,

$$\frac{t_w}{t_f} = 2\frac{V_w}{V_f} \qquad 6.35$$

multiplying and dividing by the residual volume in the cake (V_r) and defining $f=V_r/V_f$ and $n=V_w/V_r$,

$$t_w = 2\,n\,f\,t_f \qquad 6.36$$

For the washing stage of the cycle we can use,

$$r = (1-\varepsilon)^n \qquad 6.37$$

in which r is the fraction of soluble material that remains in the cake after washing; n is the ratio between volume of wash and the volume retained by the cake; and ε is the washing efficiency. The total area of the filter needed to process a given volume of suspension per unit time will be given by the sum of the areas needed for each of the different stages.

Nonconventional Filtration: Ultrafiltration/Microfiltration in Cross-Flow Geometry

Filtration has evolved into a rather sophisticated technique with the advent of membranes whose pore size and configuration can be controlled to such a degree that the filtration area is maximized while keeping the total volume of the unit small (Christy and Vermant 2002; Schu and Gautam 2001). Membrane filtration methods are classified according to the pore size of the membranes. Microfiltration membranes have pore sizes from 0.02 to 10 μm; they retain intact cells and cell debris but let pass viruses, proteins, and salts. They are used upstream to separate whole cells and cell debris from the supernatant (Reis et al. 1991; Foley 2006). Therefore, they are a potential replacement for centrifugation. Ultrafiltration membranes are operated from 0 to 10 bar, have a pore size from 0.001 to 0.3 μm, and a molecular weight cut off (MWCO) from 1000 to 10^6 Da. Ultrafiltration membranes also retain proteins but let peptides and salts pass by. Their nominal molecular weight cut off (NMWCO) is 1 kD to 1000 kD. Ultrafiltration membranes are used in bioprocesses as a means to concentrate the final product, reduce the feed volume, exchange buffers before chromatography steps, and eliminate contaminants (such as viruses). Virus filtration membranes retain viruses but let proteins and salts pass. Their NMWCO ranges 10 nm to 0.05 μm. These membranes are used to either harvest viruses or as one step in a virus reduction protocol. High-performance membranes can sort proteins of different sizes. Their NMWCO is 10 kD to 300 kD. Membranes can separate by both size (Sakhamuru, Hough, and Chaudhuri 2000) and charge (Ebersold and Zydney 2004), as they can themselves bear a charge; thus, they may be used to replace chromatography. Reverse osmosis membranes only let water pass. Their NMWCO is less than 1 kD and are composed of materials able to tolerate pressures up to 100 bar. They are commonly used to purify water, by eliminating small molecules via diffusion. As such, dialysis membranes have pore sizes between 1 and 3 μm.

Finally, nanofiltration membranes are often made by nonporous membranes such as ceramics. They are used for the separation of small molecules; for example, in the separation of hydrogen from air in the power industry.

Inorganic, mixed organic/inorganic, and organic materials are used to fabricate membranes. Typical examples of inorganic membranes are glasses, ceramics, metals, and inorganic polymers. Ion-containing polymers, such as polysiloxanes and polyphosphazenes, are examples of organic/inorganic membranes. Finally, organic membranes are normally made by a variety of natural and synthetic polymers. Examples of natural polymers used are polysaccharides (cellulose), polypeptides, and rubbers. Thermoplastic and rubbery polymers are examples of synthetic ones. Ceramics, glasses, metals, and ion-containing polymers find most of their uses in industries other than biotech. For example, most of the advances in ceramics and ion-containing polymers have been focused on the fabrication of membranes dedicated to the separation of hydrogen from air to produce raw materials for fuel cells or as the core of fuel cells.

The fabrication methods of membranes are very diverse and, of course, they depend on the materials used for their construction. A well-fabricated membrane must allow maximum flow through it, have the best selectivity, and be mechanically strong. Because flux is a function of membrane thickness, most of the efforts are focused in the synthesis of thin membranes with good mechanical properties. Membranes can be made by the arrangement of fibers, particles, or by the fabrication of a thin film. Fiber-based membranes can be manufactured by wet-lay (paper filters), dry-lay (spun-bonded olefins), wound (glass filament cartridges) or woven (polymeric and/or metal filter meshes). Particle-based membranes can be made by sol-gel (ceramic ultrafilters), compression or sintering (metal and glass filters and frits), or extrusion (alumina microfilter monoliths). Finally, film-based membranes can be prepared by extrusion of dense films (silicone films), extrusion and stretched dense films (Teflon and olefin microfilters), cast or extruded films with phase inversion step (cellulose acetate ultrafilters), nuclear-particle track etched (polycarbonate microfilters), and electrochemical deposition (homoporous alumina microfilters).

There is a variety of membranes available in the market. Millipore, one of the main producers, commercializes two basic types of membranes: regenerated cellulose and polyethersulfone-based. Regenerated cellulose membranes are very hydrophilic and, therefore, have low protein adsorption. They are good for low protein loads or when the feed is heavily fouling. However, they do not tolerate well basic conditions. On the contrary, polyethersulfone membranes can be used from pH 1 to 14. Unfortunately, they have a strong tendency to adsorb proteins because they are very hydrophobic. Some manufacturers (such as Millipore) have modified their surface to reduce protein adsorption.

Table 6.2 shows examples of adsorption levels for a series of membranes (Millipore). Low pro-

Table 6.2. Amount of Protein Adsorbed by Three Types of Membranes. The Last Two Columns Show the % Decrease in Yield for Low- and High-Protein Loads. Copyright Millipore Corporation, Billerica, MA, with permission.

Membrane	Protein Adsorption (g/m^2)	% Yield Loss	
		Low protein	High Protein
Polyethersulfone	0.5	5.0	0.10
Biomax	0.2	2.0	0.04
Regenerate cellulose	0.1	1.0	0.02

tein experiments were made by processing 1000 L of a 0.1 g/L protein solution concentrated to 2 g/L on a 10 m^2 membrane. High protein concentration experiments were done with 1000 L of a 10 g/L protein solution, which was concentrated to 200 g/L on 20 m^2 membrane.

The performance criteria for a membrane are: 1) speed 2) selectivity, and 3) stability. A good membrane should allow the fluid to move through quickly, being able to discriminate between the components of a mixture and not change its properties after repeated use. These performance criteria are associated with three membrane properties: transport through the membrane, its pore characteristics, and its surface properties. The three more important transport properties are solvent flow, solute rejection and solute diffusion. The most important pore characteristics are pore size distribution, pore shape, and pore morphology across the membrane. Finally, the most important surface properties are chemical composition, hydrophobicity or hydrophilicity, surface charges, solute-membrane affinity, and surface texture.

The solvent flow is controlled by the following experimental parameters: the permeability of the membrane, its thickness and area, the transmembrane pressure, and the viscosity of the fluid. Because the membrane will foul by cake formation, pore obstruction, or concentration polarization, the flow is not constant during filtration if the pressure is kept constant. The flow characteristics of the membrane determine the speed at which the filtration process can be done (performance criteria 1).

Performance criteria number 2 is associated with the ability of the membrane to reject a particular solute while allowing the passing of others. Solute rejection can be defined in a variety of ways and one must be aware of this before purchasing a particular membrane. The following definitions are commonly used. The observed solute rejection, R

$$R = 1 - \frac{C_{Af}}{C_{Ab}} \qquad 6.38$$

C_p and C_b are the concentration of protein in the permeate and retenate respectively. The intrinsic solute rejection, R_0

$$R_0 = 1 - \frac{C_{Af}}{C_{AW}} \qquad 6.39$$

in which C_w is the concentration at the membrane surface. The sieving coefficient S_{50} is

$$S_{50} = \frac{C_p}{C_b} \qquad 6.40$$

The sieving coefficient is measured after 50% of the feed has been filtered. And the log reduction value, LRV, is

$$LRV = -\log(S_{50}) \qquad 6.41$$

Manufacturers use either quantity to describe the performance of their product. When a membrane is used to concentrate a protein solution, it is customary to indicate the degree of concentration by using a conversion ratio and the volume concentration factor. The conversion ratio is defined as,

$$CR = \frac{Q_{Filtrate}}{Q_{feed}} \qquad 6.42$$

in which $Q_{Filtrate}$ and Q_{feed} are the volumetric flow rates of the filtrate and feed, respectively. Another important membrane parameter is the volume concentration factor, VCF, which is defined as the amount of the feed stream that has been reduced in volume from the initial volume,

$$VCF = \frac{V_{Feed}}{V_{retentate}} \qquad 6.43$$

When a membrane is used to exchange buffers, it is customary to express the extent of dialysis by the diavolume, which is equal to total buffer volume introduced to the operation during diafiltration divided by the retenate volume.

The flow (in the absence of cake formation or concentration polarization) is determined, for a given driving force, by the size and size distribution of the membrane pores. Membranes dominated by meso and macropores are used in pressure-driven liquid filtration. If the pores in a membrane would be all of the same size, then the pore size would be sufficient to determine the efficiency of a membrane to retain a particular solute. Unfortunately, a membrane with a nominal pore size has actually a distribution of pore sizes and the manufacturer only provides the average. Membranes from different manufacturers with the same MWCO can have different permeabilities, because the pore size distribution is not the same. Therefore, to know the pore size distribution is as important as to know the pore size.

Some, but not all, manufacturers indicate how the pore size and distribution have been determined, but the reported numbers may have different meanings depending on the method used. To make an informed decision we need to understand the limitations of various methods used to determine the pore size distribution. The simplest way of determining the pore size is direct visualization. Optical (low resolution) and electron microscopy (high resolution) can be used for this purpose. Optical microscopy is limited to a 0.5-μm resolution, transmission electron microscopy (TEM) is limited to 1 nm, and scanning electron microscopy (SEM) is limited to 2 to 6 nm. If the inspection of the membrane is limited to a single spot the results are not very reliable. However, we have found that SEM can yield very reliable and reproducible results by randomly targeting the spot (blind choice) and by repeating the measurement ten or more times. An idea of the pore size and distribution can be obtained by simple hydraulic measurements; that is, the flow rate across the membrane can be measured and regressed with the appropriate model. A traditional method for the characterization of membranes is to measure the bubble pressure or point. In this method, the membrane is filled with a wetting liquid and then it is displaced by increasing the pressure of an inert gas. The differential pressure needed to displace the liquid film is used to calculate the pore diameter assuming that the pores are cylindrical, which is a poor assumption for polymer-based membranes, but it may be a good approximation in ceramic membranes. To calculate the pore size distribution, one needs the contact angle (liquid/gas/membrane) and the surface tension (liquid/gas). A variation of this method is liquid/liquid porosimetry, in which an immiscible liquid replaces the gas. In gas liquid diffusion, the membrane is wetted by a liquid and the diffusion of a gas is measured. In the gas adsorption/desorption method (BET method) the pressure of a condensable gas is increased to provide a monolayer on the membrane, then the pressure continues to increase to form a multilayer on the internal surface area of the pores and it is continued until full condensation occurs. The pressure is then reduced back to the initial value. The forward and backward pathways followed by the gas to condense and evaporate are not the same (hysteresis). From that hysteresis loop, the pore volume distribution can be calculated. Polydisperse solute rejection is a very convenient method to characterize membranes for biological applications. Solutes of different molecular weights are filtrated and the membrane is rated at a particular MWCO by the mass at which 90% of the solute is rejected. A more standardized approach to obtain similar information is to measure the retention of standard particles (normally poly(styrene) latex). The "challenge" level may be diluted 10^8 to 10^{10} particles/L or concentrated, from 10^{13} to 10^{15} particles/L. Both monodisperse and polydisperse particles can be used. The measurement of latex size, size distribution, and concentration is done by illuminating the sample with a laser beam (see Chapter 4). The retention of the particle can take place inside the membrane (depth) or at the surface (surface capture). Therefore, this type of measurement sometimes yields ambiguous results because of the pore shape. Still, it is the most reliable one from a practical viewpoint. When a filter is to be used to maintain sterile conditions, it is best to test its microbial retention. In this type of test microbes are considered monodisperse. *Pseudomonus diminuta* (0.3 µm with an aspect ratio 1.5 to 1) is the standard bacteria. It is popular for filters with pore sizes between 0.22 and 0.45 µm. The bacterial challenge test for sterilizing microfilters is $1\text{x}10^7$ organisms/cm^2 with no organisms detectable in the permeate. Filters 0.22 and 0.45 µm at 4 bar have complete retention of P. *diminuta* and a LRV of 9 with 0.45-µm ones.

Surface properties of membranes are important in the retention of solutes by the membrane, in the formation of cakes and on concentration polarization. Membranes used in biotechnology must be hydrophilic (water wetting). The contact angle of air/water/solid interface is a measurement of the surface hydrophillicity. A contact angle of zero degrees corresponds to total wetting (the liquid spreads on the surface), whereas a contact angle of 90° or higher corresponds to non-wetting. Although contact angles are normally used to characterize the surface of a membrane, surface energy values are more fundamental and their value correlates with the hydrophilicity of the membrane; for example, high energy surfaces are more hydrophilic. It is important to emphasize here that either type of measurement characterizes only the surface of the membrane and not the interior. For some applications, membranes are made hydrophobic but their surface is made hydrophilic by chemical modification. Finally, streaming potential measurements can be used to determine the surface charge.

The final test of a membrane takes place when it is used for a particular process. Success of a particular ultrafiltration process is assessed by high product yield, activity, and purity or controlled bioburden and endotoxin.

Membrane Configuration

Membranes can be incorporated in a variety of housing devices. Some of them are limited to laboratory scale, whereas others are available in laboratory, pilot plant, and production scales. A common theme is the presence of a support that helps to maintain the integrity of the membrane. The simplest configuration is a funnel in which the driving force is either the liquid height (hydrostatic head) and/or vacuum. Common devices used in laboratory experiments are syringe filters, centrifuge tube microfiltration units, and dialysis bags. Figure 6.6 shows a picture of some of these devices. Syringe filters consist of a plastic housing containing the membrane. They are available from a number of vendors in a wide range of selectivities (MWCO). The driving force is the applied mechanical pressure either by hand or by using a syringe pump. A slightly different device is a cen-

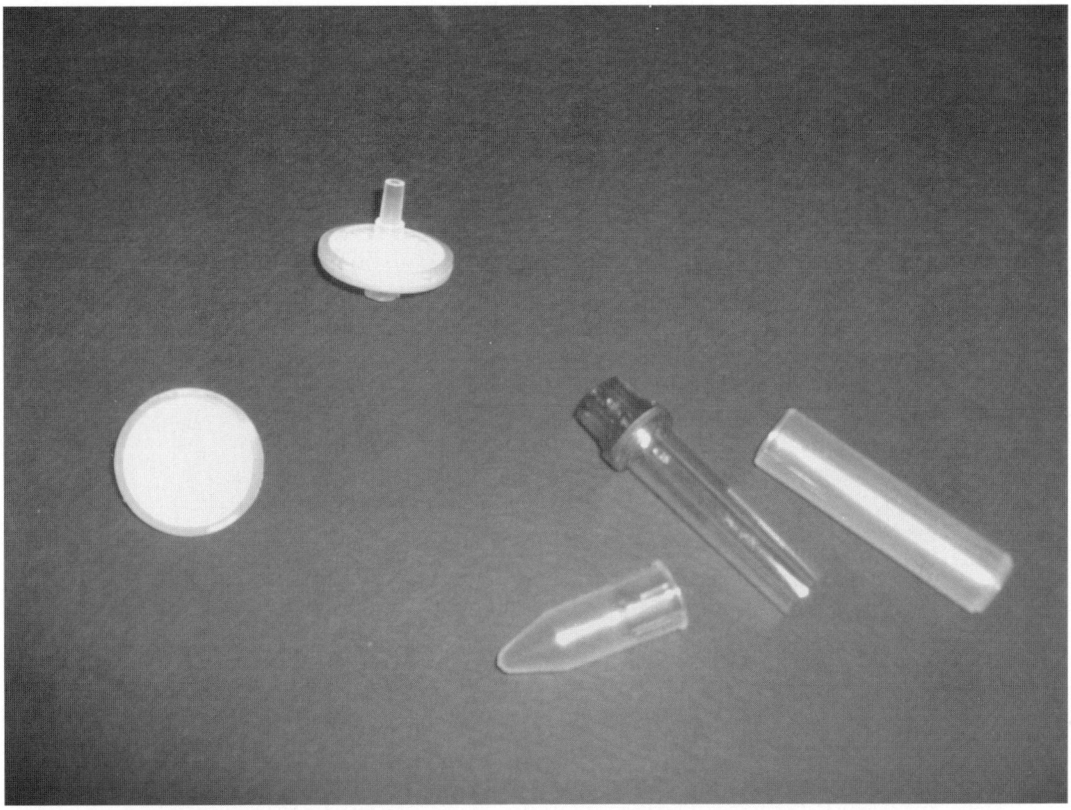

Figure 6.6. Picture of a syringe filter (left) and a centrifuge-base ultrafiltration device (right).

trifuge tube microfiltration unit. They consist of two parts: 1) a retenate section that is a cylindrical tube with one open end and a membrane at the other end, and 2) filtrate section, which is a simple test tube. The two sections are connected and the material to be filtrated is placed on the retenate section. The entire device is placed in a centrifuge and spun for a period of time (a few minutes is normally enough). These tubes can be also used in dialysis mode by reducing the volume first, adding fresh buffer to the retenate section and spinning again. When the cycle is repeated a few times the original buffer is exchanged by the new one. Dialysis bags were extremely common in laboratories, but now they have been partially displaced by tube devices or dead end filtration units. Dialysis bags are made of cellulose and have a well-controlled pore size. The liquid to be dialyzed is placed inside the bag and the bag is closed. The bag is placed in a large beaker containing the fresh buffer. Standard practice is to dialyze for 24 hours with three buffer exchanges. The driving force for this kind of device is of course diffusion.

There are two basic designs of ultrafiltration devices: dead end (stirred cells) and cross-flow. Stirred cell modules are common sights in the laboratory (Figure 6.7). They consist of a reservoir that contains the membrane at one end and a pressure-tight cap at the other end. The contents of the cell are stirred by a magnetic bar and pressurized by an inert gas (nitrogen). These devices can be operated at pressures ranging from 6 to 9 bars with operating volumes of a few μL to several liters, and can be used in either filtration or dialysis mode.

Cross-flow modules are the most efficient filtration devices and can be easily scaled up from lab-

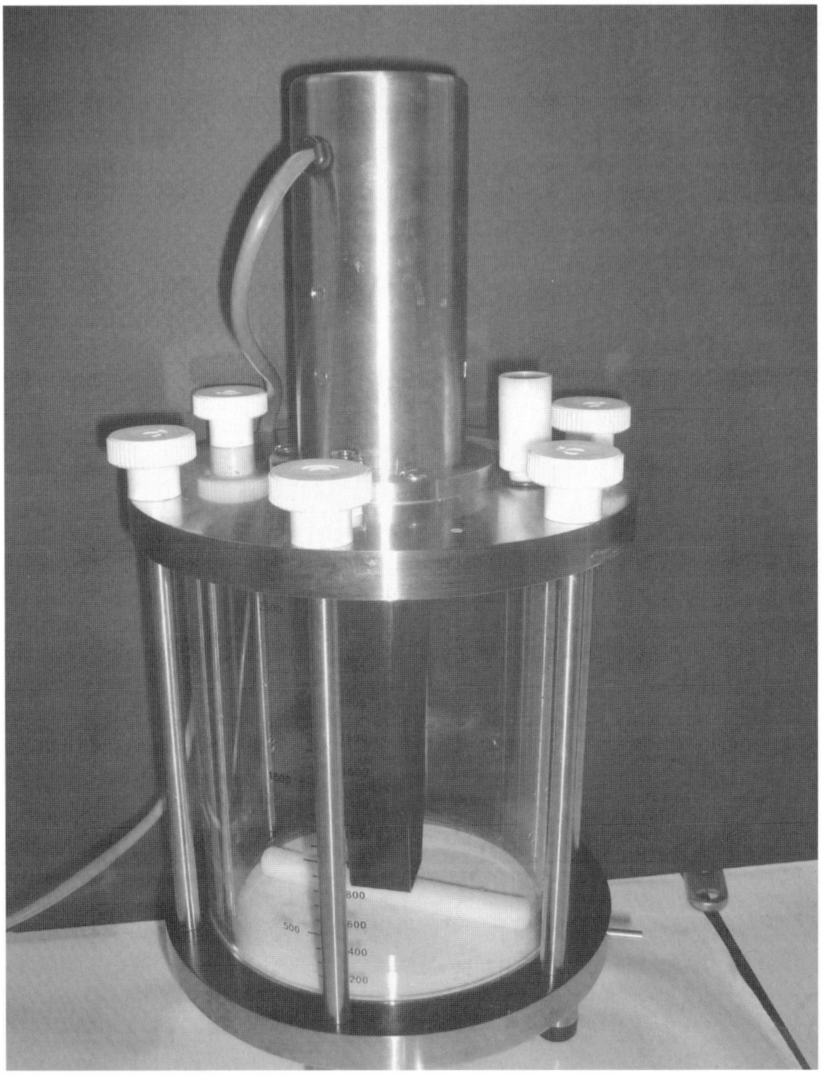

Figure 6.7. Picture of stirred cell. High pressure resistant glass is used for the main container and a motor-driven stirrer is used to minimize cake formation.

oratory (Figure 6.8) to process scale. Because of the direction of the flow, cake formation and concentration polarization are minimized. There are three different configurations: plate and frame, spiral wound, and hollow fibers. Frame and plate devices (see Figure 6.9) are a miniature version of frame and plate filters described earlier. A typical module consists of alternating layers of membranes and separator screens. The feed is added to alternating channels at one end of the cartridge and the filtrate passes through the membrane into the filtrate channels. Channels are available with and without turbulence promoters. Hollow-fiber devices (Figure 6.10) consist of a large number of narrow-diameter (typically from 0.1 to 2 mm) fibers packed in, for example, a cylindrical pattern. The feed is pumped inside the fibers and the filtrate passes through the fiber's wall into the

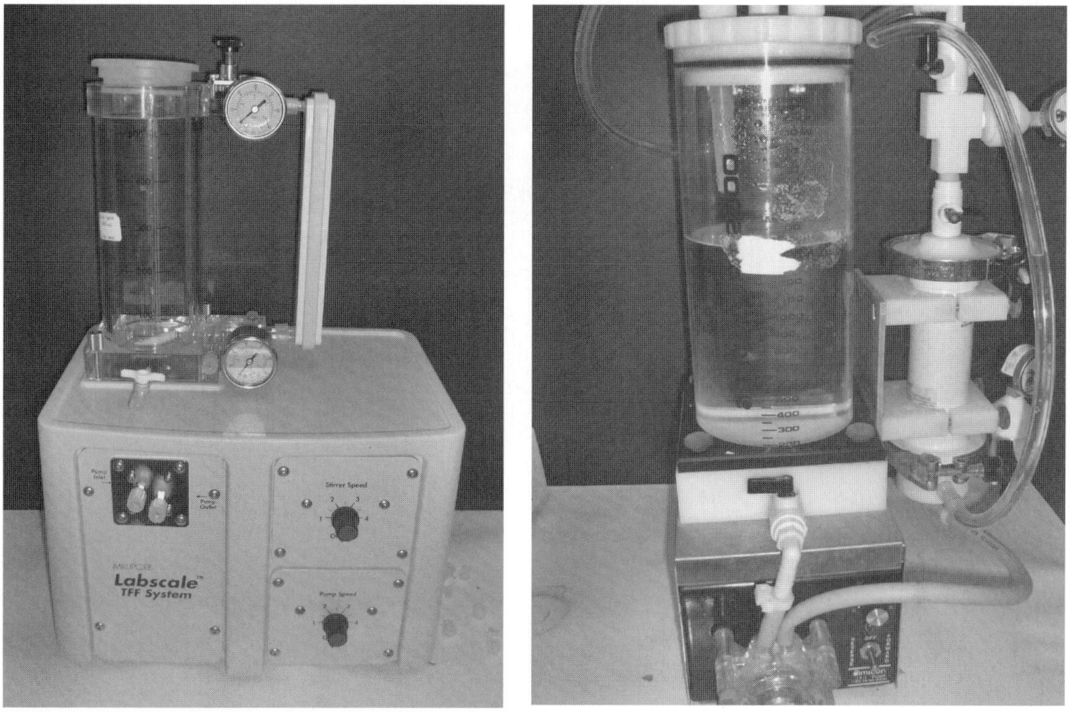

Figure 6.8. Picture of a pelican cartridge ultrafiltration instrument (left panel) and a spiral wound instrument (right panel). Both instruments are equipped with manometers to monitor the retentate and filter pressures. In the right panel, the peristaltic pump head used to run the system is visible.

shell side of the device. High pressures are needed to pump the fluid because of the narrow size of the fibers. Spiral wound cartridges (Figure 6.11) consist of a series of planar membranes stacked on top of each other and separated by screens or spacers. The entire membrane is then wound into a spiral. The fluid is pumped at one end and flows in the axial direction. The filtrate passes through the membrane and spiral into the center of the spiral where it is removed. Both frame/plate and hollow-fiber membranes are easily scalable. Spiral wound membranes, on the contrary, are more difficult to scale up because two dimensions must be changed: the cartridge length and/or the cartridge width.

Modeling of the Ultrafiltration Process

Modeling of ultrafiltration devices allows the understanding of the filtration process and provides guidelines for the design and scale up of ultrafiltration membranes. Different approaches have been followed and only a few are described below. Lately, computational fluid dynamics has been used to fully understand the flow inside the membrane modulus and a brief mention of this technique is included for completeness.

It is illustrative to consider the dependence of flux on the pressure drop across the membrane for cross flow filtration. Figure 6.12 shows the flow rate across the membrane versus the pressure drop

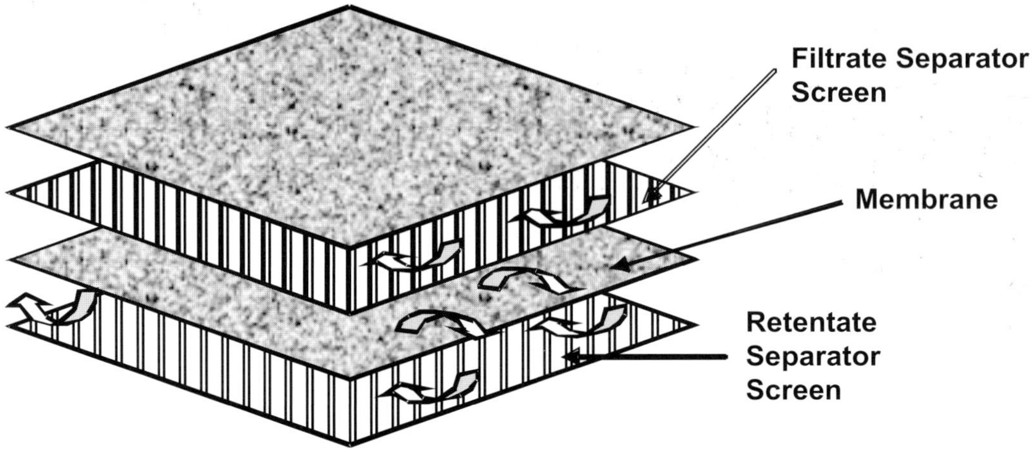

Figure 6.9. Schematic of a plate and frame microfilter

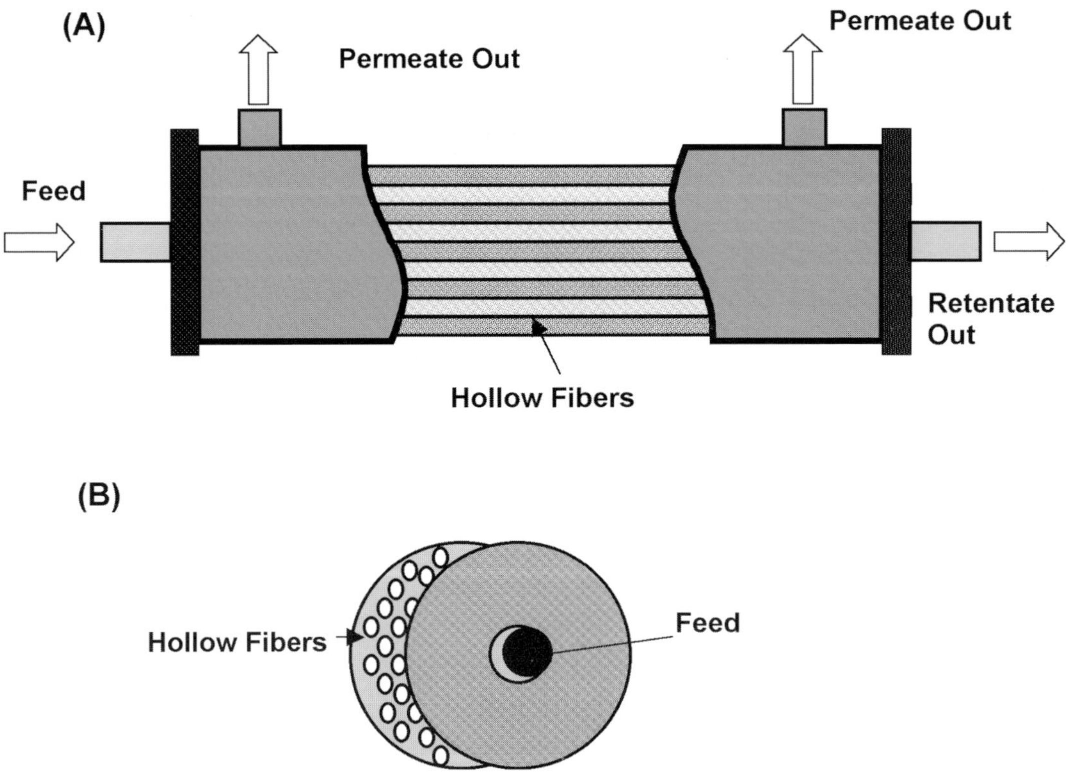

Figure 6.10. Schematic of a hollow fiber cartridge. (A) side view (the center part of the housing has been removed to show the fibers inside) and (B) front view.

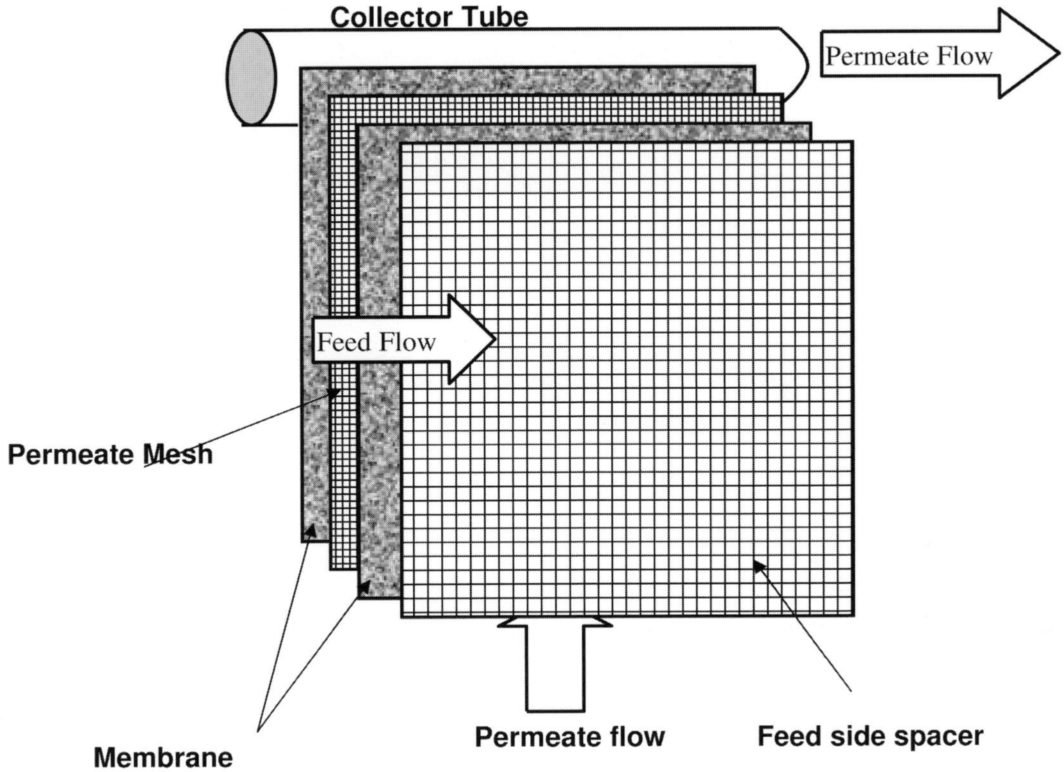

Figure 6.11. Schematic of a spiral wound device

for three cases: a saline solution, a dilute protein concentration, and a concentrated protein solution. The flow is linear for the saline solution but it becomes independent of pressure for the protein solutions. The more concentrated solution reaches the independent pressure phase earlier than the dilute protein solution. This observation is the consequence of accumulation of the protein at the protein surface, a phenomenon known as concentration polarization. Any model should capture this change in flow regime. Figure 6.13 shows a schematic of the flows and concentration profiles for cross-flow filtration. The main fluid flow is parallel to the membrane, but the filtrate moves perpendicular to it.

A solute mass balance in a controlled volume of thickness δ is:

$$JC_A = -D_{AB}\frac{dC_A}{dx} + JC_{Af} \tag{6.44}$$

in which J is the fluid flux, C_A is the concentration of solute retained by the membrane, C_{Af} is the concentration of A in the filtrate, and D_{AB} is its diffusion coefficient in the bulk. Integration between x=0, C_A=C_{AW} and X=δ, C_A=C_{Ab} yields,

$$J\delta = D_{AB}\ln\frac{(C_{Aw} - C_{Af})}{(C_{Ab} - C_{Af})} \tag{6.45}$$

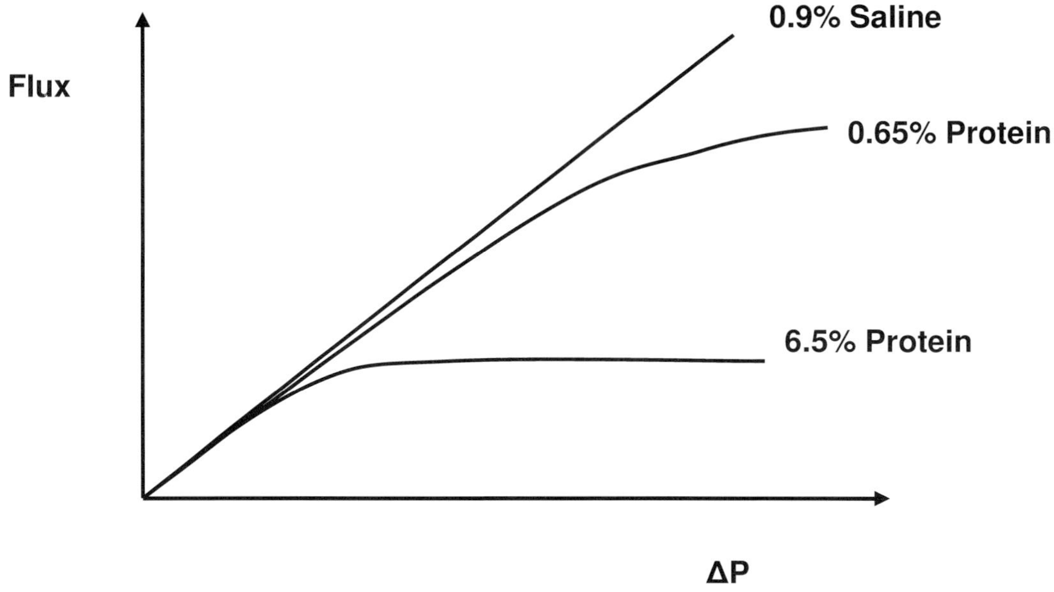

Figure 6.12. Effect of solution composition on the flux of an ultrafiltration device as a function of the pressure drop across the membrane.

or

$$C_{Aw} = C_{Af} + (C_{Ab} - C_{Af})\exp\left[\frac{J\delta}{D_{AB}}\right] \qquad 6.46$$

Equation 6.45 can be written as

$$J \approx \frac{D_{AB}}{\delta}\left(\frac{C_{Aw} - C_{Ab}}{C_{Aw} - C_{Af}}\right) = k_o\left(\frac{C_{Aw} - C_{Ab}}{C_{AW} - C_{Af}}\right) \qquad 6.47$$

in which k_o has the meaning of a mass transfer coefficient (discussed in Chapter 4). Correlations for k_0 have been presented in that chapter and will not be repeated here. Trettin and Doshi (1980) showed that for large concentration driving forces the correct equation for the flux is,

$$J = 2.046k_0\left(\frac{C_{Aw} - C_{Ab}}{C_{AW} - C_{Af}}\right)^{1/3} \qquad 6.48$$

An alternative approach to analyze mass transfer in cross-flow filtration is to treat it as a dead-end device (flux=driving force/resistance). The driving force is equal to the total pressure drop across the membrane minus a difference in osmotic pressure caused by the different concentration of solutes across the membrane multiplied by a phenomenological coefficient called the osmotic reflexion coefficient, σ, which is zero for a solute completely rejected by the membrane and one for

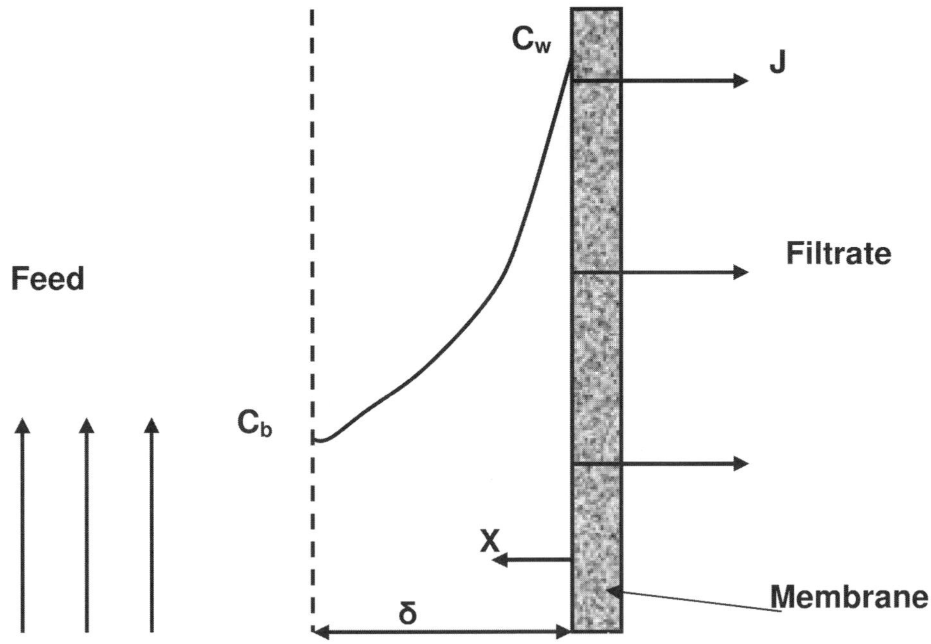

Figure 6.13. Concentration distributions in a cross-flow filtration device.

a completely permeable solute. The resistance is equal to the sum of the inverses of the permeability of the membrane and the gel layer. Therefore, the flux is equal to:

$$J = \frac{\Delta P - \sigma \Delta \Pi}{\mu (R_M + R_p)} \qquad 6.49$$

in which $\Delta\Pi$ is the osmotic pressure. For an ideal solution, $\Delta\Pi = RTc_W$. According to this equation the filtrate flux initially increases with pressure; this has been confirmed experimentally. The osmotic pressure increases as the solute accumulates at the surface of the membrane. At one point, the flux for a completely rejected solute is independent of pressure and equal to

$$J = k_0 \ln \frac{C_{Aw}}{C_{Ab}} \qquad 6.50$$

which can be used to determine the value of C_W and k_0 by regressing experimental data.

Equation 6.46 shows that high fluxes, poor stirring (large boundary layer thickness), and large molecules (small diffusion coefficient) will increase polarization of the membrane. The diffusion coefficient is sensitive to temperature, but it is not common to operate ultrafiltration devices at high temperature (that may damage the product). Instead, the flux can be kept low and the filtration time may be increased because most of the bioprocessing is in batch or semibatch mode. An alternative is to increase the membrane area, but of course that will increase the cost. Turbulent flow will decrease concentration polarization. Therefore, it is customary to include partial obstructions in the channels that increase the turbulence.

No membrane has perfect retention. Therefore, a portion of the protein to be retained will be lost into the filtrate. Because of the abundance of ultrafiltration/dialysis steps in any bioprocessing, the product losses into the filtrate may be significant. Simple calculations can be used to estimate the amount lost. A mass balance for the solute that is retained in a device used to both concentrate a product and exchange the buffer is,

$$-\left[\left(V_0+(Q-JA)t\right)\frac{dC_A}{dt}+C_A(q-JA)\right]=JAC_A(1-R) \qquad 6.51$$

in which V_0 is the initial volume, Q is the flow rate of dialyzing buffer, A is the area of filtration, C_A is the concentration of solute upstream from the membrane, and R is the retention of the membrane. In Equation 6.51, the following relation between the upstream and downstream concentrations was used,

$$R=\frac{C_A-C_{Ap}}{C_A} \qquad 6.52$$

Rearrangement of equation 6.51 yields,

$$-\left[\left(V_0+(Q-JA)t\right)\frac{dC_A}{dt}\right]=C_A(Q-JAR) \qquad 6.53$$

This equation can be integrated by separating variables,

$$-\frac{dC_A}{C_a(Q-JAR)}=\frac{dt}{V_0+\left(Q_d-JA\right)t} \qquad 6.54$$

Integrating,

$$\left(\frac{C_{Af}}{C_{Ai}}\right)=\left(\frac{V_0}{V_f}\right)^{\frac{Q-JAR}{Q-JA}} \qquad 6.55$$

Rearranging the exponent on the right hand side,

$$\left(\frac{C_{Af}}{C_{iA}}\right)=\left(\frac{V_0}{V_f}\right)^{\frac{CFN(1-R)-R(CF-1)}{1-CF}} \qquad 6.56$$

in which $N=Q/V_0$ is the number of initial volumes added for dialysis and CF is the concentration factor, $CF=V_f/V_0$. If no buffer is added, then

$$\left(\frac{C_{Sf}}{C_{Si}}\right)=\left(\frac{V_0}{V_f}\right)^{R} \qquad 6.57$$

It is important to calculate the percentage of solute lost because of imperfect retention. This amount is equal to,

$$\left(\frac{C_{A0}V_0 - C_{Af}V_f}{C_{0A}V_0}\right)X100 = \left(1 - \frac{C_{Sf}}{C_{A0}}\frac{V_f}{V_0}\right) \quad 6.58$$

Using Equation 6.56 for the ratio of concentrations,

$$\%\,\text{protein loss} = \left(1 - \left[CF\right]^{\frac{CFN(1-R)-R(CF-1)}{1-CF}-1}\right)100 \quad 6.59$$

The % protein loss is plotted in Figures 6.14, 6.15, and 6.16 for different values of N (N=0 no dialysis, N=5 moderate dialysis, N=10 normal dialysis) for various retention values. The figures clearly show the economic consequences of the use of "less than perfect" membranes particularly when multiple passes are used. For dialysis without concentration, it is straightforward to show that the percent loss is,

$$= 100\left(1 - e^{-N(1-R)}\right) \quad 6.60$$

A similar approach can be used to estimate the volume of the flush that is needed to reduce the concentration of a salt to a desired value. For a constant volume operation, the math is quite simple. A mass balance for the salt (keep in mind that for the salt R=0) is,

$$V\frac{dC_S}{dt} = JAC_S \quad 6.61$$

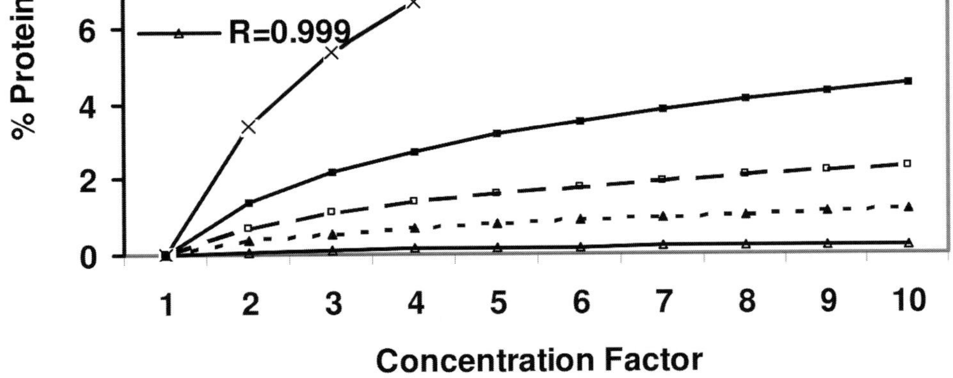

Figure 6.14. Decrease in yield vs. concentration factor for various retention factors for one pass.

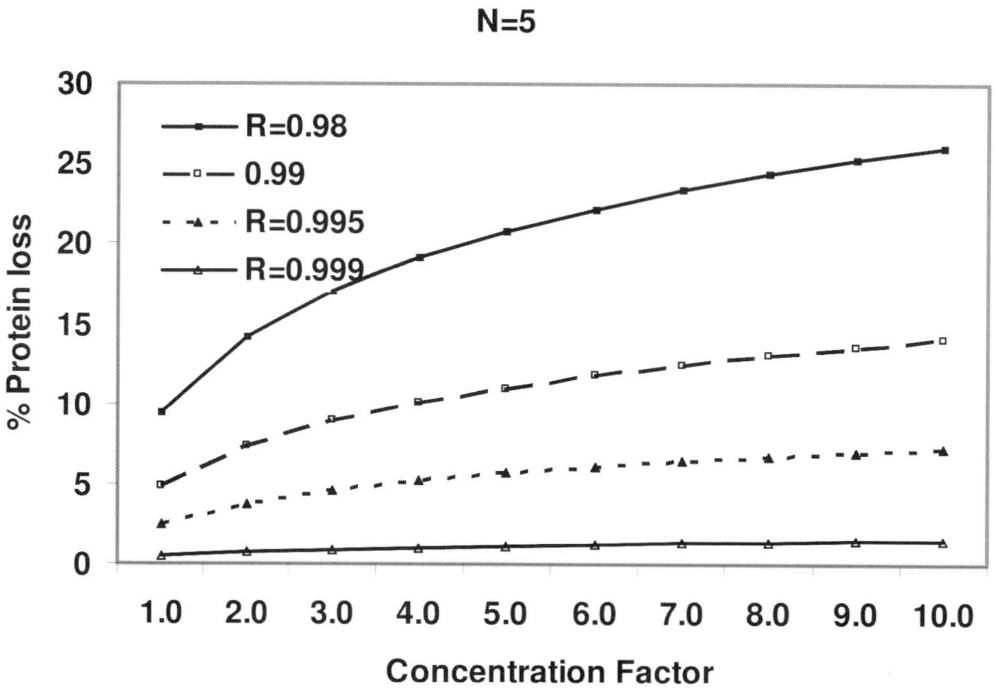

Figure 6.15 Decrease in yield vs. concentration factor for various retention factors for five passes.

Integrating between the initial and final conditions yields, or

$$C_{Sf} = C_{Si} \exp\left(\frac{-JAt}{V}\right) \tag{6.62}$$

$$C_{Sf} = C_{Si} \exp\left(\frac{-Volume\ of\ flush}{Volume\ of\ vessel}\right) \tag{6.63}$$

The critical step in the scaling up of filtration systems is the calculation of the filtration area that is needed to process a determined amount of feed per unit time. The following expressions may be used to estimate the area (for a given flow rate) for the three main types of operating modes, which are single pass; batch, feed, and bleed; and multiple-stage feed and bleed (Zeman and Zydeny 1996). For single pass (Figure 6.17),

$$A = \left[\frac{Q_0}{k_0 \ln\left(\frac{C_{AW}}{C_{Ab0}}\right)}\right]\left[\frac{1-\frac{Q}{Q_0}}{0.33+0.67\left(\frac{Q}{Q_0}\right)^n\left(1+\frac{\ln\frac{Q}{Q_0}}{\ln\frac{C_{Aw}}{C_{Ab0}}}\right)}\right] \tag{6.64}$$

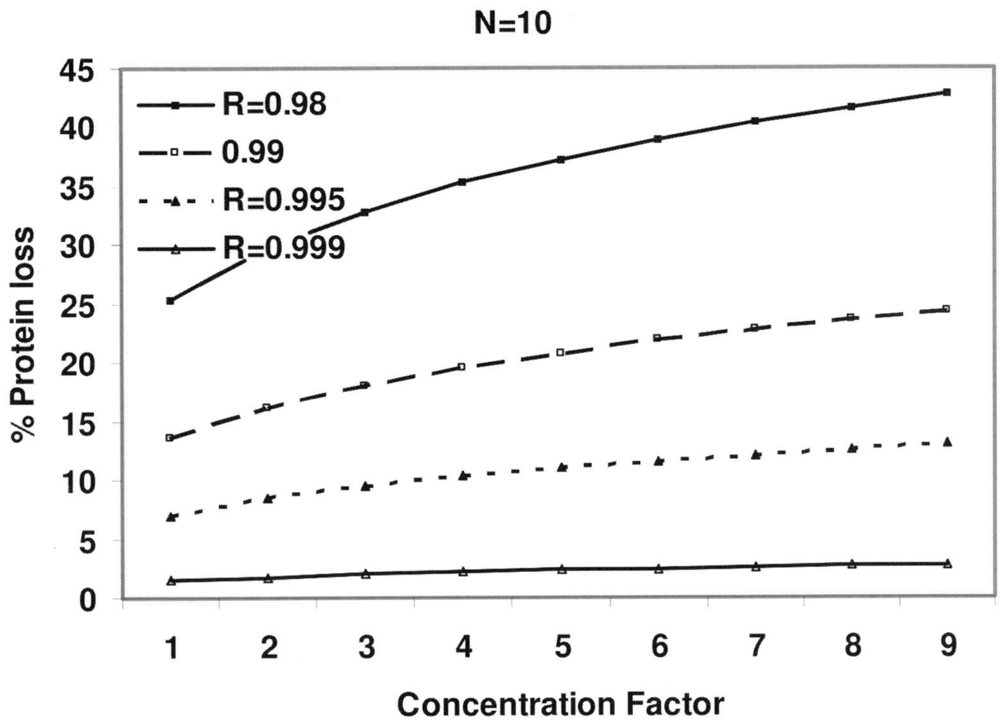

Figure 6.16. Decrease in yield vs. concentration factor for various retention factors for ten passes.

in which Q_0 is the inlet feed flow route, k_0 is the mass transfer coefficient at the inlet conditions and n is a parameter that varies from 0.5 to 0.8 for proteins and other biomacromolecules. For a batch system,

$$A = \left[\frac{V_0}{k_0 \ln\left(\frac{C_{AW}}{C_{Ab0}} \right) t} \right] \left[\frac{1 - \frac{V}{V_0}}{0.33 + 0.67 \left[1 + \frac{\ln \frac{V}{V_0}}{\ln \frac{C_{AW}}{C_{Ab0}}} \right]} \right] \quad 6.65$$

This equation predicts the area reasonably well if $V/V_0 > 0.5$ otherwise, it over predicts V/V_0 at longer times. For feed and bleed the equation is,

$$A = \left(\frac{Q_0}{k_0} \right) \frac{1 - \frac{Q}{Q_0}}{\ln \frac{C_{AW}}{C_{Ab0}} + \ln \frac{Q}{Q_0}} \quad 6.66$$

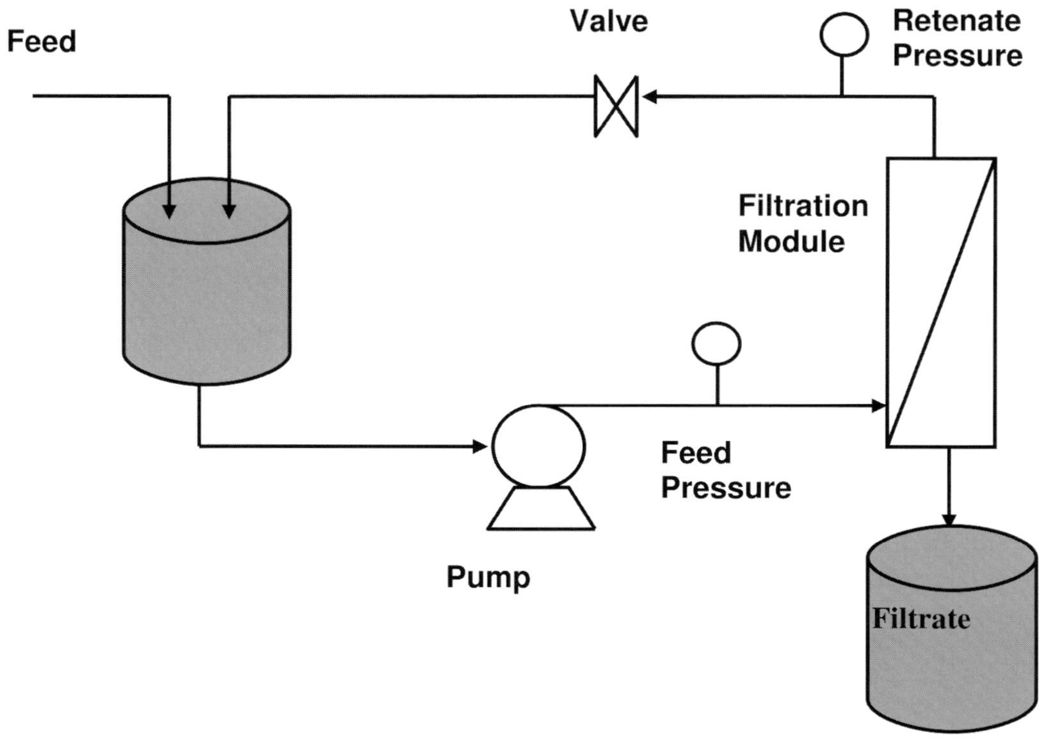

Figure 6.17 Schematic of an ultrafiltration set-up with recycle. This configuration corresponds to the devices shown in Figure 6.8.

Finally, for multiple stages and feed and bleed in between stages (Figure 6.18) the area for the ith stage is,

$$A_i = \left(\frac{Q_0}{k_0}\right)\frac{\left[\left(\frac{Q}{Q_0}\right)^{-1/N}-1\right]\left(\frac{Q}{Q_0}\right)^{i/N}}{\ln\frac{C_{AW}}{C_{Ab0}}+\frac{i}{N}\ln\frac{Q}{Q_0}} \tag{6.67}$$

in which N is the number of stages. Usually, equal areas may be chosen from practical considerations. However, the optimum is slightly off. Overall, the area required for a given volume to be treated increases in the following order: multistage, batch, single pass, and feed and bleed but the treatment time decreases in this order: single pass, feed and bleed, multistage, and batch. Therefore a choice needs to be made between minimum area and minimum time. Multistage systems require more energy to function; therefore, that must be taken into account too.

Membrane Fouling in Cross-Flow Filtration

The most important problem in any ultrafiltration device is membrane fouling. In the ultrafiltration of proteins, the problem arises mainly from the formation of a protein layer, which reduces flux, at

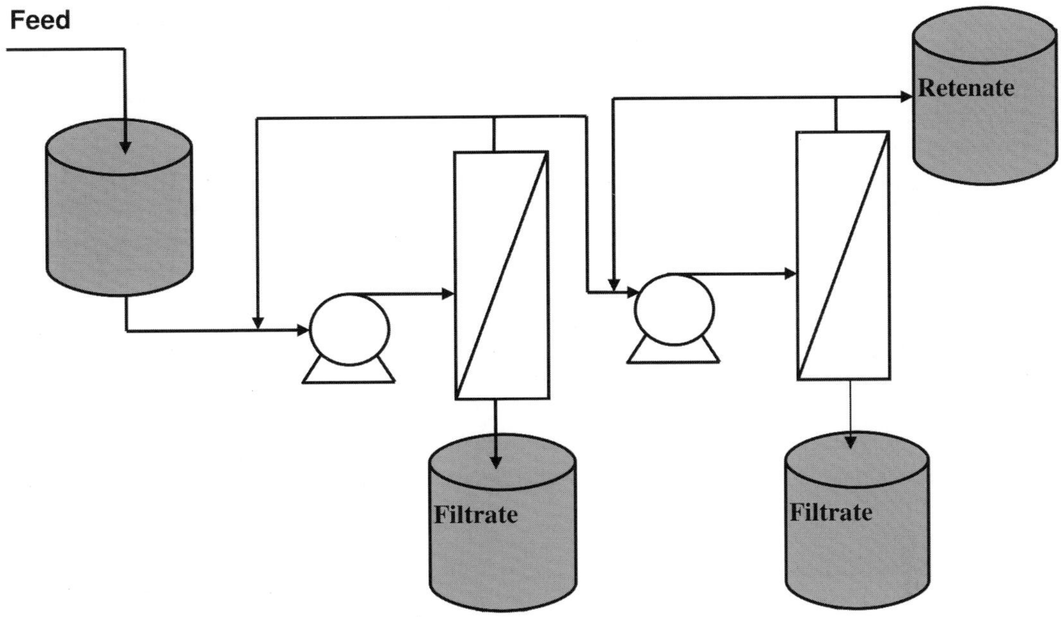

Figure 6.18. Schematic of a two-stages ultrafiltration set up with bleeding.

the surface of the membrane. The pressure is adjusted to maintain the flux because of production needs. These higher pressures increase "cake" resistance by cake compaction/collapse.

Four main remedies for this problem have been proposed and implemented: addition of filter aides, reverse filtration, backwashing, and a change in the fluid dynamics of the device. The addition of "filter aids" does not eliminate the formation of the protein layer at the membrane surface. Instead, it forms a "cake" of relatively controlled porosity and eliminates the possibility of cake collapse. The problem with this approach is that the membrane is practically replaced by a cake with its own filtration characteristics. A better option is to perform periodic reverse filtration (Mores and Davis 2003; Mallubhotla and Belfort 1996). By flowing clean solvent in the reverse direction of the filtration process the protein layer is literally washed away. The problem with this approach is that the process needs to be stopped at pre-established intervals. This needs to be built into the process to avoid discontinuities. Backwashing is a similar approach. A more permanent solution to membrane fouling is to change the fluid dynamics of the device to eliminate or minimize the formation of a deposit all together (Wu, Howell, and Field 1993; Cole and Brantley, 1996; Schutyser et al. 2002). Most of the current research on the operation of membranes is dedicated to that.

A simple protocol to clean a membrane is as follows (Chellam, Jacanelo, and Bonacquisti 1998). First the feed valve is closed and the water is drained. Next, compressed air is injected to remove the remaining fluid from the fibers and the device is pressurized. The backwash valves are opened for initial cleaning. Clean water is fed while the compressed air is maintained. This removes debris away from the membrane. The compressed air is turned off and the air is purged. The feed and filtrate valves are closed and the system is pressurized again to wet the membrane. The feed and filtrate valves are opened and the remaining air is purged. The total time elapsed is approximately 2 minutes.

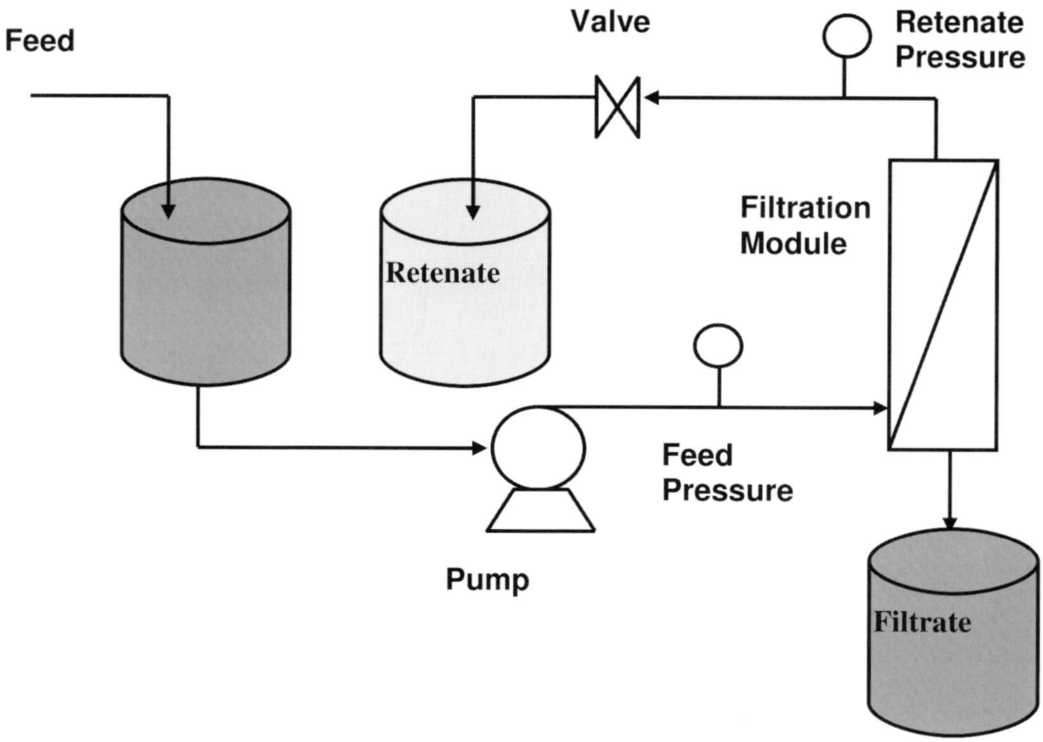

Figure 6.19. Schematic of a batch ultrafiltration set-up with no recycling.

Operating Membrane Devices

Figure 6.19 shows a simple ultrafiltration setup. A pump pumps the feed to the retenate side of the membrane module. The filtrate is collected on the filtrate side. Two manometers are used to monitor the upstream and downstream pressures. A valve is used to depressurize the outlet. Along a channel or fiber there are concentration gradients both in the axial and radial directions. The pressure drop along the length of the fiber on the filtrate side is generally small, whereas the pressure changes appreciably on the retenate side. Therefore, the transmembrane pressure drop is not constant along the fiber (it is higher at the feed point and smaller at the point in which the retenate leaves the device). Thus, the transmembrane pressure drop is given as the average pressure drop along the fiber.

The process variables in any ultrafiltration process are the cross-flow rate, the transmembrane pressure, the filtrate flow, and the membrane area. The process constraints are filtrate flow (production) and yield. High cross-flow rates are recommended to avoid membrane fouling and minimize concentration polarization. However, if the material is sheer sensitive, high cross flow rates may be detrimental because the product needs to be recycled several times. High filtrate flows reduce the membrane area needed but maximize concentration polarization. A trade-off is to work near the transition point between pressure-independent and pressure-dependent regimes. If the membrane has small pores the control of filtrate flow is not necessary because the tangential flow will be high enough to avoid concentration polarization. On the contrary if the membrane has large pores, control of the filtrate flow is needed to achieve enough tangential flow. The total filtration area is determined by process needs: total flow rate and processing time. It is customary to oversize the calculated area by 20% and choose a molecular weight cut off one order of magnitude higher than the one needed. Millipore recommends choosing a membrane with a NMWCO one-third to one-fifth the molecular weight of the molecule to be retained. In case of fractionation, the differences in molecular weight must differ by a factor of five.

A common use of membranes in bioprocesses is diafiltration, which is used to change buffers or to desalt a solution. The protein may be more stable in the dialysis buffer; therefore, it will be convenient to change the buffer as soon as possible. In addition, the protein may be more prone to aggregation in one buffer than in the other. Aggregates will increase concentration polarization and may change the filtration characteristics of the membrane (literally, the membrane pores will be changed by the "gel pores"). Therefore, direct experimentation with laboratory units that are easy to scale up is a necessity to optimize the process.

Diafiltration can be done at variable or constant volume. In the variable volume operation, the volume is first reduced and then fresh buffer is added to complete the original volume. The process is repeated as many times as needed to achieve the desirable level of buffer exchange. In a constant volume operation, the filtrate is removed at the same rate as the fresh buffer is added. The concentration at which diafiltration is performed needs to be chosen. The protein solution may be concentrated before dialysis or not. High concentrations in the feed will decrease the volume to be treated, but it will reduce the flux. Low protein concentrations increase the flux but also increase the area of membrane needed and the amount of buffer to be used. There is not a clear-cut criterion to make an informed decision. A possible strategy consists of a concentration step, followed by dialysis, and finished by a second concentration step. Because the flow through the membrane is usually a function of the buffer composition, the flow will change during the process if the pressure is controlled. Millipore recommends calculating a diafiltration optimization parameter equal to the product concentration in the feed stock times the filtrate flux at that point for the initial buffer and for the final buffer. This produces curves in a diafiltration optimization parameter-protein concentration plane that exhibit a maximum. The concentration can be chosen from a line joining the two maximums.

Different experiments can be performed to assess the performance of a membrane. An obvious one is to monitor the flux at various transmembrane pressures at several cross-flow rates and protein concentration. The number of experiments needed can be determined by following the experimental design strategy described in the appendix. During the experiments, several parameters need to be monitored; the most important ones are yield, recovery, and impurity removal.

Conditioning of the membrane before starting the process must be done according to the manufacturer guidelines. Typically, the membrane needs to be conditioned with running buffer to eliminate pirogens and any other contaminant added during installation. After operation, some of the product will remain inside the cartridge. This needs to be removed by either low-pressure air blowdown, buffer displacement, buffer flush, or buffer recirculation. Buffer displacement is a good option because the only drawback is a slight dilution of the product. The use of air may create foaming with the consequent detrimental effect on the product. Furthermore, as a result of the number of passes needed to remove the product from the membrane, buffer recirculation may also create problems of sheer induced denaturation. After the product has been recovered, the membrane needs to be cleaned and conditioned for storing. A common treatment is to use bases and acids in a cyclical or alternating fashion. Alternatively, enzymes may also be used (Arguello et al. 2002). Membranes do not respond well to dry conditions after they have been used. Therefore, they are normally stored in a buffer that inhibits microorganism growth; 5% alcohol solution will be sufficient. Otherwise, a dilute solution of sodium azide or a 0.1-M solution of sodium hydroxide can be used.

Novel Membranes

One of the areas of development in membrane filtration is the construction of composite membranes. The motivation for this approach is to resolve the paradigm of having mechanical strength

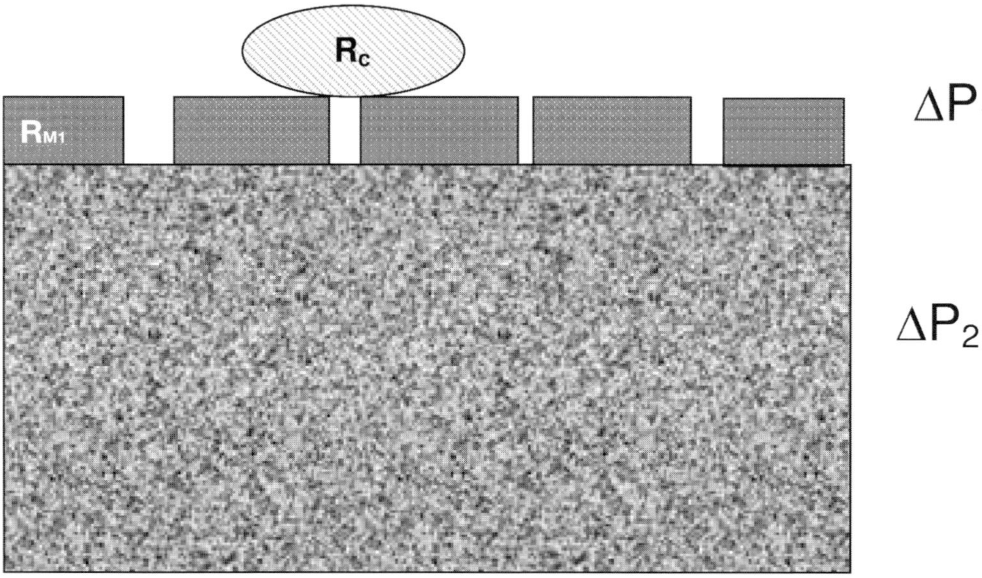

Figure 6.20. Schematic of a composite membrane consisting of a support layer and a thin controlled pore layer. A protein adsorbed at the membrane/liquid interface is shown. Adapted with permission from Ho, C-C., and A. L. Zydney. 2001. Protein fouling of asymmetric and composite microfiltration membranes. *Industrial and Engineering Chemistry Research* 40:1412-21. Copyright by the American Chemcial Society.

without losing selectivity. A common approach is to build the membrane in two layers. The filtrate-side layer is thick, mechanically strong, and has poor selectivity. The retenate-side layer is very thin, mechanically weak, and has well-controlled pore sizes. Figure 6.20 is a schematic of such a device.

It is interesting to see how the mathematical description of the system changes because of the added complexity. For example, consider the case in which the membrane is being fouled by protein deposits (see Figure 6.20). The objective is to calculate the flux decline index. In this device we have three areas of flux, flow through open pores in the upper layer, flow through blocked pores in the upper layer, and flow through the lower layer (Ho and Zydney 2001),

$$J_{open} = \frac{\Delta P_1}{\mu R_{M1}} \qquad 6.68$$

$$J_{block} = \frac{\Delta P_1}{\mu (R_{M1} + R_c)} \qquad 6.69$$

in which ΔP_1 is the pressure drop on the thin membrane, R_{M1} is the resistance of the membrane, and R_c is the resistance of the protein layer. The total flow, Q, through the upper layer is,

$$Q = J_{open} A_{open} + J_{blocked} A_{blocked} \qquad 6.70$$

in which A_{open} and A_{closed} are the areas that are free and blocked respectively and,

$$\frac{J_{open}}{J_0} = \frac{R_{M1}^2 + R_{M1}R_{M2} + R_{M1}R_c + R_{M2}R_c}{R_{M1}^2 + R_{M1}R_{M2} + R_{M1}R_c + R_{M2}R_cX} \quad 6.71$$

$$\frac{J_{blocked}}{J_0} = \frac{R_{M1}^2 + R_{M1}R_{M2}}{R_{M1}^2 + R_{M1}R_{M2} + R_{M1}R_c + R_{M2}R_cX} \quad 6.72$$

in which J_0 is the flux in the absence of obstructions. The flow of solvent through the thin membrane must be the same as the solvent flow through the macroporous support. That flow is,

$$Q = \frac{\Delta P_2 A}{\mu R_{M2}} \quad 6.73$$

The rate of change of the fraction of pores that are open (X) and of the protein resistance are,

$$\frac{dX}{dt} = -\alpha J_{open} X C_{Ab} \quad 6.74$$

in which m_p is the mass of protein deposited per unit area, f_p is the fraction of bulk proteins that contribute to the deposits, and α is a pore blockage parameter. Equations 6.74 and 6.75 can be combined to yield,

$$\frac{dR_c}{dt} = R_c \frac{dm_p}{dt} = f_p R_c J_{blocked} C_b \quad 6.75$$

$$\frac{dR_c}{dX} = -\frac{f_p R_c J_{blocked}}{\alpha J_{open} X} \quad 6.76$$

Equation 6.76 together with 6.71 and 6.72 can be solved to yield,

$$R_c = R_{M1}\sqrt{1 - \frac{2}{R_{M1}}\left[\left(\frac{f_p R_c \ln(X)}{\alpha}\right) - R_{\rho 0}\left(1 + \frac{R_{\rho 0}}{2R_{M1}}\right)\right]} - R_{M1} \quad 6.77$$

The strategy to calculate the flux decline index is as follows. Solve equation 6.71 for J_{open} as a function of X using Equation 6.77. Use Equation 6.74 to calculate X(t). Then, Q is calculated from Equation 6.70 with Equations 6.71 and 6.72 and the variation of the total flow, Q, over time can then be calculated to obtain the flux decline index from,

$$K = -\frac{1}{Q}\left(\frac{dQ}{dt}\right) \quad 6.78$$

Ho and Zydney (2001) found that K shows a maximum for several membranes that they tested except for a PCTE-based membrane. This is an important consideration in choosing the area and operating conditions for a given process using this type of membranes.

Two other novel uses of ultrafiltration are the integration of membrane filtration and chromatography in a single unit and the use of affinity filtration membranes. A hybrid filtration/chromatography process has been developed by (Dai et al. 1999, 2000; Xu et al. 2005) who included ion exchange beads in the shell side of an ultrafiltration membrane and used the device for the separation of various binary protein mixtures. It is not clear how this type of configuration can be scaled up and still having a device that can be cleaned in place and kept sterile. Affinity filtration membranes have been developed as an alternative to affinity chromatography (He, Dong, and Sun 1998; Galaev and Mattiasson 2001). However, it is questionable if they can compete with monolithic chromatography packing materials (see Chapter 7).

Liquid/Liquid Separations

A solution of two hydrophilic polymers or one hydrophilic polymer and a salt form two equilibrated liquid phases above certain concentrations. These phases are called aqueous two-phase systems (ATPS). Chapter 2 showed a typical phase diagram for an ATPS. Phase diagrams are a representation of phase composition dependence on phase-forming species concentrations for a given temperature, pressure, and molecular weight of the polymers (Bamberger et al. 1985).

ATPS is a technique that can be used as a protein purification process and for the separation and characterization of cells. Although it is ideally suitable for the removal of cell debris and simultaneous concentration of a target protein, it can be used as a very selective procedure at later stages of the purification process by adding affinity ligands to one of the phase-forming polymers. For either industrial or recombinant proteins the method offers notable advantages. For industrial enzymes, the method's utility stems from the ease with which it can be adapted to continuous production and scaled up to meet industrial needs. For recombinant proteins, the method's utility is based on its attractiveness as a gentle first step or on its selectivity in small scale by using affinity ligands. It is also a very powerful technique to fractionate cell populations in the laboratory and on a larger scale.

The partitioning behavior of any material in an ATPS is determined by the properties of the polymers and/or salt used to make the systems, by the properties of the material to be partitioned, and by the addition of additives, such as affinity ligands or additional salts. Once the appropriate phase-forming species have been chosen, phase diagrams can be used to determine the polymer/polymer or polymer/salt concentrations needed.

The partitioning behavior of cells is expressed in terms of a partition ratio, which is the quantity of cells in the top phase divided by that at the interface plus the bottom phase. The distribution of cells in an ATPS can be measured in several ways, such as by particle counting under the light microscope, by turbidity measurements, by radio labeling, or by cell plating. ATPS have been used to separate a wide variety of cells both in batch and in thin layer counter-current distribution (CCD) modes with and without the addition of affinity ligands. For example, Stendahl, Tagesson, and Edebo (1973) showed that the virulent parent strain of *Salmonella typhimurium* 395 partitions mostly into the top phase of a PEG/Dx system, whereas the nonvirulent mutant partitions into the bottom phase. This difference in partitioning correlates well with their susceptibility to undergo phagocytosis by polymorphonuclear leukocytes. In spite of the simplicity of batch partitioning, better separations are achieved by CCD. The technique has been copiously used to separate plant, bacteria, and mammalian cell populations. For example, the cellular slime mold *Dictyostelium discoideum* has been studied by thin layer counter-current distribution to identify the earliest stages of differentiation into pre-spore and pre-stalk cells (Sharpe 1994). Still, the most efficient way of affecting the partition of whole cells is by the addition of affinity ligands. These can be specific antibodies conjugated with PEG, charged polymers attached to a ligand specific for a particular antigen present on the cell surface, or fatty acids attached to PEG. Obviously, the use of specific antibodies is recommended for very selective separations, whereas the use of a fatty acid-PEG derivative is normally used in less selective separations.

Generally, cells partition between the interface and the PEG-rich phase or into the PEG-rich phase exclusively. Although PEG is nonionic, the addition of some salts creates a potential difference between the phases. This potential has been compared to the well-known Donnan potential that develops across membranes. The ions of some salts have different affinities for each phase, whereas others have an equal affinity for each phase. When a salt is used whose ions do not have the same affinity for each phase (for example sodium phosphate), a Donnan potential between the two phases is created that affects the partitioning behavior of charged particles, such as whole cells. We would like to emphasize here that, of course, each phase remains electrically neutral. Such a system is called a "charged sensitive ATPS" and it is the manipulation of the potential difference that permits us to separate cells that differ in their surface charge and to determine their charge. The surface charge measured by an ATPS partitioning experiment is not necessarily the same as the one measured by electrophoresis. Electrophoresis measures electrophoretic mobility, which is the ratio of surface charge to particle size, whereas cell partitioning measures charges that are not limited to the surface, but are present deep in the polyglycan layers (Walter and Widen 1994). Walter and Widen performed a series of experiments in which electrophoresis was done in the presence of the phase-forming species rather than in the presence of just buffer to compare ATPS results with electrophoresis ones. They concluded that the differences in cell populations partitioning behavior, up to that point thought to just reflect differences in surface charge exclusively, are actually indicators of more complex phenomena. This is not surprising because the polymers used in ATPS may interact with regions on the cell surface, like hydrophobic pockets. We have shown, for example, that block co-polymers of PEG and PPO (poly(propylene)glycol oxide) adsorb on hydrophobic surfaces extending several molecular diameters into the bulk (Huang and Forciniti 2002). It has been argued that ATPS partitioning has the added advantage of maintaining the viability of the cells; however, considering that PEG increases the fluidity of cell membranes, the cells will stay viable only for a limited time depending on the toxicity of additives.

There are several relatively recent reviews (Albertsson 1986; Hatti-Kaul 2000; Kimura and Kobayashi 1996) that cover various aspects of cell partitioning and the reader is referred to them for a complete collection of publications as well as to the ATPS home page for a comprehensive list of references. General features of cell partitioning are described in the monograph by Albertsson (1986) and in the book edited by Walter, Brooks, and Fisher (1985). For those interested in the physics behind the partitioning of cells the chapter by Brooks in Walter et al. (1985) is an excellent starting point. Practical aspects of the technique as well as numerous examples can be found in Walter and Johansson (1993) and Hatti-Kaul (2000).

One difference that needs to be highlighted between the partitioning of soluble molecules and particulates (cells) is the way in which the distribution coefficient is defined. Although some authors prefer to calculate the partition coefficient as the ratio between the concentration of cells in the top and bottom phases and then to calculate the amount of cells at the interface by that difference, it is the amount of cells in the top phase that is more meaningful. This is true because it has been shown (Albertsson 1986) that the amount of cells in the top phase is independent of the total amount of cells added. The total amount of cells increases the amount of material accumulated at the interface also but not the amount of cells in the bottom phase. This is contradictory to what is thought of as a true partitioning (from a thermodynamic equilibrium point of view), in which an increase in the total amount of cells should result in a proportional increase of the cell concentration in each phase.

The partition behavior of soluble substances can be readily explained from thermodynamic principles, in which the partition coefficient is defined as the ratio between the chemical potential of that particular solute between the two phases. The chemical potential is a complex but well-defined function of the chemical composition of the phases and on temperature. Gravity, of course, does not play a role in the partitioning of small molecules. However, equilibrium thermodynamic arguments alone cannot be used to explain the partition behavior of particulates because of some important transient phenomena occurring at the interface between the two phases during phase separation. However, a

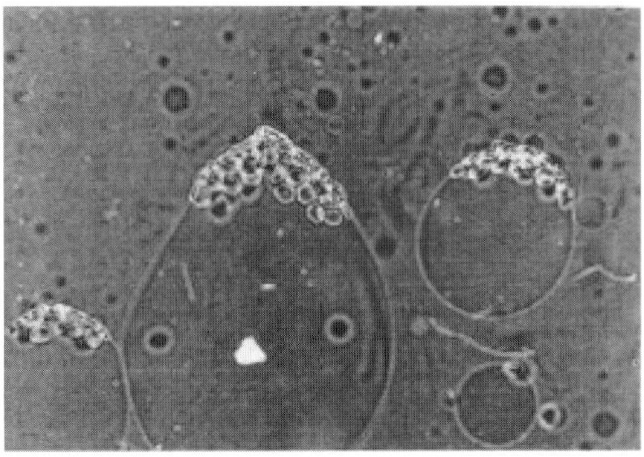

Figure 6.21. Red blood cells adsorbed at the interface of PEG-rich drops in a continuous dextran phase (from Van Alstine, 2000). Copyright by Humana Press; reproduced with permission.

thin line separates the partition behavior of solutes and particles. Recently, Kimura and Kobayashi (1996) have observed that large RNA molecules (although soluble) behave like particles in ATPS. This is an important observation because it allows the study of solutes and particles without changing the chemical nature of the material to be partitioned. An intriguing aspect of cell partitioning is associated with the dynamic changes that occur before an emulsion of two phases separate into their distinct components. At this point, the cells accumulate mostly at the interface between one phase and the other. Figure 6.21 illustrates this by showing an emulsion of a Dx-rich phase in a continuous PEG phase in which erythrocytes accumulate mostly at the continuous/dispersed phase interface and in the PEG continuous phase (Van Alstine 2000). As the emulsion coalesces, the surface area decreases and the cells are forced to move into one phase or the other. Obviously, these phenomena are absent in the partitioning of soluble substances in which thermodynamic equilibrium is reached between the material inside the small droplets and the bulk (continuous) phase.

At the interface between two immiscible liquids (one made up by molecules A and the other by molecules B), there are interactions between molecules A and B. An interfacial tension arises due to the differences in attractive forces between the molecules in the different fluids—these attractive forces between the molecules inspired the work of van der Waals on the continuity of the states. Interfacial tension is a force per unit length, which is equal to the energy per surface area. The cartoon in Figure 6.22 illustrates the different forces present at the interface between two liquid interfaces in which a particle has been deposited. The energy barrier produced by interfacial tension prevents one liquid from becoming emulsified into the other. To form an emulsion, the surface free energy must be lowered by adding a third component that accumulates at the interface. Thus, the adsorption of cells at the liquid/liquid interface occurs as the system seeks a minimum energy. It is this adsorption of particles at the interface that distinguishes particle partitioning from solute partitioning. This distinction arises from the fact that when a particle is located in the interface the area of the interface is reduced by an amount equal to the particle cross-section. Because the formation of any interface requires work, the loss of area produces a proportionate decrease in free energy, an effect that can be seen from Gibbs equation.

Some simple thermodynamic arguments can be made to rationalize the effect that interfacial tension has on particle partitioning. Although particles may or may not distribute between the phases by thermal motion, we can start with Boltzmann's equation, which assumes that the particles diffuse freely (Van Alstine 2000),

$$K_p = \frac{n_1}{n_2} = \frac{c_1}{c_2} = \exp(\Delta E / k_B T) \quad 6.79$$

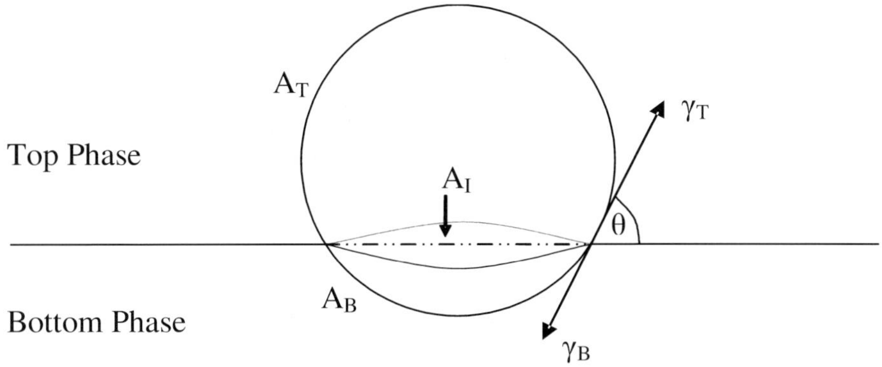

Figure 6.22. A particle deposited at the interface of two partially miscible phases. A_I is the area of the interface occupied by the particle.

in which K is the partition coefficient, n_1 and n_2 are the number of cells in two compartments (they can be either top and bottom phase or one bulk phase and the interface), c_1 and c_2 are the corresponding concentrations, ΔE is the difference in energy between the two phases, T is the absolute temperature, and k_B is the Boltzmann constant. The exponential indicates that the partition behavior is a stochastic process, or that a particle has a certain probability of being in a particular phase. Now, we can consider an ideal spherical particle in a phase system characterized by the interfacial tension γ_{TB} with an interfacial free energy of γ_T or γ_B. The interfacial free energies γ_T or γ_B represent the energies associated with the formation of a unit of surface area for the particle when it is either in the top or bottom phase respectively (see Figure 6.22). By estimating the work required to move a particle from the interface to the top phase and by using this estimate in Boltzmann's equation, we get,

$$\ln K_p = \left(\frac{-\gamma_{TB} \pi a_p^2}{k_B T} \right) (1 - \cos\theta_c)^2 \qquad 6.80$$

in which θ_c is the contact angel between the particle and the top phase and a_p is the particle surface area. From this we can see that partitioning of a particle will only occur if there are some chemical or physical interactions between the phase-forming species and the particles such that the particles are pulled away from the interface and into one of the phases. Although there is experimental proof that this equation breaks down, probably as an indication of the non-stochastic nature of the partitioning phenomenon, it is still useful to demonstrate the interrelationship between cell partitioning, the surface free energies, and the wetting of a particle by the individual phases.

The partition coefficient of soluble molecules reaches a steady state value within minutes after the appearance of the macroscopic interface, in spite of the fact that a dispersion of one phase into the other may take several hours to coalesce. Therefore, equilibrium is quickly attained. The situation is quite different in the partition of particles. It has been observed that the amount of cells in the PEG-rich phase decreases with time, whereas the amount of cells at the interface increases. It has also been observed that the rate at which these changes are noticed is faster at longer tie-line lengths. Experimental evidence shows that a fine emulsion of the Dx-rich phase persists in the top phase for hours (if the phases are not separated by centrifugation) and that, depending on the type

Table 6.3 Range of the Exponents in the Correlations Between Interfacial Tension and Tie-Line Length (Equations 6.81, 6.82, and 6.83).

	−a	b	−e	f	−g	h
$\gamma = a(TLL)^b$	4.7–1.1	4.9–2.5				
$\gamma = e(\Delta Dx)^f$			6.3–1.1	6.2–2.7		
$\gamma = g(\Delta PEG)^h$					4.9–(−0.09)	6.5–2.1

of cells, the cells accumulate at the interface between the Dx droplets and the continuous PEG phase. Over time, these droplets coalesce and move into the interface where they "deliver" their cell load. If the concentration of cells is high, cells at the interface will eventually sediment into the bottom phase. These non-equilibrium aspects of particle partitioning must be taken into consideration as we interpret experimental results.

The interfacial tension of aqueous two-phase systems depends on polymer molecular weight, tie-line length, and temperature. A few correlations have been used that relate the interfacial tension with the length of the tie line,

$$\gamma = a\left(TLL\right)^b \qquad 6.81$$

$$\gamma = e\left(\Delta Dx\right)^f \qquad 6.82$$

$$\gamma = g\left(\Delta PEG\right)^h \qquad 6.83$$

The parameters a, b, e, f, g, and h depend on temperature and molecular weight of the polymers used. Their range at 25°C is summarized in Table 6.3 (Bermudez and Forciniti 2003).

In the processing industry, ATPS can be used for the separation of cell debris from the target protein; some preliminary purification is achieved as well because some impurities remain in the bottom phase. A disc stack centrifuge can easily separate the two liquid phases. Normally, the centrifuge is set such as the entire bottom phase and the entire interface are separated from the top phase. This is necessary because of the tendency of cells and particulates to accumulate at the interface between the two phases.

Experiments

Two experiments are described: dead end and cross-flow filtration of lysozyme.

Dead End Filtration

In this experiment, measurements of retention time are used to determine the specific resistance of the membrane and the resistance of the polarization layer. This data can then be used to scale up the process.

Materials

A 1 mg/mL solution of hen lysozyme. 10 cm diameter, 10,000 molecular weight cut off membrane (YM10 Amicon). A dead end ultrafiltration cell. Lysozyme enzymatic activity kit. Bradford kit.

Methods

Place the membrane in a beaker and soak it in Nanopure^{TM} water. Change water every 15 minutes for 1 hour. Be careful when handling the membrane, oils will clog the pores. Assemble the ultrafiltration cell (for example, Millipore stirred cell unit) according to manufacturer instructions. Add sample to the unit (Figure 6.7). Insert magnetic stir bar and screw lid on tight. Place unit inside anchoring device on stir plate. Make sure the buffer reservoir (if connected) is switched to "gas" and turn on gas flow to unit to begin filtering. **Warning: DO NOT EXCEED 80 psi.** Turn on stir plate and set to desired stir rate. Connect a tube to the opening in the bottom of the cell and collect the filtrate in a beaker place on a top loading balance. Allow filtration to continue until the desired volume is reached. Shut off the gas flow. Purge the unit using the valve on the top. Remove the unit from the stir plate. Unscrew lid and collect material inside for further use. Measure the concentration of the protein and the enzymatic activity in the filtrate and in the retentate. Repeat the experiment for 5 pressures and 2 pHs (recommended: 4.5 and 10). Plot the flux versus the pressure drop across the membrane and calculate from the slope of the liner part of the curve the overall resistance of the membrane. Calculate the amount of enzyme recovered and the purification fold.

Cross-Flow Filtration

The following description applies to any cross-flow ultrafiltration module. The results of the experiments can be used to calculate mass transfer coefficients.

Materials

Solutions at various concentrations of lysozyme in phosphate pH 8 buffer. Lysozyme enzymatic activity kit. Bradford kit.

Instrument

Any cross flow ultrafiltration system; for example Labscale TFF system by Millipore (Figure 6.8, left panel). A cross-filtration membrane modulus with a molecular weight cut off of 10,000.

Methods

Condition the membrane modulus according to the manufacturer specifications. Select a recirculation rate (this fixes the filtrate pressure). Place 100 mL of a 1 mg/mL protein solution into the sample reservoir, start stirring and begin pumping of the protein solution. Start a stopwatch. Collect the filtrate in a beaker placed on a top loading balance until half the sample volume (approximately 50 mL) has filtrate. Measure the concentration and activity of lysozyme in filtrate and retenate. Clean the membrane with a warm 0.1 N NaOH solution and repeat the process for increasing pressure drops (higher recirculation rates). A plot of flux versus pressure drop should look like the lower curves in Figure 6.12. Select a pressure drop, such as the flux that is independent of pressure, and repeat the experiment for five or six higher protein concentrations. Remember to clean the membrane in between runs. Use Equation 6.50 to calculate the protein concentration at the surface of the membrane and the mass transfer coefficient.

Summary

Centrifugation and filtration are competing methods in the early stages of any bioprocess. Following the first separation between cells and supernatant or in between cell debris and supernatant ultrafiltration is used several times mostly to exchange buffers or to condition streams for chromatographic separations. Both techniques are mature and equipment is available in laboratory, pilot plant, and process scale.

Ultrafiltration continues to evolve into a more selective separation technique as new materials are being developed by the leading industries. Whether or not ultrafiltration may replace some chromatography steps in the future remains to be seen. Liquid/liquid extraction may become a more viable alternative if the scale of bioprocesses increases because of market demands.

Nomenclature

$\vec{a}$ acceleration due to the force field (m/sec^2)
a_p particle surface area (m^2)
A_{open} Area available for filtration (m^2)
A_{closed} Blocked area (m^2)
A area of filtration (m^2)
C_i concentration of species i
C_{ib} concentration of species i in the bulk
C_{if} concentration of species i in the filtrate
C_{iw} concentration of species i at the membrane
C_{AW} concentration at the membrane surface (mol/L)
CF concentration factor, $CF=V_f/V_0$.
D_{AB} diffusion coefficient in the bulk (cm^2/sec)
d diameter of a particle (m)
f_p fraction of bulk proteins that contribute to fouling
F_B bouyancy force (N)
F_D drag force
$\vec{g}$ acceleration of gravity (9.8066 m/sec^2)
J_0 flux in the absence of obstructions (mol/m^2 sec)
J flux of filtrate (mol/m^2 sec)
K_p partition coefficient
k_B Boltzmann's constant (1.3807×10^{-23} J/K)
k_D Darcy's Law permeability of the bed (m^2)
k_0 mass transfer coefficient (mol/m^2 sec)
$L_{1,2}$ length (m)
m_p mass of protein deposited per unit area (kg)
n number of discs
N number of stages.
Q Volumetric flow (m^3/sec)
r_i radial distance (m)
r fraction of soluble material that remains in the cake
R_0 intrinsic membrane retention
R_M resistance of the filter medium
R_C resistance of the cake
R retention of the membrane.
S_{50} sieving coefficient
t time (sec)
T absolute temperature (K)
v_g sedimentation velocity under gravity

v_p	velocity of the particle (m/sec)
v_w	sedimentation velocity under a centrifugal force (m/s)
v	velocity of liquid (m/sec)
V	volume to be filtrated (m^3)
V_0	initial volume (m^3)
X	fraction of pores that are open
Z	accelerator factor
ΔE	is the difference in energy between the two phases (J)
ΔP	is the pressure drop across a bed of thickness l (Pa)

Greek Letters

ρ_s	density of the particle (kg/m^3)
ρ	density of the solvent (kg/m^3)
μ	viscosity of the solvent (kg/cm sec^2)
ω	rotational speed (radian/sec)
Σ	"Sigma Value"
η	efficiency of the centrifuge
θ	angle of the discs
α	specific cake resistance
ε	washing efficiency
$\Delta\Pi$	osmotic pressure (Pa)
θ	contact angel between the particle and the top phase
θ_D	inclination angle of the discs in a disc-stack centrifuge
γ_{TB}	interfacial tension (Pa m)
γ_i	interfacial free energy (Pa m)

References

Albertsson, P. A. 1986. *Partition of Cell Particles and Macromolecules,* 3rd ed. New York: Wiley Interscience.

Arguello, M. A., S. Alvarez, F. A. Riera, and R. Alavarez. 2002. Enzymatic cleaning of inorganic ultrafiltration membranes fouled by whey proteins. *Journal of Agricultural and Food Chemistry* 50:1951-1958.

Axelsson, H., and B. Madsen. 2006. *Centrifuges, Sedimenting. Ullman's Encyclopedia*. New York: John Wiley & Sons, pp. 89-117.

Bamberger, S., D. E. Brooks, K. A. Sharp, J. M. Van Alstine, and T. J. Webber. 1985. Preparation of phase systems and measurement of their physicochemical properties. In *Partitioning in Aqueous Two-Phase Systems: Theory, Methods, Uses, and Application to Biotechnology*, edited by H. Walter, D. E. Brooks and D. Fisher. Orlando, FL: Academic Press.

Bermudez, O., and D. Forciniti. 2003. Distribution of cells at interfaces-analytical and health related issues. *Recent Research Developments in Biotechnology and Bioengineering* 5:243-61.

Boychyn, M., S. S. S. Yim, P. Ayazi Shamlou, M. Bulmer, J. More, and M. Hoare. 2001. Characterization of flow intensity in continuous centrifuges for the development of laboratory mimics. *Chemical Engineering Science* 56:4759-70.

Chellam, S., J. G. Jacanelo, and T. P. Bonacquisti. 1998. Modeling and experimental verification of pilot-scale hollow fiber direct flow microfiltration with periodic backwashing. *Environmental Science and Technology* 32:75-81.

Cole, J., and J. D. Brantley. 1996. Dynamic membrane filtration in cell and protein production. *BioPharm* 66:68-71.

Christy, C., and S. Vermant. 2002. The state-of-the-art of filtration in recovery processes for biopharmaceutical production. *Desalination* 147:1-4.

Dai, X-P., R. G. Luo, and K. K. Sirkar. 1999. An integrated process for biomolecule isolation and purification. *Biotechnology Progress* 15:1095-105.

Dai, X-P., R. G. Luo, and K. K. Sirkar. 2000. Pressure and flux profiles in bead-filled ultrafiltration/microfiltration hollow fiber membrane modules. *Biotechnology Progress* 16:1044-54.

Ebersold, M. F., and A. L. Zydney. 2004. Separation of protein charge variants by ultrafiltration. *Biotechnology Progress* 20:543-9.

Foley, G. 2006. A review of factors affecting filter cake properties in dead-end microfiltration of microbial suspensions. *Journal of Membrane Science* 274:38-46.

Galaev, I. Y., and B. Mattiasson. 2001. Protein purification by affinity ultrafiltration. *Biotechnology and Bioprocessing* 26:243-62.

Hatti-Kaul, R. (ed.) 2000. *Methods in Biotechnology: Aqueous Two-Phase Systems-Methods and Protocols,* 11th ed. Totowa, NJ: Humana Press.

He, L.-Z., X.-Y. Dong, and Y. Sun. 1998. Modeling and analysis of the affinity filtration process, including broth feeding, washing, and elution steps. *Biotechnology Progress* 14:594-600.

Ho, C-C., and A. L. Zydney. 2001. Protein fouling of asymmetric and composite microfiltration membranes. *Industrial and Engineering Chemistry Research* 40:1412-1421.

Huang, Y., and D. Forciniti. 2002. Ethylene oxide and propylene oxide random copolymer/sodium chloride aqueous two-phase systems: Wetting and adsorption on dodecyl-agarose and polystyrene. *Biotechnology and Bioengineering* 77:786-95.

Kimura, K., and H. Kobayashi. 1996. RNA partitioning accompanied by adsorption: high-molecular-mass RNA adsorbed at the interface like a particle. *Journal of Chromatography B* 680:213-9.

Mallubhotla H., and G. Belfort. 1996. Semiempirical modeling of cross-flow microfiltration with periodic reverse filtration. *Industrial and Engineering Chemistry Research* 35:2920-8.

Mores, W. D., and R. H. Davis. 2003. Yeast-fouling effects in cross-flow microfiltration with periodic reverse filtration. *Industrial and Engineering Chemistry Research* 42:130-9.

O'Mahony, K., R. Freitag, B. Dhote, F. Hilbrig, P. Muller, and I. Schumacher. 2006. Capture of bacteria from fermentation broth by body feed filtration: a solved problem? *Biotechnology Progress* 22:471-83.

Reis, R. V., L. C. Leonard, C. C. Hsu, and S. E. Builder. 1991. Industrial scale harvest of proteins from mammalian cell culture by tangential flow filtration. *Biotechnology and Bioengineering* 38:413-22.

Sakhamuru, K., D. W. Hough, and J. B. Chaudhuri. 2000. Protein purification by ultrafiltration using a β-galactosidase fusion tag. *Biotechnology Progress* 16:296-8.

Schu, P., and M. Gautam. 2001. Ultrafiltration membranes in the vaccine industry. *Biotechnology and Bioprocessing* 26:225-41.

Schutyser, M., R. Rupp, J. Wideman, and G. Belfort. 2002. Dean vortex membrane microfiltration and diafiltration of rBDNF *E. coli* inclusion bodies. *Biotechnology Progress* 18:322-9.

Sharpe, P. T. 1993. Surface changes during growth and development of slime molds. In *Methods in Enzymology,* edited by H. Walter and G. Johansson, pp. 334-44. Orlando: Academic Press.

Stendahl, O., C. Tagesson, and M. Edebo. 1973. Partition of salmonella typhimurium in a two-polymer aqueous phase system in relation to liability to phagocytosis. *Infection and Immunity* 8:36-50.

Trettin, D. R., and M. R. Doshi. 1980. Limiting flux in ultrafiltration of macromolecular solutions. *Chemical Engineering Communications* 4:507.

Van Alstine, J. M. 2000. Eukaryotic cell partition: experimental considerations. In *Methods in Biotechnology*, edited by R. Hatti-Kaul, Chapter 11. Totowa, NJ: Humana Press.

Walter, H., and K. E. Widen. 1994. Cell partitioning in two-polymer aqueous phase systems and cell electrophoresis in aqueous polymer solutions red blood cells from different species. *Journal of Chromatography* A 668:185.

Walter, H., and G. Johansson. eds. 1993. *Methods in Enzymology.* Orlando: Academic Press, p. 228.

Walter, H., D. E. Brooks, and D. Fisher. eds. 1985. *Partitioning in Aqueous Two-Phase Systems: Theory, Methods, Uses and Applications to Biotechnology.* Orlando: Academic Press.

Wu, D., J. A. Howell, and R. W. Field. 1993. Pulsatile flow filtration of yeast cell debris: Influence of preincubation on performance. *Biotechnology and Bioengineering* 41:998-1002.

Xu, Y., K. K. Sirkar, X-P. Dai, and R. G. Luo. 2005. A new integrated membrane filtration and chromatographic device. *Biotechnology Progress* 21:590-7.

Further Reading

Zeman, L. J., and A. L. Zydeny. 1996. *Microfiltration and Ultrafiltration: Principles and Applications.* New York: Marcel Dekker Inc.

Chapter 7
Chromatography

Introduction

Chromatography is a separation technique based on the uneven distribution of analytes between a stationary (usually a solid) and a mobile phase (either a liquid or a gas). Chromatography can be done in two or three dimensions. Two-dimensional chromatography is done on a thin layer of a porous material, such as cellulose or silica, supported on a glass plate. Sometimes two-dimensional chromatography is called paper chromatography because the use of paper (supported or not) was the dominant technique for many years. Paper chromatography is used solely for analytical purposes. The solid phase in paper chromatography is a strip of paper on which the sample is seeded as a small drop. One end of the paper is immersed in a container whose bottom is filled with a solvent that moves upward by capillary action. The solvent then carries the solutes upward different distances depending on their relative affinity for the solid and liquid phases. Column chromatography is by far the most common form of chromatography used for analytical purposes and the only one used in processing. Although the mobile phase may be a gas (gas chromatography) or a liquid (liquid chromatography), only liquid chromatography is used in bioprocessing. The columns may be operated at low (LPLC) or high pressures (HPLC); they are both found in bioprocessing.

Figure 7.1 shows a typical and simplified chromatography set-up. The main element of the set-up is the chromatography column. One or more detectors are used to monitor the presence of various solutes exiting the column and process variables such as pH and conductivity. Ultraviolet, conductivity, and refractive index detectors are frequently used. The use of two detectors in series is also common. Further downstream, fractions of the different samples are collected. Upstream from the column there is a valve that is used to switch from the buffer used to elute the column to the "injection loop" that is used to seed the sample into it. Upstream from the injection valve there is a pump/mixer device that is used to pump the buffers that carry the sample through the column; up to four buffer reservoirs are commonly used. There are several operating modes available for chromatographic separations. The bed can be fixed, expanded, or simulated moving. Fixed beds are the most common, whereas expanded beds are an option to treat "dirty" feeds, and simulated moving beds have not found their niche in bioprocessing yet. Two possible modes of operation are available: 1) the column may be saturated with the sample first and then eluted, or 2) the sample may be added as a pulse and then eluted. The first mode of operation is dominant in affinity chromatography,

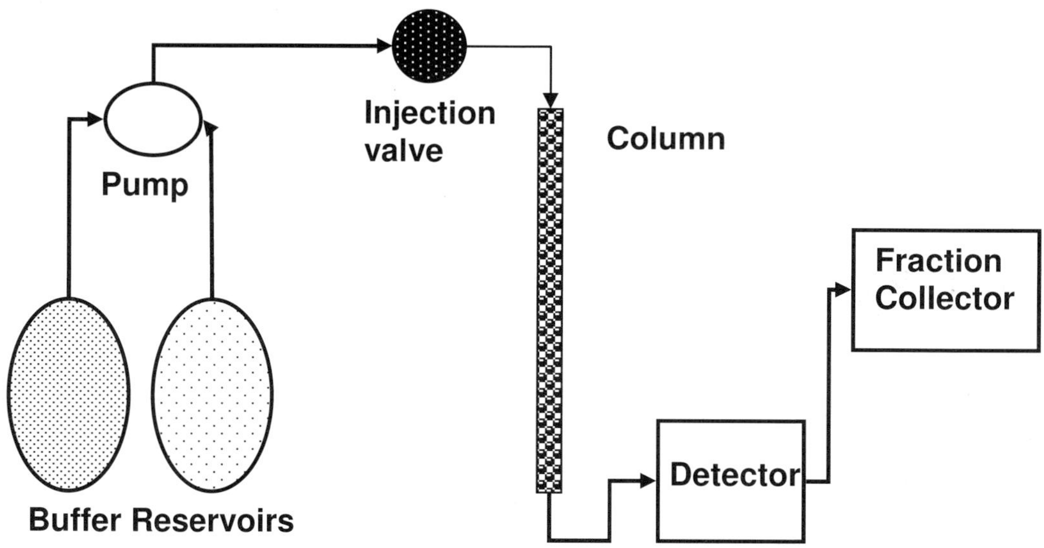

Figure 7.1. Typical liquid chromatography set-up

whereas the second mode is more common in the other types of chromatography. The mobile phase, which moves the sample and the impurities through the column, may either be a buffer of constant composition or one made by the combination of two or more solutions (gradient).

At the core of the chromatography system is the column containing the stationary (solid) phase. Because of high operating pressures, HPLC columns are normally built in stainless steel. Figure 7.2 shows a picture of two chromatography columns: laboratory and process scale. Notice the differences in size and the differences in construction materials. Different types of stationary phases can be used depending on the kind of mixture to be separated. In size exclusion chromatography, the gel is a filtration media that sorts solutes by size. If there is a difference in the charge of the analytes at some particular pH, an ion exchange gel (either cationic or anionic) may be used. If the analytes differ in their hydrophobicity, a hydrophobic chromatography gel could be used. Finally, affinity chromatography gels, which have functional groups that bind a single species in a mixture, are used for very high resolutions. Figure 7.3 shows these various options.

We start our discussion describing the different types of solid phases. Then, we describe the different types of chromatography associated with the various solid phases. This is followed by several definitions that allow the characterization of the chromatographic process. The chapter continues with an analysis of adsorption and elution in fixed beds. Finally, recent developments in expanded and simulated moving beds are described.

Characteristics of the Stationary Phase

The key of any chromatography technique is the uneven distribution of solute particles between a stationary and mobile phase. Therefore, large contact areas between both phases improve separation by minimizing mass transfer limitations and by increasing the capacity. Because of this, the solid

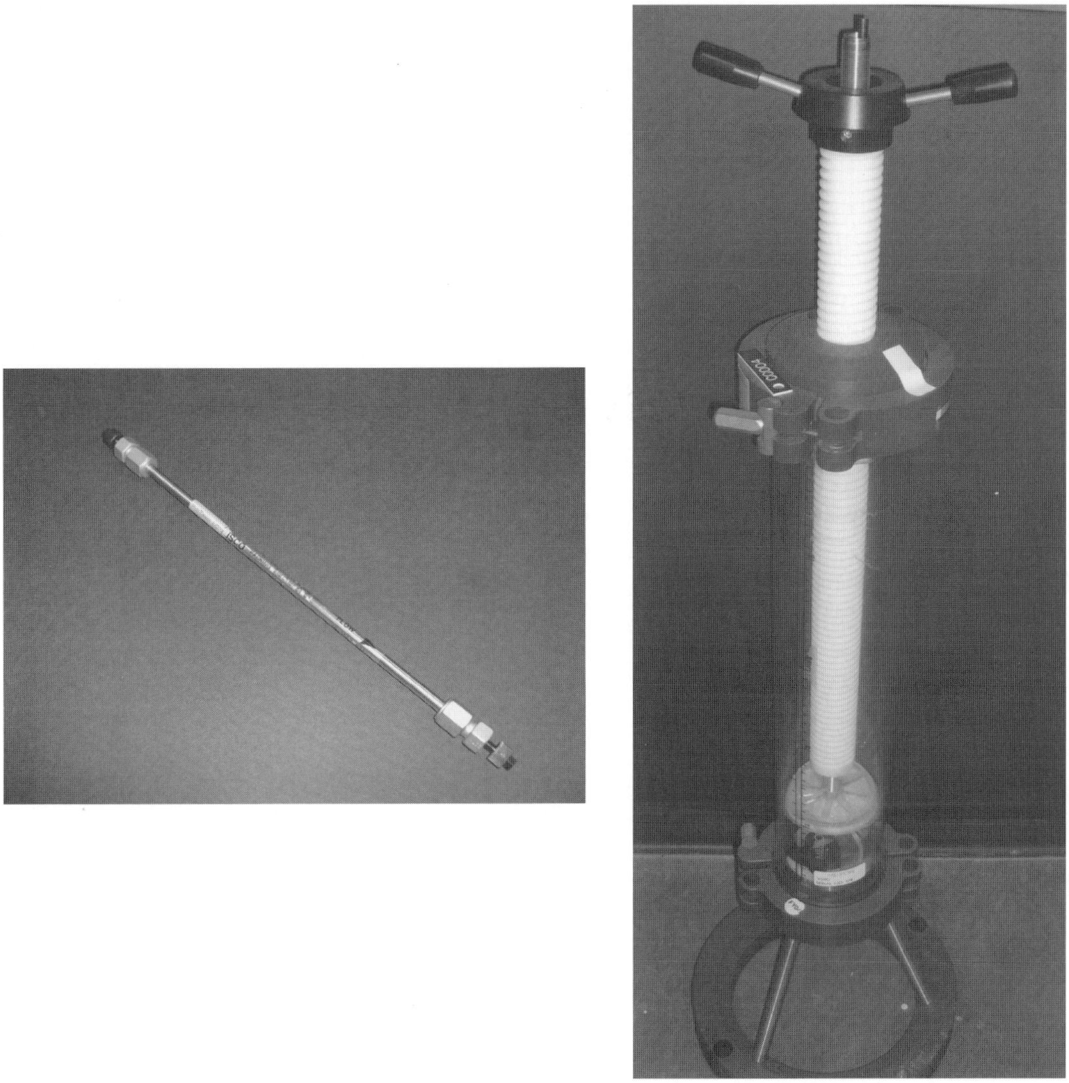

Figure 7.2. Picture of chromatography columns at different scales. Left panel: laboratory (HPLC) scale. Right panel: process scale (1 liter).

support is a highly porous material compacted in the form of small beads (between 10 to 100 μm in diameter). This highly porous material must be chemically and mechanically stable. Currently a limitation of the industrial standard material, beads, is that they are compressible (they deform upon stresses). Because of this operational problem, new materials in the form of monoliths have recently been developed.

Solid phases are made by a variety of polymeric materials, but dextran and agarose somehow dominate the market. Dextran is a polymer of glucose, whereas agarose is a polymer of glucose and 3,6 and hydro L-galactose (Figure 7.4). The hydroxyl groups in the sugars are reactive and, there-

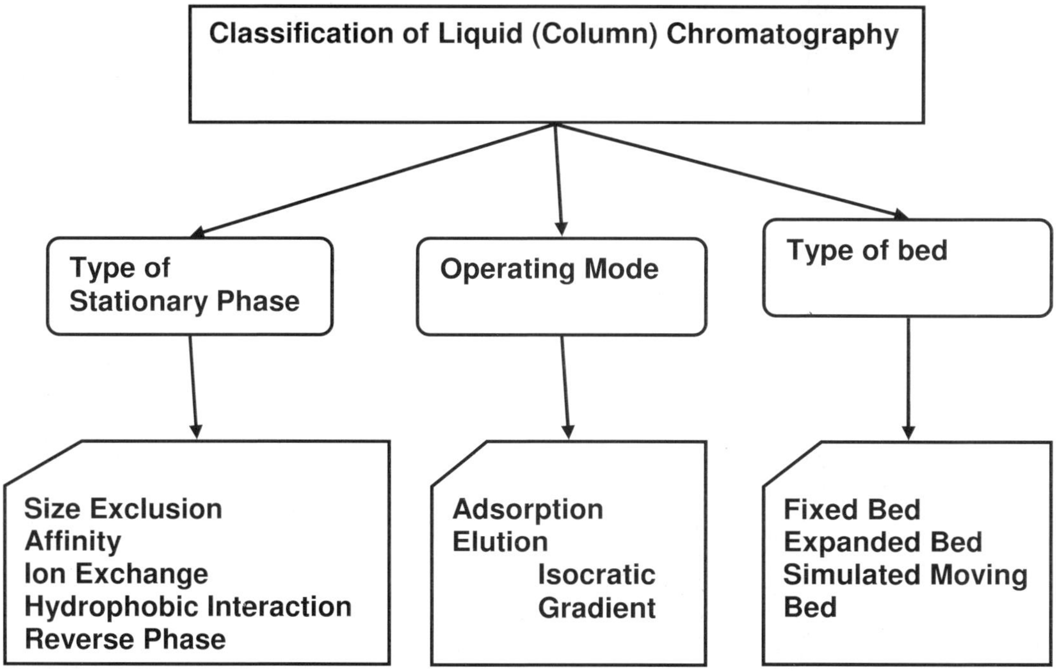

Figure 7.3. Classification of chromatography processes

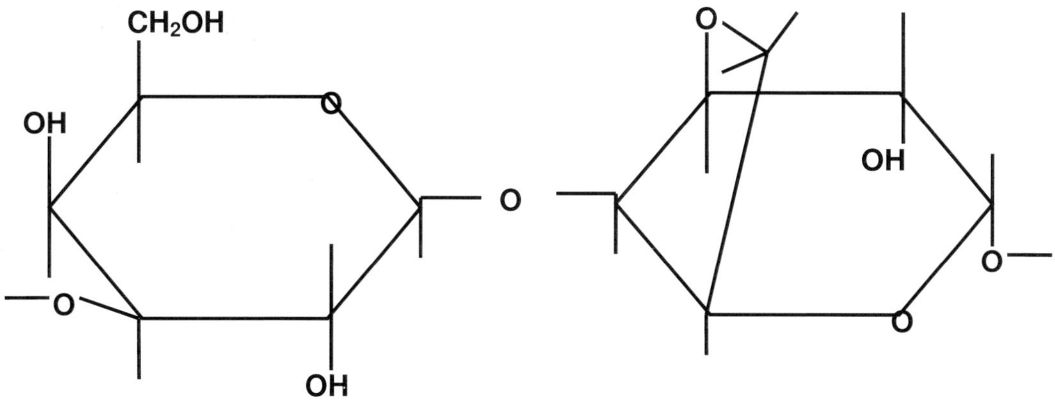

Figure 7.4. Repeating unit of agarose

fore, several modifications (like the addition of ionizable or short hydrocarbons) are possible, which lead to several different chromatography modes. For example Superdex (Pharmacia) is a gel permeation media consisting of cross-linked dextran and agarose. This combination makes the gel have the filtration properties of dextran and the physical stability of cross-linked agarose. Sephacryl HR is made of cross-linking allyl dextran with N,N′-methylene bisacrylamide, which creates a hy-

Table 7.1. Carbon Leakage in Some Common Matrices (adapted from Sofer, G., and L. Hagel. 1997. *Handbook of Process Chromatography: A Guide to Optimization Scale-up and Validation.* New York: Academic Press, with permission). Copyright by Academic Press.

Matrix	Carbon content in dried gel (%)	Carbon Leakage (%)	
		pH 2	pH 14
Agarose	44.3	0.2	0.8
Styrene-divinylbenzene-1	70.6	0.8	4.0
Styrene-divinylbenzene-2	70.9	0.3	3.5
Styrene-divinylbenzene-3	64.9	2.0	18
Methacrylate	51.4	0.1	140
Polystyrene-silica	12.6/46.0% silica	3.0	3.2

drophilic matrix of high mechanical strength. Sephadex is an ion exchange media made by cross-linking dextran with epichlorohydrin. An example of a polyacrylamide-based gel is Bio-Gel P-6DG commercialized by BioRad. This gel is only stable at a narrow pH range (from 5.5 to 6.5). Despite this drawback, this gel is still compatible with dilute organic acids, 8 M urea, 6 M guandine HCl, and detergents, and is ideal for desalting protein solutions.

Solid phase materials are sold as powders that must be hydrated before use. The swollen material forms a gel that is then packed into the column. These gels are not rigid; therefore, they are deformable upon the action of external forces like the pressure drop imposed along the column to induce flow.

One of the problems intrinsic to any chromatography media is leaching; that is, the release of ligands, surface groups, silica, or sugars into the eluent. Leaching may take place during the storage and elution stages of column operation, but is more commonly seen during sanitation or cleaning of the media because of the harsh conditions used in these processes (for example a concentrated NaOH solution). The leachable material may be toxic and maximum tolerances may be established by the corresponding regulatory agency. Because of its relatively uncontrollable occurrence, leaching is particularly important during the validation of the process. Before a chromatography media is chosen for process separations, the leaching is estimated by subjecting the media to extreme pHs and by identifying the leachable products. Then, analytical methods are optimized to reach the detection limit imposed by regulatory agencies. After that, the method with the optimized parameters is tested with the column.

The kind of material that is leached out depends on the chromatography media that is used. For example, in a media containing a phenyl group (like phenyl Sepharose) the following products may leach: 1) the phenyl group plus the spacer plus a piece of the polysaccharide, 2) the phenyl group plus the spacer, 3) the phenyl group plus fragments of the spacer, 4) phenol, and 5) benzene. Polysaccharides are common supports in various types of chromatography—for example, Sepharose (Pharmacia). Sofer and Hagel (1997) studied the release of sugars from Sepharose upon storage in various solutions. They found that the gel was stable upon storage for four weeks in acetic acid, sodium sulfate, ammonium hydroxide, and sodium hydroxide. However, the gel leaked sugars upon storage in sulfuric acid. Table 7.1 summarizes some common compounds often found as the result of leaching (Sofer and Hagel 1997).

Another example of media leaching is the release of protein A in the affinity chromatography of antibodies. Often, protein A will leach as protein A-antibody complexes. The ordinary solution for this problem is to follow the affinity column with an ion exchange column to remove protein A. Because there are very sensitive methods to detect traces of protein A in the preparation, the severity of having the product contaminated by protein A is attenuated. Unfortunately, the same kind of

Table 7.2. Effectiveness of NaOH to Destroy Several Common Viruses (adapted from Sofer, G., and L. Hagel. 1997. *Handbook of Process Chromatography: A Guide to Optimization Scale-up and Validation.* New York: Academic Press, with permission). Copyright by Academic Press.

NaOH	HIV	BVD	CPV	BHV	POL	SV-40	MLV	ADV
				Inactivation (log_{10})				
0.1M	>3.5	2.5	5.6	8.2	5.5	3.9	>3.8	>6.0
0.8M	>3.5	>4.7	6.1	>8.1	5.1	6.2	>3.7	>7

Abbreviations: HIV (human immunodeficiency virus type I 2.0 10^6), BVD (bovine viral diarrhea virus, 9.5 10^6), CPV (canine parvovirus, 2.0 10^9), BHV (bovine herpes virus type I, 6.9 10^9), POL (human polio virus type 2 (7.1 10^8), SV-40 (simian virus 40, 1.7 10^8), MLV (murine leukemia virus, 2.6 10^5), ADV (human adenovirus type 2, 2.2 10^8).

analytical protocol may not be available for other leaching materials. Because ion exchange is a very common bioseparation process it is worthwhile to mention the detection methods used to detect ion exchange gels leakage. For anion exchangers (amine), gas chromatography and smell are used, whereas for cation exchangers, (sulfite) ion chromatography, elemental analysis, and GC-MS may be used.

Cleaning and sanitation of a chromatographic system are two periodical maintenance procedures. Cleaning in situ is a necessity for having an economically viable process. This is done every one to 10 cycles depending on the process, and eliminates material that is not removed during the regeneration of the column. It is accomplished normally by operating the column in reverse flow. The purpose of sanitation is to inactivate vegetative cells. Sanitation is performed between batches and often the use of a 0.5- to 1.0-M NaOH solution with a contact time of 30 to 60 minutes is sufficient (Sofer and Hagel 1997).

The kind of cleaning agent needed depends on the contaminant, whose identity depends on the type of raw material used in the process. For example, high lipid content is a characteristic of feeds originated from tissue or plasma, although they also come from the cell membrane of microorganisms. Phenolic compounds are the main contaminants when plants are used. Mechanically disrupted cell feeds will contain high lipids, proteases and endotoxins for cytoplasmatic proteins and high lipid and protein content, proteases, and endotoxins for periplasmatic ones. The amount of lipids and proteins will be low if osmotic shock is used. Finally, if the protein is extracellular, the feed may contain proteases. Endotoxins are a major concern when *E. coli* is used as the host system, whereas proteases are of concern when yeast is utilized. The most common cleaning agents are NaOH, NaCl, nonionic detergents, and organic solvents. NaOH is efficient for the removal of endotoxins, nucleic acids, viruses, precipitated proteins, and hydrophobic proteins. NaCl is used to remove soluble and precipitated proteins as well as nucleic acids. Nonionic detergents—ethanol, isopropanol, or acetonitrile—are used to remove lipids. Table 7.2 shows the inactivation efficiency of two different concentrations of NaOH on eight different viruses. The manufacturers of the chromatographic media normally recommend a type of cleaning as well as cleaning materials to avoid because they may jeopardize the stability of the material.

The design parameters (flow rate, operating temperature, bed height, and buffers) are chosen based on laboratory experiments performed with columns whose diameters are 3 cm or less. Scaling up to process scale is done by increasing the diameter of the column, which causes the compressible gel to be "less supported" by the column's walls. One can imagine that beads within the bed are keeping each other together by friction. The beads next to the column wall are kept in that position by friction against a stationary object. As we move from the column wall into its center the supporting effect of the wall weakens. In small diameter laboratory columns, that is not a problem, but it becomes important as the column diameter becomes larger or when the aspect ratio (D/L) is large. This loss of wall support increases the compression of the gel and consequently increases the pressure drop along the bed. The increase in pressure drop may cause the column not to work in large scale; therefore, modifications of bed height or flow rate may be needed after the column has been built with the consequent change in production capacity. From a practical viewpoint, it is important to have a simple correlation based on parameters available in the laboratory that would allow the designer to estimate the increase in pressure drop during the scaling up of the column.

The pressure drop along the column increases as the flow rate increases. Sticker and Fotopoulus (2001) observed that after a critical flow rate is reached the pressure drop does not increase any further. This flow rate corresponds to a critical superficial velocity, which depends on the type of gel. Less rigid gels have lower critical velocity. Thus, this critical velocity can be used to characterize a gel. Before the critical velocity is reached, the pressure drop increases with flow rate because of both an increase in the fluid velocity and a decrease in the permeability as the gel compresses. Beyond the critical velocity, the increase in pressure drop is the result only of a decrease in bed porosity. They also observed that compression is linear with flow velocity below the critical velocity. They propose to use the following equation to determine the critical velocity, $v_{c,}$

$$v_c L_0 = m\left(\frac{L_0}{D}\right) + b \qquad 7.1$$

in which L_0 is the initial bead height, D is the diameter of the column, L/D_0 is the aspect ratio, and m and b are empirical constants that must be determined experimentally. This equation is used to fit experimental data obtained from a set of columns with aspect ratios between 2 and 15. The compression of the bed, λ, is calculated from,

$$\lambda = \lambda_c \frac{v}{v_c} \qquad 7.2$$

in which v is the superficial velocity and λ is,

$$\lambda = \frac{L_0 - L}{L_0} \qquad 7.3$$

in which L is the actual bead height. Sticker and Fotopoulus found that λ_c does not change significantly with the aspect ratio. Therefore, it is taken as a material specific parameter. λ_c can be determined by averaging the values of λ obtained in the experiments done to determine m and b (Equation 7.1). The value of λ is used to calculate the actual length of the bed (Equation 7.3) and the porosity of the bed, ε, from,

$$\varepsilon = \frac{\varepsilon_0 - \lambda}{1 - \lambda} \qquad 7.4$$

ε_0 is the settle bed porosity and can be obtained from the manufacturer or estimated by seeding and eluting a blue dextran solution. Finally, the pressure drop in the column can then be calculated using the Blake-Kozeny equation,

$$\Delta P = \mu \frac{K_0}{d_p^2} \frac{(1-\varepsilon)^2}{\varepsilon^3} Lv \qquad 7.5$$

in which d_p is the average diameter of the beads that should be supplied by the manufacturer or may be determined by scanning electron microscopy, μ is the viscosity of the buffer that can be measured or estimated from tabulated values, and K_0 is an empirical constant equal to 150 (Geankoplis 1993). The authors illustrate the use of this approach to scale up a process based on protein A Sepharose 4FF. Laboratory experiments show that a column 20 cm in bed height operated at 100 cm/h at 6^0C is needed. Binding studies in laboratory scale suggest that a 100-cm diameter column will be needed in process scale. Following the above methodology, a plot of pressure drop versus superficial velocity for various column diameters and bed heights and at two temperatures was generated (Figure 7.5). The straight lines shown in the figures correspond to the post-packing pressure drop across a bed of any cross-sectional area assuming a degree of compression equal to λ_c. It is clear from the upper panel that a superficial velocity of 100 cm/h is not reachable with a 20-cm bed and 100-cm diameter column. The options are either to increase the temperature to 22^0C or to decrease the height of the bed to 15 cm. It is important to notice that to generate these plots the characteristic length chosen was that of the compressed length. This can be found by using experimental data or by choosing a "reasonable" value that approximates the compressed length. The authors recommend λ_c because this amount of compression is below physical damage and because further compression during operation does not occur unless the critical velocity is reached.

The approach described above relies on direct experimental data of the deformation of the bed. This is very convenient for design purposes, but adds little to our understanding of bed compression. One of the main problems in developing theoretical models of mechanical deformation of compressible chromatography gels is the lack of good models for solid deformation. Keener et al. (2002) solved a tension balance for the gel, and in combination with a careful analysis of the mechanical compression of the gel (the compression that arises after the gel has been compacted by flow), they found an expression for the stress profile along the column. That expression is,

$$p = \exp(\frac{x}{\eta}) \qquad 7.6$$

with

$$\eta = \left(\frac{1-\nu}{4\nu \tan\phi}\right)\left(\frac{D}{L_0 - h}\right) \qquad 7.7$$

in which $p=\sigma(Z)/\sigma_0$, $x=(z-h)/(L-h)$, h is the extent of column compression, z is the axial coordinate, σ_0 is the applied area-average axial stress, $\sigma(z)$ is the stress profile, ν is the Poisson ratio for the material, and ϕ is the angle of wall friction. These equations can be used to determine the stress profile throughout the column. Furthermore, they show that if η is greater than unity, mechanical forces will compress the gel uniformly. This information can be used to pack the bed minimizing the risk of the formation of channels and other irregularities.

About 20 years ago, the first report on monoliths "gels" was published (Hjerten, Liao, and Zhang 1989). Monolithic columns consist of a single piece of solid material with interconnected skeletons and pores. Monolithic columns are not compressible, have longer stability, and higher flow rates can be used. They are made by a variety of materials including acrylamide, acrylate, methacrylates, and silica. Silica-based columns are most common in analytical laboratories. For example, Merck commercializes a series of monolithic columns under the trade name Chromolith, which have a binodal pore size distribution. Mesopores are about 12 nm in diameter, whereas macropores are about 3 µm in size. Their use has been very popular in the separation of small molecules but there are few reports on the use of monolithic columns for protein separations. Still, monolithic chromatography

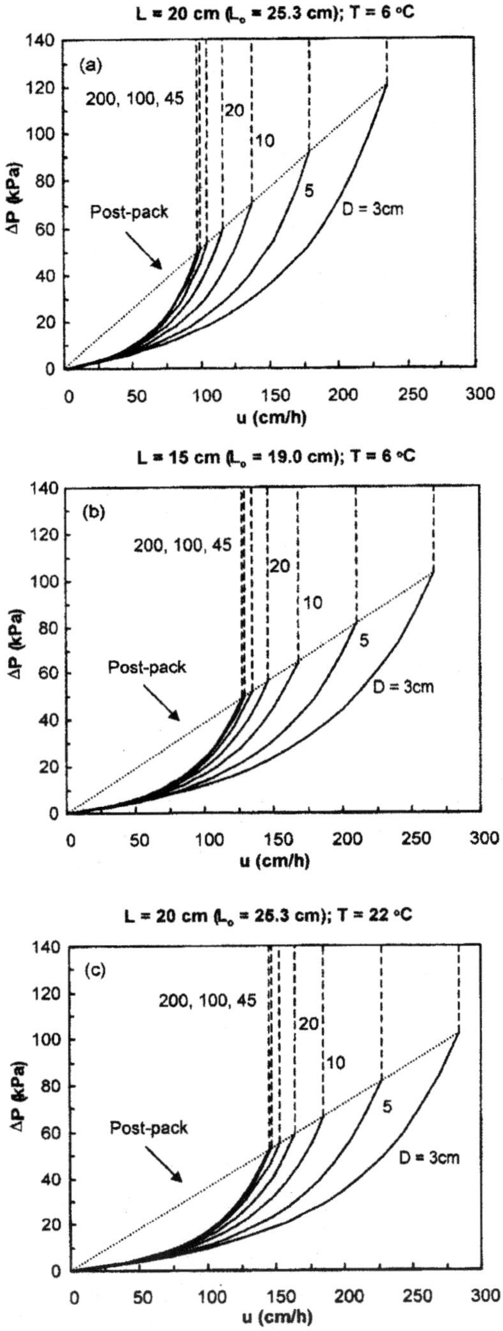

Figure 7.5. Experimental and predicted pressure flow profiles for packing of compressible media in PBS at 6°C. Loss of wall support with increasing diameter is emphasized by normalizing for bed height differences in the abscissa. From Sticker, J. J., and A. Fotopoulus. 2001. Pressure-flow relationships for packed beds of compressible chromatography media at laboratory and production scale. *Biotechnology Progress* 17:744-51, with permission). Copyright by the American Chemical Society.

based on polymer blocks has reached process scale. Bia Separations (Slovenia) represented in the United States by Iris Technologies International (GA, USA) produces a series of columns from 8 mL to 800 mL using a matrix of poly(glycidyl methacrylate-co-ethylene glycol dimethacrylate) or poly(styrene-co-divinylbenzene). A somehow unique feature of this device is that the flow is in the radial direction. This is achieved by a series of distributors on the outside of the monolith that distribute the eluent (or sample) on the outer surface. The flow is directed inward, where a collecting tube directs the fluid to the outlet. The 800-mL column can be operated at flow rates up to 2 L/min and the pressure drop does not exceed 1.0 MPa at a flow rate of 250 mL/min. Columns are available with a variety of ligands including quaternary amine, diethylamine, butyl, sulfonyl, carboxymethyl, reversed phase (C-18 or C-22), protein A, and protein G. They also produce disks of the same material that can be staked into a single housing (up to four disks), which opens the possibility of performing multidimensional chromatography (more than one separating mode) in a single column.

Membranes have been used as a solid support in chromatographic separations. This technique, known as membrane chromatography, was developed approximately 15 years ago (Brandt et al. 1988; Fulton et al. 1992). Ion exchange, affinity, reverse-phase, and hydrophobic interaction chromatography membranes are available (Ghosh 2002) and they are available in flat beds, hollow fibers, and radial flow. The use of membrane-based chromatography devices is attractive because the predominant flow mechanism is convection. This added to the ability of enclosing a very large membrane area in a small volume reduces both equipment size and processing time. Membrane-based devices are more appropriate for adsorption than elution chromatography. Membrane chromatography has become quite popular in laboratory and pilot scale and it is used in industrial scale for virus removal and polishing (after protein A treatment) in the purification of human antibodies (Zhou and Tressel 2006). The main concerns when using this technique are significant backpressures with viscous solutions and non-uniform flow. The potential problems with cleaning and sterilization have been eliminated by the fabrication of single-use (disposable) devices.

Types of Chromatography

Chromatography is used in bioprocessing in the capture, purification, and polishing steps. The type of chromatography used as well as the specific gel for that mode depends on the purpose of the step (capture, purification and polishing). The following sections describe each type of chromatography as well as the most common commercial gels.

Size Exclusion Chromatography (SEC)

Size exclusion chromatography (SEC) is used to separate molecules according to size. SEC gels consist of very small porous particles whose surfaces (ideally) do not interact with the analytes. In a SEC gel, small particles penetrate the pores media, whereas large ones do not. Thus, small molecules begin to "lag behind the largest ones" as the column is eluted. This phenomenon of different resident times for differently sized molecules in an SEC media is what allows SEC to separate molecules according to their size. The column is normally eluted with a single buffer and the separations are relatively coarse. Gels are made of inert, cross-linked polymers. Different degrees of cross-linking yields different pore sizes and different molecular sieving capacities or separation ranges.

Typical matrices are cross-linked dextran (Sephadex), cross-linked polyacrylamide (Biogel), and agarose. Table 7.3 summarizes some gels that are available commercially.

Table 7.3. Several Agarose-Based Gel Permeation Matrices.

Gel	% agarose	Bead size (μm)	Fractionation range for globular proteins
Sepharose 6B	6% agarose	45-165	10000-4000000
Sepharose 4B	4% agarose	45-165	60000-20000000
Sepharose 2B	2% agarose	60-200	70000-40000000
Superdex 30		34-44 μm	<10000
Superdex 75			3000-70000
Superdex 200			10000-600000
Superdex 6	6% agarose	>13 μm	5000-5000000

Ion Exchange Chromatography (IEC)

Ion exchange chromatography (IEC) separates molecules based on their charge density and is a variation of adsorption chromatography in which the solid adsorbent has charged groups chemically linked to an inert support. Ions are electrostatically bound to the charged groups; these ions may be exchanged for ions in an aqueous solution. Because charged molecules generally bind to an ion exchanger reversibly, molecules can be bound or eluted by changing the ionic strength or the pH of the eluting solvent.

The selection of the appropriate ionic exchange resin depends, as in any other type of chromatography, on the stage in the process in which the chromatography step is going to be used (capture, intermediate purification, polishing). Vendors will provide enough information for a rational choice. Notice that different gels are needed depending on the stage of the process because of purity requirements (highest in polishing), capacity requirement (highest in capture), and in the potential presence of particulates that may clog the resin (highest in capture).

The gel consists of very small porous particles whose surface (inner and outer) is covered by ionic groups (basic or acidic) such as sulfate, carboxylic, and amine. The operating pH is selected so that both the ionic groups on the surface of the gel and the ionizable groups on the analytes are (at least) partially dissociated. Under those conditions, the analytes bind through Columbic forces to the solid support. The strength of these forces depends on the dissociation constants of the various groups as well as on the concentration of electrolyte in the buffer. The sample is commonly added in a low ion strength buffer to allow the analytes to bind to the gel. Then, they are eluted, for example, with a salt gradient. Because proteins are separated in ion-exchange chromatography according to their net surface charge, it is imperative to consider how the charge of a protein varies as the pH is varied. Figure 7.6 shows this. At a particular pH the protein will have a net charge of zero; that pH is known as the protein's isoelectric point. At lower pHs the protein will be positively charged and it will have an affinity for a cation exchange resin. At pHs higher than the isoelectric point the protein will be negatively charged and it will have an affinity for an anion exchange resin.

Ion exchangers are classified into anion and cation exchangers depending on the charge present and into weak and strong according to the value of the dissociation constant of the acid or the base. Table 7.4 shows the most common ligands found in commercial gels.

If the isoelectric point of the protein is not known and it cannot be estimated by isoelectric focusing, then it is better to choose strong ion exchangers, because they remain charged in a broad pH range. The main problem with strong ion exchangers is that the adsorptive force may be so strong that the protein will denature or it will not be eluted. Thus, weak ion exchangers, under the right pH conditions, should be preferred because they are gentle with biological products.

An example of an ion exchange resin is the Sephadex family, which is commercialized by Pharmacia, and includes the following resins: mixed weak and strong anion exchanger (DEAE

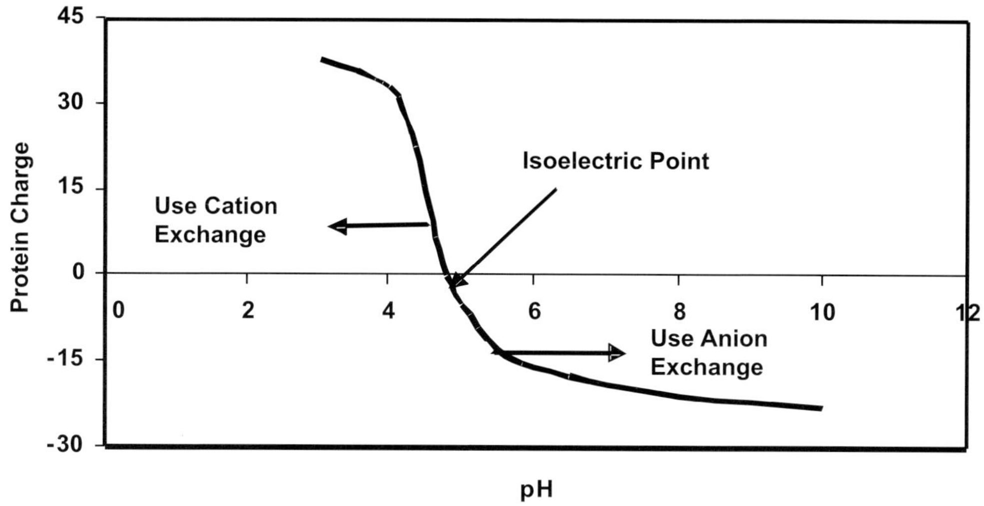

Figure 7.6. Titration curve for a protein. The arrows show the pH regions where cation and anion exchange resins may be used.

Sephadex), strong anion exchanger (QAE-Sephadex), weak cation exchanger (CM Sephadex), and strong cation exchanger (SP-Sephadex). Most ion exchange gels change their volume as the pH and/or the salt concentration is changed. This is a problem because changes in pH and salt concentration are used to elute the column. For example, some Sephadex products (QAE-Sephadex) may change its volume from approximately 55 mL per gram of dried gel to 20 mL per gram of dried gel as the ionic strength is changed from 0 to 0.5 (NaCl). The decrease in volume with ion strength is exponential; therefore, the volume reduces to half as the ionic strength is increased from 0 to 0.1 (a value normally used to elute a column). An increase in pH may decrease, increase, or not affect the volume depending on the gel. For example, QAE-sephadex volume is not a function of pH, whereas CM sephadex C-50 nearly doubles its volume as the pH is increased from 4 to 10. On the other hand, the volume of DEAE Sephadex decreases by 50% as the pH is increased from 5 to 12. These changes in volume are not 100% reversible. Therefore, the pressure drop in the gels tends to increase after several cycles.

Table 7.4. Different Types of Ion Exchange Gels.

Gel	Nature	pH Range
Anion Exchangers		
$X\text{-}CH_2N\text{+}(CH_3)_3$	Strong	2-11
$X\text{-}CH_2NH\text{+}(CH_3)_2$	Intermediate	2-7
$X\text{-}(CH_2)2NH^+(CH_3CH_2)_2$	Weak	3-6
Diethylamino-ethyl (DEAE)		
Cation Exchangers		
$X\text{-}SO_3^-$	Strong	2-11
$X\text{-}COO^-$	Intermediate	6-10
$X\text{-}CH_2COO_-$	Weak	7-10
Carboxymethyl (CM)		

Table 7.5. Common Target Protein/Ligand Pairs.

Target	Bound Ligand
Enzyme	Substrate, inhibitor, substrate mimic
Antigen	Antibody
Antibody	Antigen
Hormone	Receptor
Receptor	Effector

Affinity Chromatography

Affinity chromatography gels' surfaces are covered with molecules (ligand) that bind a particular molecule or a family of molecules in the sample. An example of the former is protein-A that is used for the purification of antibodies, whereas an example of the latter is Procion Red (a dye) that is used for the isolation of dehydrogenases from complex mixtures. The difference between these two ligands is in their specificity. Protein-A has a high specificity for IgG, whereas Procion Red is a promiscuous ligand that can bind to any NAD dependent dehydrogenase. Many other different types of ligands can be used; therefore, this type of chromatography has the largest amount of gels available and specific gels may be developed for a particular separation. Table 7.5 summarizes some general types of ligands used in affinity chromatography.

In affinity chromatography, the sample is injected in a buffer that favors the binding of the target protein to the gel. Then, the column is washed with a series of buffers to eliminate all impurities. Finally, a buffer is chosen that weakens the interaction of the target protein with the gel.

Some antibodies like IgG3, rat IgGs, and mouse IgG1 do not bind (or have very low affinity) for protein-A columns. An alternative for the purification of antibodies is to use protein-G as a ligand. Protein G columns may also be used to remove contaminating antibodies, for example bovine IgG from fetal calf serum, to very low levels. Protein G binds to the constant region of IgG. In one version of this gel (Millipore Prosep-G) the region of native protein G responsible for binding albumin has been eliminated, which increases the capacity of the gel by eliminating non-specific interactions. Millipore protein-G gel has no affinity for IgA, IgM, IgGE, IgGD, and albumin but has affinity for IgG of human, mouse, and other species. In the case of Millipore, protein G is bonded to porous glass beads with a capacity from 5 to 24 mg of antibody per mL of gel depending on the antibody. The antibodies are adsorbed in a pH range from 4 to 8 and eluted in a pH range from 3 to 2. This very low pH may create some problems because proteins are not stable at extreme pH values. This particular gel may be sanitized with 0.3% HCl, 1% phosphoric acid, or 6 M guanidine HCl and cleaned with 80% ethanol/0.5% acetic acid. The gel may be used in packed or fluidized beds.

Hydroxyapatite or $[Ca_{10}(PO_4)_6(OH)_2]_2$ (commercialized by Pharmacia) is unique because it is both the support and the ligand. It consists of pairs of charged crystal calcium ions and clusters of six oxygen ions from the phosphate. The resins have very high capacity and are resistant to hydroxide, detergents, and organic solvents. However, the crystal is destroyed below pH 5 and it is dissolved by chelating agents such as EDTA. Binding to proteins takes place by both non-specific electrostatic attractions and specific interactions between carboxyl groups in the protein and the calcium "loci." Elution is done by screening electrostatic interactions (non-specific) or by displacement (specific).

Hydroxyapatite columns (HAC) are operated at the lowest possible pH (alkaline pHs decrease binding) and low ionic strength to maximize interactions between the proteins and the gel. It is common to add organic solvents such as ethanol or ethylene glycol to keep hydrophobic proteins in solution. Basic proteins are eluted with 0.1 to 0.3 M phosphate fluoride, chloride, or thiocyanate or

Table 7.6. Operational Parameters and Responses Associated With Various Stages of a Hydroxyapatite Chromatography Run (adapted from BioProcess Guide. 2006. *Hydroxyapatite Chromatography*. GE Healthcare, pp 1-9, with permission).

Operation	Parameters	Responses
Equilibration	pH, conductivity, temperature, redox potential, flow	Back pressure
Sample	Volume, pH, conductivity, temperature, redox potential, protein concentration, holding time	
Sample application	pH, conductivity, temperature, redox potential, flow, application time	Back pressure, UV profile
Wash	pH, conductivity, temperature, redox potential, flow	Back pressure, UV profile
Elution	pH, conductivity, temperature, redox potential, flow	Back pressure, UV profile
Eluate	Holding time	Volume, pH, conductivity, temperature, redox potential, protein concentration, retention time
CIP	pH, conductivity, temperature, redox potential, flow	Back pressure

CIP: cleaning in place

with less than 0.003 M Ca^{2+} or Mg^{2+}. Acidic proteins elute with phosphate but they will not elute with Ca^{2+} or Cl^-. Neutral proteins will elute with phosphate, fluoride, and chloride, but do not elute with Ca^{2+} and thiocyanate.

Proteins can be eluted isocratically, with a stepwise protocol, which can be accomplished either by a serial application of several isocratic elutions or by a gradient elution. In HAC non-binding molecules will be fractionated according to size. According to Pharmacia, HAC can be used during capture, intermediary purification, and polishing operations because of both its high capacity and its capabilities of operating in a mixed purification mode (i.e., both gel permeation and affinity chromatography in combination) that imparts unusual selectivity. However, it is known that HACs may work great for a particular mixture but fail completely for others without a reasonable explanation. Table 7.6 summarizes some important features of HAC gels.

Some natural materials, like alginate, have been used as affinity ligands. Alginate is a non-toxic polysaccharide consisting of guluronic and mannuronic acids. Alginate has been used in laboratory scale and in fixed and expanded beds for the purification of starch degrading enzymes (Sharma, Sharma, and Gupta 2000; Mondal, Sharma, and Gupta 2003), lipases (Sharma, Roy, and Gupta 2001), pullulanase (Roy and Gupta, 2002), and pectinase (Roy et al. 2004).

A common type of affinity chromatography consists of the use of dyes as ligands, which are normally attached to cross-linked dextran gels such as Sepharose. Common dyes include Procion Blue MX-3G, Procion Blue H-B, Procion Red H-3B, Procion Red HE-3B, and Procion Yellow H-A. Most dyes are promiscuous ligands, that is, they bind a family of proteins. For example, Procion Red binds dehydrogenases, whereas Cibacron Blue binds nucleotide dependent enzymes. At least in the latter case, the affinity is because of the fact that the dye mimics the structure of NADH.

Chelating groups, such as iminodiacetic acid, may be used to adsorb metal ions onto the gel, which then recognize some amino acids like histidine, cysteine, or tryptophan that are present on the surface of the protein. This affinity chromatography mode is known as immobilized metal affinity chromatography (IMAC). IMAC has been used for the purification of recombinant proteins tagged with a polyhistidine tail (Belew and Porath 1990), human growth hormone (Maisano, Testpro, and Grandi 1989), tissue plasminogen activator (Furlong et al. 1988), human factor XII (Pixley and Coleman 1986), and others (Wong, Albright, and Wang 1991). The technique can selectively separate proteins differing in as little as one histine residue on their surface. Its main drawback is the leaking of metal ions into the preparation.

Hydrophobic Interaction Chromatography

Hydrophobic interaction chromatography (HIC) separates solutes by exploiting differences in their hydrophobicity. In the particular case of proteins, those differences in hydrophobicity are caused by different distributions of amino acids on the surface of the protein. The gel consists of a solid (porous) matrix support to which hydrophobic groups are attached.

One matrix that is used is cross-link beaded agarose (Pharmacia). Pharmacia commercializes butyl Sepharose, octyl Sepharose, and phenyl Sepharose. Phenyl ligand densities vary from 20 μmol/mL to 40 μmol/mL of gel, whereas butyl densities are approximately 50 μmol/mL of gel and octyl densities are approximately 5 μmol/mL of gel. The maximum flow rate is 4 mL/min. The gels are stable in a wide range of pH values (from pH 3 to 13), and to exposure to 1 M NaOH, 70% ethanol, 30% isopropanol, and 6 M GuHCl. Phenyl Sepharose is also stable in 8 M urea, 1 M acetic acid, and 3 M ammonium sulfate. Butyl and octyl are stable in 1 mM HCl. Regeneration of the columns is achieved with distilled water. Precipitated proteins can be removed by washing with alkali or ethanol/isopropanol. Elution is done with salt gradient of ammonium sulfate or sodium sulfate but other "weaker" salts such as NaCl may be used. The quality of the separation depends strongly on the type of salt used.

A special type of hydrophobic interaction chromatography is called Ca binding protein HIC (Chao, Bu, and Yiu 1990; Gopalakrishna and Anderson 1985). Some proteins change their conformation so drastically in the presence of Ca^{2+} or other ions like Zn^{2+} that their surface hydrophobicity increases several folds. This property allows the protein to bind to HIC gels (such as phenyl Sepharose) in the absence of high salt concentrations. Therefore, other hydrophobic proteins are washed out, whereas the Ca-dependent protein is retained by the column. The elution is performed in the presence of chelate agents that remove Ca^{2+}. A variety of buffers may be used; the only precaution is to stay away from buffers, which form insoluble complexes with Ca, like phosphate.

For HIC an initial salt concentration just below the amount needed is used to salt out the protein. Hydrophobic interactions are weaker at cold temperatures; therefore, columns are normally run at room temperatures. Elution in HIC is done by either 1) reducing the concentration of salting-out ions in the buffer with a decreasing salt gradient, 2) increasing the concentration of chaotropic ions in the buffer with an increasing gradient, 3) eluting with a polarity-reducing organic solvent (ethylene glycol added to the buffer), or 4) eluting with detergent added to buffer composition. Increasing the salting-out effect strengthens hydrophobic interactions while increasing the chaotropic effect weakens hydrophobic interactions. Figure 7.7 can be used to select the appropriate buffer.

Reverse-Phase Chromatography

The principles of reverse-phase chromatography (RPC) are similar to those of hydrophobic interaction chromatography. The difference is that in RPC the solid support is highly hydrophobic (C8 and higher), which allows the mobile phase to be aqueous. However, in order to have the best resolution, a gradient elution with organic solvents is needed. This creates a limitation for its use in the separation of proteins because the use of organic solvents in a mobile phase may denature them.

Although RPC has been used for proteins up to 100,000 Da, it is better suited for the purification of peptides. It has a very high resolution. For example, RPC is able to separate rabbit insulin from human insulin that differs only at a single amino acid (threonine in rabbit vs. serine in human) (Rivier and McClintock 1983). Elution is done with mobile phases that contain an ion pairing agent and an organic modifier. The most common ion pairing compounds are 1) acids: acetic, formic, perchloric, phosphoric, trifluoroacetic (most common), heptafluorobutyric (good for basic proteins), and 2) bases: triethylamine, tetramethylammonium, tetrabutylammonium, and triethylammonium acetate (unique selectivity for some peptides). The two most common organic modifiers are

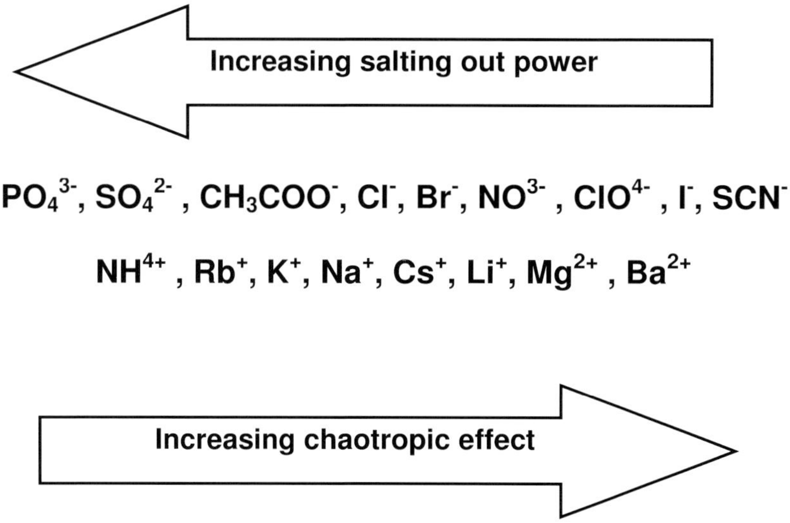

Figure 7.7. Guidelines to choose buffers in reverse phase chromatography.

1) acetonitrile (ACN), which has low viscosity and has low absorbance in the ultraviolet (UV) range, and 2) isopropanol, which can be used alone or in combination with ACN and is used for large or very hydrophobic proteins. Ethanol is used in large-scale process because it is more environmentally friendly. Most peptides will elute at 70% organic; therefore, gradients up to 40% are normally a good first try. Large or hydrophobic proteins may require up to 85% organic to elute.

General Principles

If chromatography is used for analytical purposes, the technique must be selective, stable, and reproducible to satisfy the "well-characterized biotechnology product" requirement. The protocol must be able to resolve minor impurities, such as de-amidation products and oxidized methionine (selectivity). It must be stable in that its selectivity must be constant over hundreds of injections. Finally, it must be reproducible. Selectivity and resolution must remain the same for the same column and new columns for different batches. Vendors will normally provide proof of this that must be collected for compliance to regulatory agencies.

Chromatographic theory is helpful in understanding the effect of changes in process parameters on resolution and capacity but it is rarely used as the only design tool. It is useful to define some key parameters that are important in all kinds of chromatographic separations and that are used by vendors in the commercialization of their products. The number of theoretical plates is a measure of the separation power of a column. It is defined by,

$$N = \frac{L}{H} \qquad 7.8$$

in which L is the length of the column and H is the distance needed for the streams exiting from a section of the column to be at thermodynamic equilibrium. One stream is the eluent moving downstream and the other is the solid "moving" upstream. The retention factor is defined by,

$$k = \frac{N_S}{N_M} \tag{7.9}$$

in which N_s and N_M are the number of moles of the compound of interest in the stationary and mobile phases respectively. The retention volume, V_R, is defined by,

$$V_R = V_M + kV_M \tag{7.10}$$

in which V_M is the mobile phase volume. The partition coefficient, K_D, is related to the retention factor by,

$$k = \frac{K_D V_S}{V_M} \tag{7.11}$$

in which V_S is the volume of the stationary phase. The selectivity, α, of any chromatographic separation is given by the ratio of the retention factors of any two species. The resolution is defined by,

$$R_S = \frac{(V_{R2} - V_{R1})}{\frac{w_2 + w_1}{2}} \tag{7.12}$$

in which V_{R1} and V_{R2} are the retention volumes of species 1 and 2 and w_1 and w_2 are their peak width in units of volume. Assuming that the mobile phase volumes and the plate numbers are the same for both species, and defining $k_2+k_1=2k$, one gets (Sofer and Hagel 1997):

$$R_S = \frac{1}{4}\frac{\alpha - 1}{\alpha}\frac{k_2}{1+k}\sqrt{N} \tag{7.13}$$

Equation 7.13 consists of three terms. The term involving α accounts for the effect of selectivity on resolution, the term involving k accounts for the effect of the retention factor, and the last term involving N accounts for the efficiency of the column. What follows is an analysis of retention, zone broadening, resolution, and experimental parameters for various chromatographic techniques.

Gel Filtration

The mobile phase volume in gel filtration is equal to the void volume, V_o, and the stationary phase volume, V_i, which is equal to the volume occupied by the intraparticle fluid. Therefore, the retention volume is equal to,

$$V_R = V_0 + K_D V_i \tag{7.14}$$

The slope of a curve dK/dlogr is called the selectivity curve (r is the radius of the molecule) and indicates the sensitivity of a particular gel to changes in the size of the analytes.

The resolution for gel permeation chromatography for two solutes of radii r_{p1} and r_{p2} is given by (Sofer and Hagel 1997),

$$R_S = \frac{1}{4}\log\frac{r_{p2}}{r_{p1}}\frac{\frac{dK}{d\log r}}{\frac{V_0}{V_1}+K}\frac{\sqrt{L}}{\sqrt{H}} \qquad 7.15$$

This equation indicates that the resolution may be improved by either having a steeper selectivity curve, increasing the length of the column, decreasing the height of a theoretical plate, making the particles smaller (decreasing V_0/V_1), or any combination of the above. According to this equation, the resolution is not affected by the chemical nature of the elution buffer. In practice, however, salts (typically NaCl) are added to the eluent to minimize non-specific (ionic) interactions. It has been recommended for an optimum flow rate (Hagel, 1989) of,

$$Q = A\varepsilon 65K\frac{D_M}{d_p}60 \qquad 7.16$$

in which A is the cross-sectional area of the column, D_m is the molecular diffusivity, and d_p is the average particle diameter.

Ion Exchange, Affinity, and Hydrophobic Interaction Chromatography

Most work associated with modeling chromatographic process is focused on the interaction of the target molecule with the gel and with the mass transfer limitations arising from the flow through a porous medium. The modulator (the element in the buffer that is weakening the interaction between the protein and the gel) may act in more than one way. For example, the modulator may compete with the protein for the adsorption site, but may have a negligible effect on the capacity of the gel to interact with the protein, or the modulator may affect the capacity of the gel to interact with the protein, but may not compete for the adsorption sites, or, finally, the modulator may affect both. Because for proteins the effect of the eluent is not limited to a simple mass action phenomenon but, instead, affects the adsorption isotherm constants, empirical correlations have been used to calculate the retention factor,

$$\log(k) = a_1 - a_2\log(C_m) + a_3 C_m \qquad 7.17$$

in which C_m is the concentration of the eluent and α, β, and γ are the empirical constant. The retention in ion exchange, reverse phase chromatography, and hydrophobic interaction chromatography is approximately given by,

$$k = k_0 b^{-a} \qquad 7.18$$

For ion exchange chromatography k_0 is a measure of the ion exchange capacity of the column, a is the charge of the solute, and b is the concentration of salt in the elution buffer. For reverse phase chromatography k_0 is a measure of solvent and stationary phases constants as well as the phase ratio, b is equal to 10 and

$$a = mc \qquad 7.19$$

Here m is the ratio of the area occupied by a molecule of solute to the area occupied by one molecule of the solvent in the adsorbed state, and c is the concentration of the organic modifier. For hydrophobic interaction chromatography b=10 and

$$a = -m\,c \qquad 7.20$$

in which here m is related to the hydrophobic contact area of the solute and the molar surface tension increment of the salt and c is the salt concentration. These models are somehow simplistic because they assume that there is only one type of interaction for each type of chromatography. This may not be necessarily true. For example, in ion exchange chromatography, although the main interaction is a result of the electrostatic interactions between the protein and the solid matrix, residual hydrophobic interactions may also be encountered.

Retention in affinity chromatography is given by the association constant of the reaction between the ligand and the solute. As a rule of thumb the association constant should be in the range of 10^5 to 10^{11} M^{-1}; otherwise, the solute will only be retarded in the column.

Zone broadening for ion exchange, gel permeation, reverse phase, hydrophobic interaction, and affinity chromatography is caused by axial (Taylor-Aris) dispersion, slow mass transfer, and by eddy dispersion. Zone broadening in ion exchange and reverse phase chromatography may be affected by the slow kinetics of adsorption. In addition, there is a self-sharpening effect of the zone caused by the fact that solutes moving at the edge of the front see a lower ion strength or concentration of the organic modifier (in gradient elution) and therefore, they do not move as fast as the solutes behind the edge of the front. Part of the zone broadening observed in gel permeation chromatography is caused by injection volume. It has been proposed (Sofer and Hagel 1997) that the optimum injection volume should be,

$$V_{injection} \approx \left(\frac{V_{feed} K_{injector} V_c V_i d_p^2}{15 D_M} \right)^{\frac{1}{3}} \qquad 7.21$$

in which V_{feed} is the volume of the feed, V_c is the volume of the column, V_i is the interstitial volume, and $K_{injector}$ is a constant approximately equal to 5 for laboratory units and 12 for industrial devices. This equation is consistent with the common practice of using a sample volume of 2 to 6% of the column volume and cycle times of 1 to 5 hours. For ion exchange, reverse phase and hydrophobic interaction chromatography the rule of thumb is to use a sample load below 30% of the total capacity of the column. This, of course, should be modified according to the kind and concentration of impurities. The sample load for affinity chromatography columns is normally determined by frontal elution (5 to 10% breakthrough). In other words, the entire column is "almost" saturated by the sample.

Most parameters needed to estimate the performance of a chromatography column may be either determined experimentally or estimated using available correlations. The simplest parameters to determine are the elution volume and the zone spread. The elution volume is the distance in volume from the middle of the injection time to the middle of the elution front. If the peak is not symmetric (not Gaussian) then the following expression may be used,

$$V_R = \frac{\sum_{i=1}^{n} C_i V_i \Delta V}{\sum_{i=1}^{n} C_i \Delta V} \qquad 7.22$$

in which the summation is over the entire width of the peak. The zone spread can be read directly from the chromatogram, if the peak is Gaussian, as the width of the peak at the base is four times the standard deviation. If the peak is not Gaussian, the standard deviation of the peak can be calculated from the second moment of the distribution,

$$\sigma^2 \approx \frac{\sum_{i=1}^{n} C_i V_i^2 \Delta V}{\sum_{i=1}^{n} C_i \Delta V} - V_R^2 \qquad 7.23$$

Analysis of Chromatography and Scaling up Principles

Column chromatography can be implemented in different modes. Elution chromatography consists of the seeding of a "pulse" of the sample followed by one or more elution buffers. Only a small fraction of the gel is, at anytime, interacting with the analytes. In adsorption chromatography, the sample is seeded as a step and the feeding continues until the entire column is saturated with the target molecule (displacement chromatography) or with the impurities (frontal mode). Frontal mode is common in the separation of DNA (impurity) from proteins in ion exchange. The benefit of frontal mode is the large capacity of the column because the impurity is very dilute. Both adsorption and elution chromatography can be done on fixed beds or expanded beds. Fixed bed chromatography is the most common mode. Expanded bed chromatography (either in adsorption or elution mode) is gaining popularity particularly when it is used at early process stages and the feed has suspended solids (for example cells or cell debris). Finally, simulated moving bed chromatography (that tries to emulate counter-current contact between the solid phase and the mobile phase without physically moving the solid phase) is an interesting alternative to separate binary mixtures and achieves a truly continuous mode. In elution chromatography of any kind, the target protein needs to be displaced from the solid phase with the appropriate buffer. With the exception of SEC, the composition of the eluent is constantly changing (linearly or not). This is called gradient elution chromatography. The formation and scaling up of the gradient are crucial for the successful design of the process scale equipment.

Gradient elution chromatography (Truei et al. 1992) is the most common approach to operate a chromatography column. In gradient elution chromatography the interactions between the proteins and the gel are weakened in a programmed manner by changing the composition of the buffer. The result of this gradual change in the interactions between the solutes and the stationary phase is an improved resolution and greater capacity. This change in buffer composition is realized by mixing two or more buffers before they enter the column. Different types of gradients may be done. The most common one is a linear gradient in which the concentration of one buffer component (buffer A) is changed linearly with time. Non-linear gradients, concave or convex, are also found. It is also common to use stepwise gradients and segmented ones. Stepwise gradients consist of feeding the column with one buffer first for a period of time. Then the concentration of the buffer is increased by a step and the elution continues for another period of time, the steps are continued until the final concentration of buffer is reached. In a segmented gradient the slope of the buffer is changed at programmed periods of time. More complex patterns may be used too. For example, in the separation of γ-crystallins isolated from calf lenses, we have improved the resolution and recovery of the proteins by using a series of alternated segmented and steps gradients. Figure 7.8 illustrates all these gradients.

As mentioned before, the main advantage of gradient chromatography is an increase in resolution and recovery of proteins. The main disadvantage is the added complexity of the chromatographic separation because the making of the gradient is quite complex and difficult to reproduce.

In laboratory scale, gradients are prepared with gradient mixers that can be placed upstream from

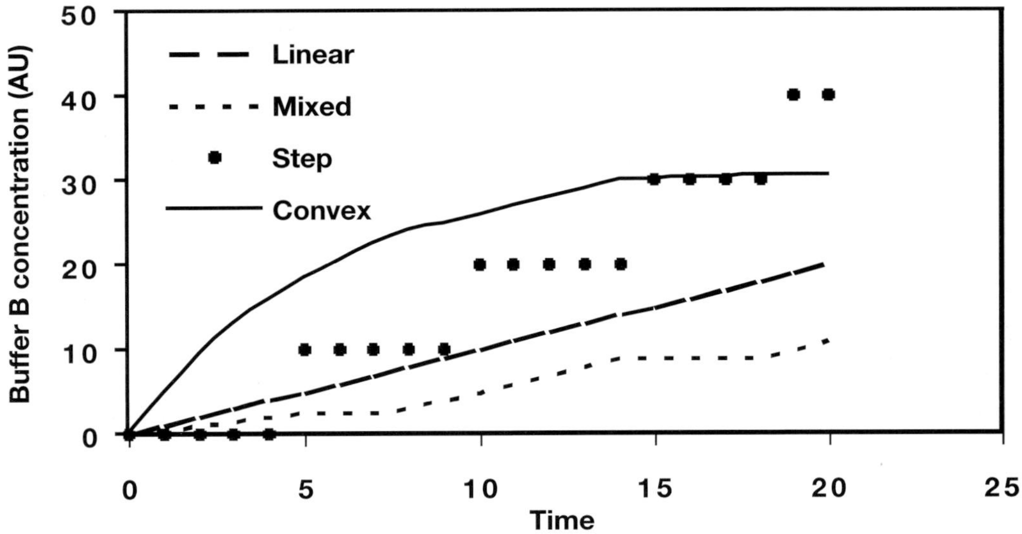

Figure 7.8. Different types of gradients use in elution chromatography. The concentration of buffer in arbitrary units is plotted versus the elution time (in arbitrary units).

the pump (in low pressure chromatography) or downstream from the pump (in high pressure chromatography). The simplest gradient consists of two containers—one containing buffer A and the other containing buffer B—joined by a capillary tube. Buffer is removed from container A with a pump and used in the column. Buffer B flows continuously into container A making the gradient (from 100% A to the desired concentration of B). Surprisingly, despite the simplicity of the device, the gradient prepared in this fashion is free of fluctuations. Instead of having two containers interconnected, one may have a multiple channel pump with one of the channels pumping buffer B into the container having buffer A. Another channel of the pump is used to pump the mixture into the column. Because the two channels work at the same speed, the gradient is also free of fluctuations. Figure 7.9 shows a picture of a multichannel pump. In high pressure chromatography the mixing is done in a mixing chamber downstream from the pump. A series of valves regulate the composition of the buffer. Very complex gradient shapes can be generated in this way. Either approach has been used in industrial separations.

Often, the expected and actual gradients are quite different. A possible reason for these differences is incomplete mixing and/or a change in the residence time in the mixer upon a scale change. It has been shown by Truei et al. (1992) that ignoring the differences between the expected gradient and the actual gradient leads to large differences in the predicted resolution of ion exchange chromatography.

The resolution of a given separation is a direct function of the shape and length of the gradient, and not a function of scale (i.e., physical size of the column). Therefore, the resolution should not change upon a scale change.

Adsorption Chromatography in Fixed Beds

In adsorption chromatography, the column is loaded until is "saturated" by the sample. The column then is washed to eliminate the remaining impurities and finally the sample is displaced from the

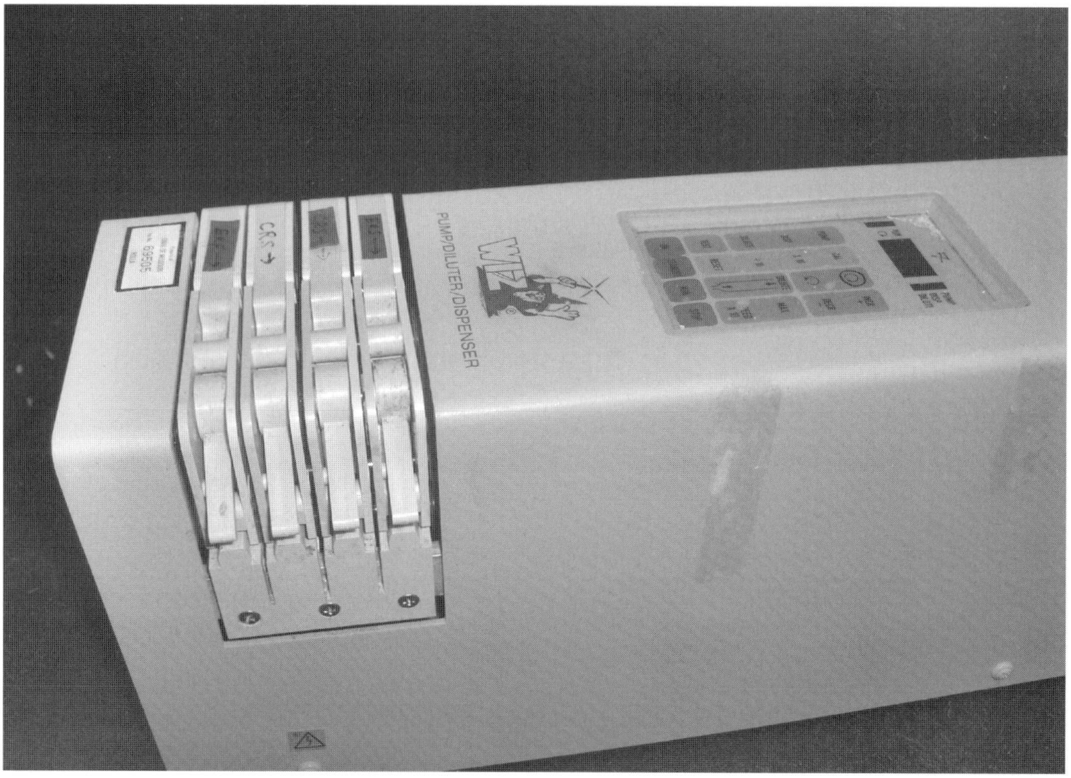

Figure 7.9. Multi-channel pump. This particular pump has four channels.

column. The key here is to know when to stop loading the column. In an ideal situation the target (what is being adsorbed to the gel in the column) will advance as a front. Figure 7.10 shows a sketch of what the loading profile will look like.

We can use the principles of mass transfer introduced in Chapter 3 to analyze this particular type of chromatography mode. The separand enters and leaves the control volume by convection (around the particles) and diffusion. The mass balance in the volume element becomes,

$$\begin{aligned} &A\varepsilon\Delta t\left(\frac{v}{\varepsilon}c_i - D_{eff}\frac{\partial c_i}{\partial z}\right)_{z,t} - A\varepsilon\Delta t\left(\frac{v}{\varepsilon}c_i - D_{eff}\frac{\partial c_i}{\partial z}\right)_{z+\Delta z,t} \\ &= A\varepsilon\Delta z\left(c_i|_{t+\Delta t} - c_i|_t\right) + A(1-\varepsilon)(q_i|_{t+\Delta t} - q_i|_t)\Delta z \end{aligned} \qquad 7.24$$

in which v is the mobile phase superficial velocity (flow rate/cross-sectional area), c_i is the concentration of separand i in the mobile phase, v/ε is the interstitial velocity, D_{eff} is the effective dispersivity of the separand, t is time, and q_i is the concentration of separand i in the stationary phase averaged over an adsorbent particle. Dividing by AεΔzΔt while letting Δz and Δt both approach infinitesimally small values, and using the definition of partial derivatives one gets:

$$D_{eff}\frac{\partial^2 c_i}{\partial z^2} - \frac{v}{\varepsilon}\frac{\partial c_i}{\partial z} = \frac{\partial c_i}{\partial t} + (1-\varepsilon)\frac{\partial q_i}{\partial t} \qquad 7.25$$

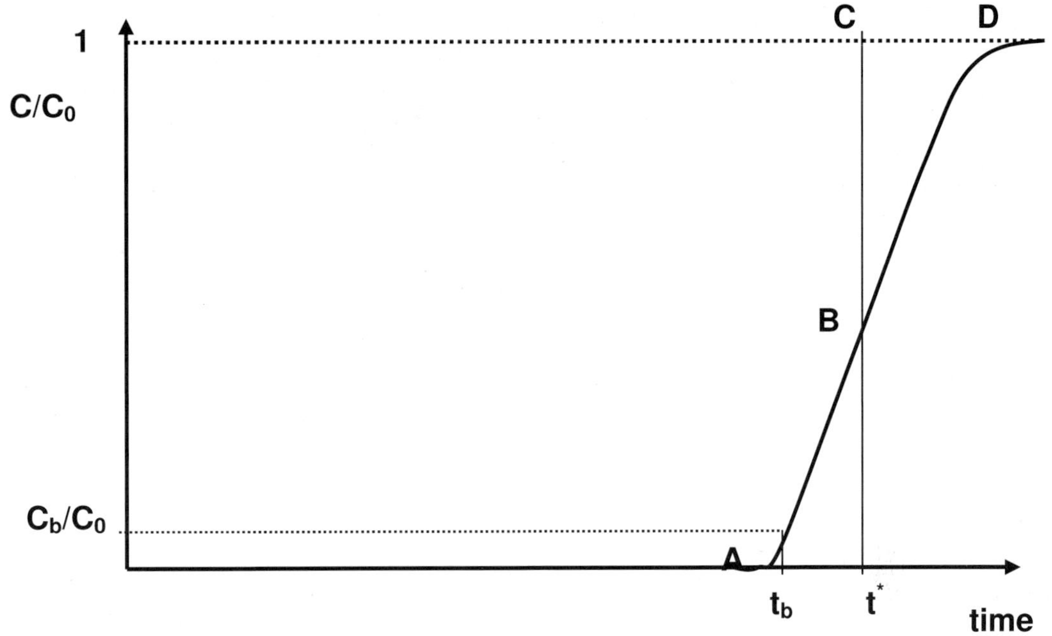

Figure 7.10. Breakthrough curve. t_b is the breakthrough time that corresponds to a concentration c_b. t* is the ideal breakthrough time (the time at which the areas delimited by points A, B, and t* is equal to area delimited by points t*, C, and D).

with

$$\frac{\partial q_i}{\partial t} = f(C_i, C_j, \ldots\ldots\ldots\ldots q_i, q_j \ldots\ldots) \qquad 7.26$$

for example,

$$\frac{\partial q_i}{\partial t} = K_T(C_i - C_i^*) \qquad 7.27$$

in which C_i^* is the liquid phase concentration that would exist at equilibrium with q_i. Although Equation 7.25 can be solved, insight into the concentration profile inside the column can be gained by a few simplifications (Harrison et al. 2003). Assuming that dispersion is negligible and that conditions of local equilibrium are maintained Equation 7.25 becomes,

$$\frac{v}{\varepsilon + (1-\varepsilon)q'_{i(Ci)}} \frac{\partial c_i}{\partial z} + \frac{\partial c_i}{\partial t} = 0 \qquad 7.28$$

This equation can be simplified further by defining an effective velocity of component i, v_i,

$$\frac{v}{\varepsilon + (1-\varepsilon)q'_{i(Ci)}} = v_i \qquad 7.29$$

Using this definition, Equation 7.28 becomes,

$$v_i \frac{\partial c_i}{\partial z} + \frac{\partial c_i}{\partial t} = 0 \qquad 7.30$$

If the equilibrium is linear, then $q'_i = K_{eq,\,i}$ and, therefore, the effective velocity of solute i is constant (the front advances as a shock). However, if the Langmuir model applies (see Chapter 2), q'_i decreases as c_i increases. Furthermore, a decrease in q'_i causes v_i to increase (Equation 7.29). Three effects need to be considered. Low concentrations of solute in the mobile phase ahead of the wave will result in higher values of q'_i and therefore, lower v_i (the front is retarded). High concentrations within the wave give lower values of q'_i and therefore higher values of v_i. Both effects result in a sharpening of the front. Low concentrations that trail the front will increase q'_i and therefore lower v_i. This will broaden the tail behind the front.

The most common scaling up principle for adsorption chromatography is to use the concept of length of unused bed (LUB). If the solute advances through the column as a front, a stream with a concentration of solute equal to that of the feed will elute from the column. We have seen that the solute advances not as a front but as a sigmoid curve. Therefore, as we monitor the concentration of the sample in the eluent we will see a mirror image of the concentration profile inside the column. We can, of course, wait until the concentration of solute in the eluent becomes equal to the concentration in the feed. This would be very non-economical because we would waste a substantial amount of the product. Therefore, it is customary to stop seeding the sample when C/C_0 is from 0.05 to 0.1. Therefore, there is a portion of the bed that is not used for adsorption. This amount of bed that is not used is called the length of unused bed. It can be shown that the length of unused bed is equal to,

$$LUB = \left(1 - \frac{q_{i,tb}}{q_{i,sat}}\right) L = \left(1 - \frac{t_b}{t^*}\right) L \qquad 7.31$$

in which t* is the ideal adsorption time (the retention time of a solute that moves as a front), $q_{i,sat}$ is the average adsorbent-phase concentration of solute i at equilibrium with fluid concentration $c_{i,o}$, based on adsorbent weight, and t_b and $q_{i,tb}$ are the time and adsorbent-phase concentration at breakthrough. The times can be calculated from

$$t^* = \int_0^\infty \left(1 - \frac{c_i}{c_{i,o}}\right) dt \qquad 7.32$$

and

$$t_b = \int_0^{Cb} \left(1 - \frac{c_i}{c_{i,o}}\right) dt \qquad 7.33$$

The total length of the bed is equal to the sum of the LUB plus the length calculated by assuming local equilibrium, with a shock wave concentration. The column is scaled up by keeping constant the LUB, which has in it the implicit assumption that the LUB in the mass transfer zone does not change as the bed height is changed.

Elution Chromatography

In elution chromatography the column is first equilibrated with the running buffer. Then, a small sample (a complex mixture to be separated) is seeded into the column. The buffer is switched back

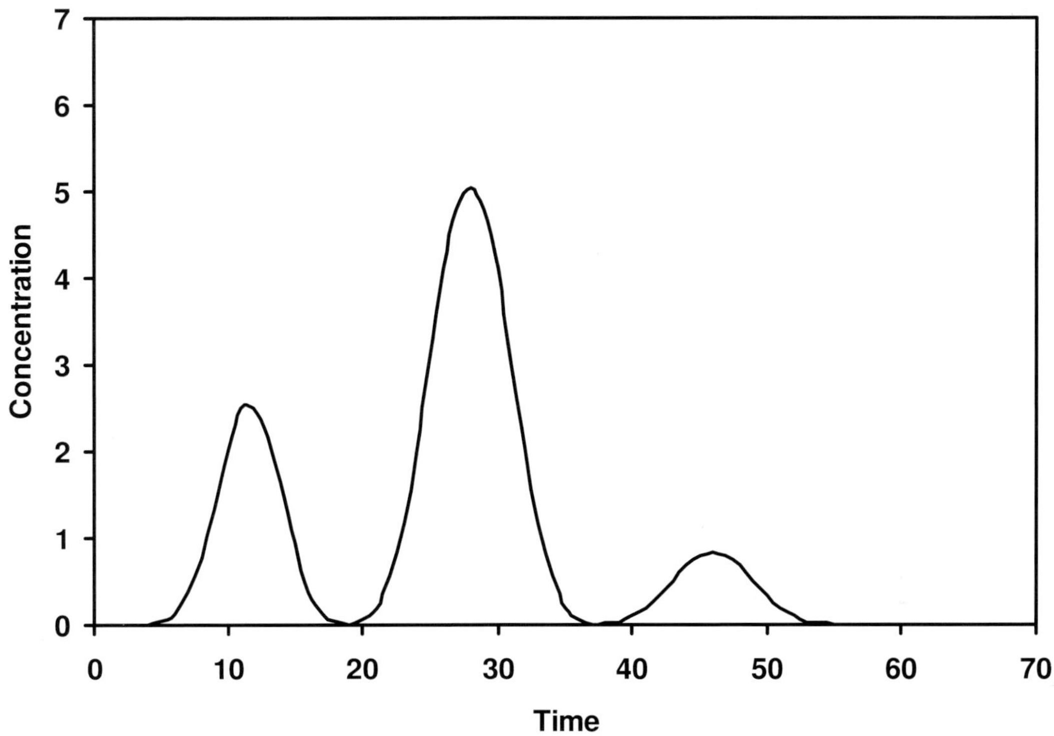

Figure 7.11. Typical elution chromatography output showing good resolution between the three peaks.

to the running one and the fluid exiting the column is monitored. Figure 7.11 shows a typical chromatogram.

The objective is to have peaks well separated from each other. It is also obvious that sharp peaks will be better separated than broad ones (less overlap). Several models are available to study the performance of an elution chromatography column.

Size exclusion chromatography (SEC) is treated separately because it is, somehow, unique. SEC is in essence a filtration process that takes place inside a column filled with the filtration material. The total volume, V_t, of the column may be broken up into three components,

$$v_t = v_0 + v_i + V_g \tag{7.34}$$

V_0 is the liquid volume between the gel beads (void volume), V_i is the pore volume inside the beads, and V_g is the volume of the gel matrix. As molecules travel through the column they partition between the pores (stationary phase) and the mobile phase. The "affinity" of a solute for either phase is dictated by its size. We can define a partition coefficient by

$$K_D = \frac{V_R - V_0}{V_i} \tag{7.35}$$

For very large molecules that cannot penetrate the gel, V_o is equal to V_R and therefore the partition coefficient is equal to zero. For very small molecules the elution volume is equal to the sum of the matrix volume and the pore volume,

$$K_D = \frac{V_R - V_0}{V_i} = \frac{V_0 + V_i - V_0}{V_i} = 1 \tag{7.36}$$

Because the volume inside the particles accessible to the solute is difficult to determine, we define a practical partition coefficient by,

$$K_{av} = \frac{V_R - V_0}{V_t - V_0} \tag{7.37}$$

The "void" volume (V_0) can be easily determined by measuring the elution volume of a molecule that is totally excluded from the pore volume (a convenient molecule is Blue Dextran that, in addition, can be used to qualitatively assess the quality of the packing by visually inspecting the movement of the dye through the column). A plot of K_{av} versus the log (molecular weight [MW]) is called a "selectivity curve." It is also used to calibrate the column. The selectivity of a column is affected by column load, flow rate, and the type of gel.

The above approach is convenient to determine experimentally the partition coefficient, and therefore, to characterize a column but it offers little insight into the details of the separation. A better choice is to use thermodynamic principles to calculate the partition coefficient. In gel permeation chromatography there are two phases—a solid and liquid phase. The chemical potential can be written as (Wills et al., 1995),

$$\mu_{i,L} = \mu_{i,L}^0 + RT(\ln c + 2B_{ii}c_i^e + 2B_{ij}c_i^e) \tag{7.38}$$

$$\mu_{i,S} = \mu_{i,S}^0 + RT(\ln q_i + 2B_{ii}q_i + 2B_{ij}q_j \tag{7.39}$$

in which $\mu_{i,L}{}^0$,and $\mu_{i,S}{}^0$ are the standard chemical potentials of species i in the liquid and solid phase, and $c_i{}^e$ is the concentration of the liquid phase (excluding the pores inside the beads). B_{ii} and B_{ij} are the second virial coefficients for equal and dissimilar species respectively. At equilibrium, the chemical potentials are equal,

$$\frac{\mu_{i,L}^0 - \mu_{i,S}^0}{RT} = \ln\frac{q_i}{c_i^e} + 2B_{ii}(q_i - c_i^e) + 2B_{ij}(q_j - c_i^e) \tag{7.40}$$

The term q_i over c_i yields the infinite dilution distribution coefficient which is a function of size only. According to Ogston (1958),

$$\ln\frac{q_i}{c_i^e} = \ln K_i^\infty = -\phi_f\left[1 + \frac{r_p}{r_f}\right]^2 \tag{7.41}$$

in which ϕ_f is the volume fraction of the fibers, r_p is the radii of the particle, and r_f is the radii of the fiber. Using this definition of the Ogston distribution coefficient, one obtains

$$\ln\frac{K_i}{K_i^\infty} = 2B_{ii}c_i^e(1 - K_i) + 2B_{ij}c_i^e(1 - K_j) \tag{7.42}$$

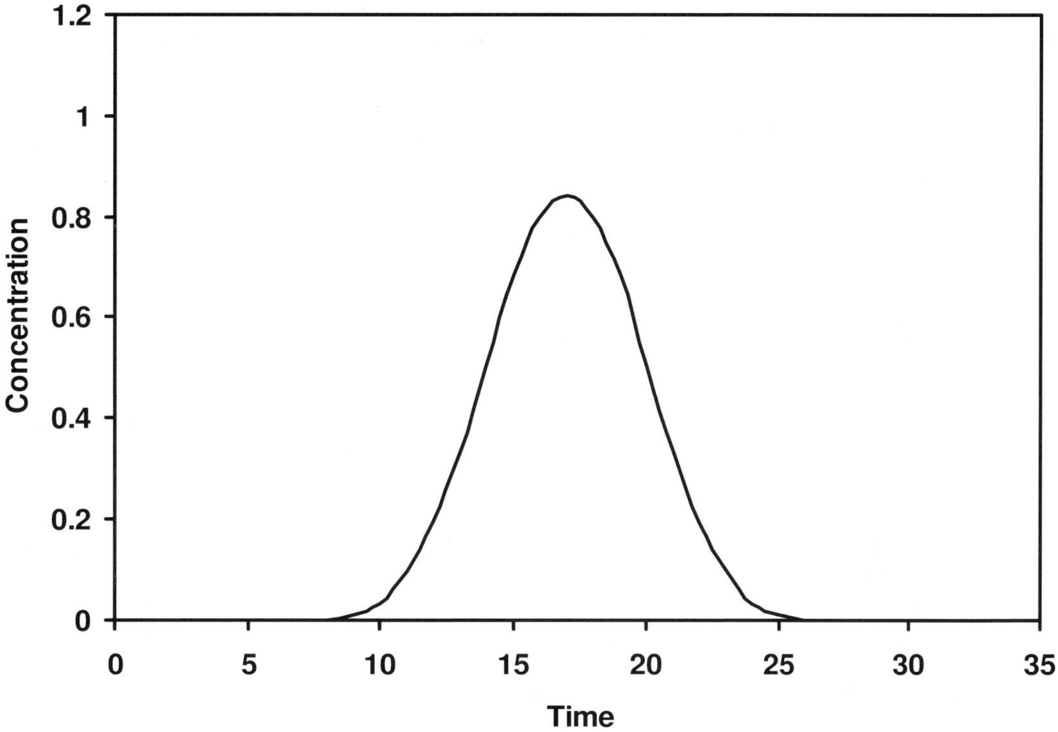

Figure 7.12. Chromatogram showing an ideal Gaussian peak.

in which

$$K_i = \frac{q_i}{c_i^e} \qquad 7.43$$

Chromatographic models can be conveniently classified into two types: 1) rate models and 2) plate models. Rate models are in theory exact and their predictive capabilities depend fundamentally on the goodness of the adsorption model and the model chosen for the porous matrix. An example of a rate model was explained in adsorption chromatography.

One of the most popular models is the plate model. In this model band broadening is explained by approximating a chromatograph as a series of well-mixed tanks at equilibrium. Plate models are considered empirical, because they do not allow the independent study of the effect of parameters such as molecular diffusion and mass transfer coefficients. This is because in plate models these parameters are lumped together into dimensionless parameters. In the following discussion it is assumed that the main cause for peak broadening is mass transfer limitations. It must be remembered, however, that slow adsorption rates broaden the peaks too. Martin and Syhnge (1941) show that the concentration profiles could be described by a Gaussian curve (see Figure 7.12). The height equivalent of a theoretical plate (H) is,

$$H = \frac{\sigma^2}{z} \qquad 7.44$$

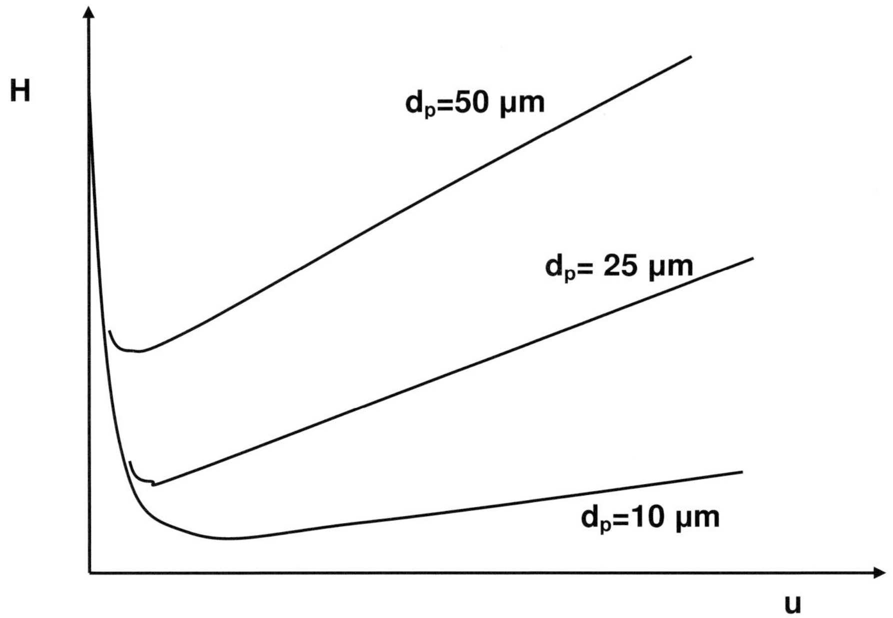

Figure 7.13. Dependence of the height of a theoretical plate with superficial velocity for different particle sizes.

in which σ^2 is the variance of the zone and z is the distance traveled. It can be written as,

$$H = \frac{L}{\left(\frac{V_R}{\sigma}\right)^2} \quad 7.45$$

For a Gaussian peak, $N = t_R^2/\sigma^2$. The problem with this definition of the height of a theoretical plate is that it applies only to Gaussian curves, which are not common. A more general definition was given by Guiochon (2006),

$$H = \frac{\mu_2 L}{t_R^2} \quad 7.46$$

in which μ_2 is the second central moment of the elution peak,

$$\mu_2 = \frac{\int_0^\infty C(t,L)(t-t_R)^2\,dt}{\int_0^\infty C(t,L)\,dt} \quad 7.47$$

A popular approach to calculate the height of a theoretical plate is known as the van Deemter Equation (van Deemter, Zuiderweg, and Klinkenberg 1956),

$$H = A + \frac{B}{v} + Cv \qquad 7.48$$

in which A accounts for the effect of eddie dispersion, B for the molecular diffusion in the mobile phase, and C for mass transfer limitations and interparticle diffusion, or written in its expanded form (Velayudhan and Ladisch 1993),

$$H = 2\lambda d_p + \frac{2\left[0.6D_m + D_S\left(\frac{V_R}{V_0} - 1\right)\right]}{v} + \frac{\frac{V_0}{V_R}\left(1 - \frac{V_0}{V_R}\right)d_p^2}{30D_p}v \qquad 7.49$$

Van Deemter presented this equation in 1956. λ is a coefficient close to one, D_m is the diffusivity of the solute in the mobile phase, D_s is the diffusivity in the pores, V_0 is the void volume, and V_R is the retention volume. Figure 7.13 shows the dependence of H versus v, the superficial velocity, according to the van Deemter type of equation as a function of particle size. At very low flow rates the second term dominates and the height of a theoretical plate decreases with superficial velocity. At normal operating conditions, the height will increase with flow rate because the last term dominates. There is an optimum (minimum height of the theoretical plate). Since 1956, new versions have been published. One because of Katz, Ogan, and Scott (1983) is,

$$H \approx 2\lambda_T d_p + +2\gamma\frac{D_m}{v} + \left[\frac{d_p^2(1 + bk + ck^2}{24(1+k)^2 D_m}\right]v \qquad 7.50$$

in which λ_T is a tortuosity factor and γ is an obstructive factor. Giddings (1965) proposed the following expression for H,

$$\frac{H}{d_p} = 2\lambda_T\frac{D_m}{vd_p} + \sum\frac{2\lambda_i}{1 + \frac{\varpi_i D_m}{vd_p}} + C\frac{vd_p}{D_m} \qquad 7.51$$

in which γ is the tortuosity of the packed bed, λ_i and ω_i are geometrical parameters, and C accounts for mass transfer phenomenon in the bulk and inside the particles and for the kinetics of adsorption. An empirical version of Giddings equation was proposed by Bristow and Knox (1977),

$$\frac{H}{d_p} = A\left(\frac{vd_p}{D_m}\right)^{1/3} + B\frac{D_m}{vd_p} + C\frac{vd_p}{D_m} \qquad 7.52$$

Finally, the following expression for monolithic columns has been proposed (Hahn and Jungbauer 2000),

$$H = \frac{2D_m}{v} + 2A + \frac{k}{30}\frac{1}{(1-k)^2}\left[\frac{vd_1^2}{D_p + d_1 v/2}\right] \qquad 7.53$$

in which D_p is the diffusivity in the pores and d_1 is a characteristic diffusional length that needs to be determined. Two other plate models that deserve a few comments here are the Craig model and the continuous flow model. Both have been recently reviewed (Velayudhan and Ladisch 1993). After comparing the results of these two models with rate models, they concluded (properly) that these two plate models are reliable and can be used for scaling up purposes. The working equations for the Craig model are:

$$\frac{c(i,N_i,t)}{c_0} = I'_{pi'}(N_i,t-N_i) - I_{pi'}(N_i,t-N_i,M) \qquad 7.54$$

I_{pi}' is an incomplete β function, and M is the impulse width. The probability that species i is in the mobile phase, p'_I, is calculated from,

$$p_i^{'} = \frac{1}{1+k_i} \qquad 7.55$$

k_i is the retention factor of species i. The phase ratio is calculated from,

$$\phi = \frac{(1-\varepsilon_z)(1-\varepsilon_y)\varepsilon_x}{\varepsilon_z + (1-\varepsilon_z)\varepsilon_y + (1-\varepsilon_z)(1-\varepsilon_y)\varepsilon_x} \qquad 7.56$$

Equation 7.54 is solved for the concentration profile along the column. The number of Craig segments is related to the plate count by,

$$N = N_{Craig}\left(\frac{k_i}{1+k_i}\right) \qquad 7.57$$

in which k_i is calculated from $k_i=\phi K_d$. The plate count, N, can be calculated from equation 7.49 (or similar). The continuous flow plate is done as follows. The retention time is calculated from,

$$t_R = \frac{L}{V}\left[1+\phi K_D\right] \qquad 7.58$$

The number of continuous segments is calculated from $N=N_{plate}$ then the concentration profile is calculated from,

$$\frac{c(i,J,t)}{c_{i0}} = e^{-\alpha i(t-\tau)} - \sum_{J=0}^{J-1}\frac{(\alpha_i)^j(t-\tau)^j}{j!} - e^{-\alpha it}\sum_{j=0}^{J-1}\frac{(\alpha_i)^j t^j}{j1} \qquad 7.59$$

α_i is equal to $N_i/t_{R,i}$ and τ is the injection time. J is an integral. For any J,

$$\frac{c(i,J,t)}{c_{i0}} = \Gamma\left(J_i,\frac{J_i t}{t_R}\right) - \Gamma\left(J_i,J_i\left(\frac{t-\tau}{t_R}\right)\right) \qquad 7.60$$

in which Γ is the incomplete Gamma Function.

Rate models are more detailed but also much more complex. Because plate count is standard in chromatography, rate models have been used to determine H. For example, Lapidus and Amundson

(1952) used a rigorous mathematical model (but assuming linear isotherms, local equilibrium, and a semi-infinite column) to show that the height of a theoretical plate can be written as:

$$H(x,t) = f(D_L, v, \tau, K_{eq}(1-\varepsilon)/\varepsilon) \qquad 7.61$$

in which D_L is the axial dispersion coefficient defined by,

$$D_L = D_m + E \qquad 7.62$$

in which E is the Taylor dispersion coefficient. D_L is 3 or 4 orders of magnitude larger than D_m. They proved that the critical parameter is v/D_L (i.e., the minimum number of plates is achieved for low values of this parameter). This result is consistent with very simple mass transfer considerations that show that the height of a theoretical plate is proportional to,

$$H \propto \frac{d_p^2}{D_m} \frac{1}{t_R} \qquad 7.63$$

Then, smaller diameter beads, high diffusivities (or low dispersion), and long retention times (low superficial velocities) would increase resolution.

Expanded Bed Chromatography

Expanded bed chromatography has gained popularity in the pharmaceutical sector. One of its main advantages is that filtration of the feed stream to remove solids is not necessary, which leads to the elimination of one purification step and the increase in yield. In expanded bed chromatography, the feed is fed to the bottom of the column and the solids are partly fluidized. The size of the equipment to obtain reliable data makes laboratory experimentation expensive. Therefore, good theoretical models are used to determine the impact of various parameters in the performance of the unit.

It has been recognized (Menkhaus and Glatz 2005) that both the type of sparging system and solid matrix used determine the efficiency of an expanded bed. Metal screens, glass beads, and mixing devices are used as inlet mechanisms. One unique feature of expanded bed chromatography is that it is necessary to have a size distribution of solids to assure stable fluidization and minimize back mixing. A few absorbents that have been used are quartz-DEAE (size range: 100 to 300 μm, density 1.1 to 1.3 g/mL); quartz-SP (size range: 100-300 μm, density: 1.1 to 1.3 g/mL) (Amersham); Steel-SP (40-200 μm, density 2.5-2.7 g/mL) (UpFront); and Zirconium dioxide-CM (45-105 μm, average density 3.2 g/mL) (BioSepra). Inlet designs may be evaluated by determining the number of theoretical plates, axial dispersion, and Peclet (Pe) number by injecting a step input of 0.25% acetone. The axial dispersion is calculated from,

$$D_L = \frac{vL_e}{2N} \qquad 7.64$$

and the Pe number by,

$$Pe = 2N \qquad 7.65$$

in which N is the number of theoretical plates and L_e is the height of the expanded bed. Pe numbers higher than 40 are needed to insure that back mixing will not limit adsorption.

To develop a model we must recognize that there are two phases: a solid phase and a liquid phase. Therefore, mass balances must be written for each (Kaczmarski and Bellot 2004). A mass balance for the liquid phase is,

$$\varepsilon(z)\frac{\partial C_f}{\partial t}+v\frac{\partial C_f}{\partial z}=\frac{\partial}{\partial z}\left(\varepsilon(z)D_L\frac{\partial C_f}{\partial z}\right)-\left(1-\varepsilon(z)\right)\frac{3}{R(z)}k_{ext}(z)\left(C_f-C_p\right) \quad 7.66$$

Subjected to:

$$C_f(z,0)=0 \quad \text{for } t=0 \quad 7.67$$

$$C_f=C_0\frac{D_L\varepsilon}{v}\frac{\partial C_f}{\partial z}=0 \qquad z=0 \quad 7.68$$

$$\frac{\partial C_f}{\partial z}=0 \qquad z=H_e \quad 7.69$$

For the solid phase,

$$\left(1-\varepsilon(z)\right)\frac{\partial q_i}{\partial t}=\frac{\partial}{\partial z}\left(D_s\frac{\partial q_i}{\partial z}\right)+\left(1-\varepsilon(z)\right)\frac{3}{R(z)}k_{ext}(z)\left(C_f-C_p\right) \quad 7.70$$

Subjected to,

$$q_i(z,0)=0 \qquad \text{for } t=0 \quad 7.71$$

$$\frac{\partial q_i}{\partial z}=0 \qquad \text{for } z=0 \text{ and for } z=H_e \quad 7.72$$

For each particle,

$$\frac{\partial(\varepsilon_p C_p)}{\partial t}+\frac{\partial q_i}{\partial t}=\frac{1}{r^2}\frac{\partial}{\partial r}\left(r^2\frac{D_m\varepsilon_p}{\lambda_p}\frac{\partial C_p}{\partial r}\right) \quad 7.73$$

Subjected to,

$$q_i(z,r,0) \qquad C_p(z,r,0)=0 \qquad t=0 \quad 7.74$$

$$\frac{\partial C_p}{\partial r}=0 \qquad r=0 \quad 7.75$$

$$\frac{\varepsilon_p D_m}{\lambda_p}\frac{\partial C_p}{\partial r}=\frac{R(z)}{3}\frac{1}{1-\varepsilon(z)}\frac{\partial}{\partial z}\left(D_s\frac{\partial q_i}{\partial z}\right)+k_{ext}(z)\left(C_f-C_p\right) \qquad r=R(z) \quad 7.76$$

With C_f and C_p the concentrations of solute in the percolating and stagnant liquid phase, C_0 is the inlet concentration, q is the particle average solute concentration, r is the distance from particle center, R(z) is the particle radius, H_e is the expanded bed height, ε_p is the particle porosity, D_s is the solid phase axial dispersion coefficients, and k_{ext} is the mass transfer coefficient from the bulk phase to the external surface of the particle.

The model presented above produces very good breakthrough curves but they are only marginally better than breakthrough curves obtained with simpler models (models that do not account for changes in bed porosity and bed size distribution in the axial direction of the bed). Parametric studies showed that the dynamic capacity strongly depends on internal mass transfer resistances. The external mass transfer resistance, axial liquid-phase dispersion, axial solid-phase dispersion have minor effects as well as particle size distribution and bed porosity. If the last two parameters are not taken into account, the breakthrough curves are broader. Overall, the dynamic capacity increases with settle bead height, as observed by others. One of the conclusions of this work is that tall beds make the model less sensitive to perturbations. Whether or not the same would be valid for experimental expanded beds remains to be seen.

Two potential problems with expanded beds are plugging and bubbling. Although a well-maintained bed should not have plugging problems, its use with unfiltered broths may result in heavy clogging. Bubbling may take place because of poor degassing of the feed, which will be the case when using unfiltered broth. Bubbling may cause turbulence that will mix the bed and increase the size of a theoretical plate; it will also affect the performance of on-line detectors like UV or, particularly, refractive index detectors.

Simulated Moving Bed

The reference for a simulated moving bed (SMB) is the true moving bed, which is not realizable because it is not possible to physically move the solid phase in a counter-current fashion with respect to the mobile phase. Still, if that would be feasible it is obvious that the separation between components that have different affinity for the solid phase would improve. Figure 7.14 illustrates the working principle of a simulated moving bed. If a sample is injected in the mid-section of a long column, the solutes will move in the direction of the mobile phase. If we move the solid phase in the opposite direction and at a speed in between the speeds at which each of the two solutes move, then the two solutes will move in opposite directions. If the sample is added continuously, the two solutes will continue being separated by moving in opposite directions. Of course, it is physically impractical to move the solids. Therefore, the movement of the solid is simulated by a cleaver connection of a series of short columns and a wise addition and removal of the feed.

The advantages of simulated moving beds are high productivity, lower power consumption, and high product concentration. The main disadvantages are that it is expensive in laboratory scale, highly complex, limited to the separation of binary mixtures, and difficult to optimize. Figure 7.15 shows an snap shot of a five-column simulated moving bed. There are four main streams: 1) the feed, 2) the mobile phase, 3) the raffinate, and 4) the extract. At each joint, a valve allows the manipulation of the direction of the flow. A pump circulates the fluid from the last column into the first one. The head of any given column is connected to the bottom of the previous one. Therefore, the system functions as a long column with feeding and extractions at various heights. At the process time depicted in the figure, the process feed is being added to column 4 the "dotted product" is being removed from column 5 and the "gray product" is being removed from column 2.

The key in the development of SMB technology is to choose the "right" ratio of mobile phase and solid-phase flows. Of course, the solid phase does not move but the choice in the selection of flows by opening and closing valves achieves the illusion of a moving solid. Work has been done to rationally design SMB using triangle theory. Still, it is not robust enough and it cannot be used for the "quick" design of the system. The interested reader is referred to the work of Houwing (2003).

SMB was first introduced by Broughton and Gerhold (1961) and then commercialized by

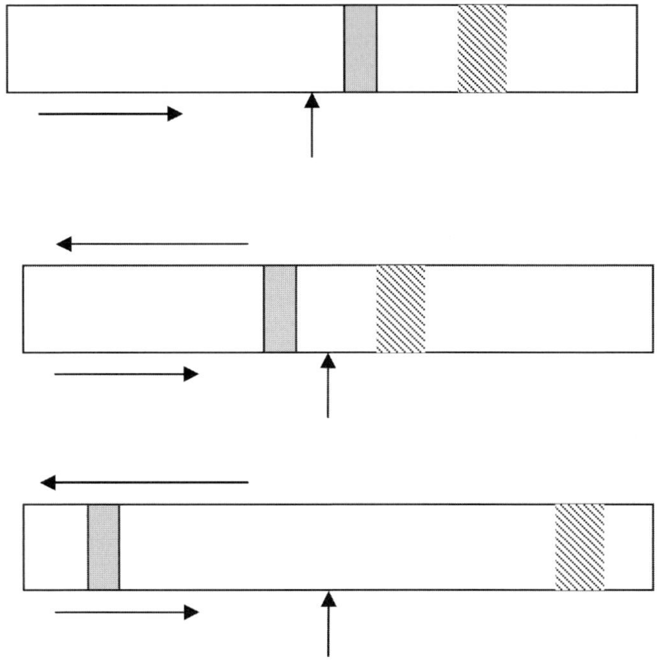

Figure 7.14. Schematic showing the principles underlying simulated moving bed technology. The vertical arrows correspond to the injection point. There are two components: gray and dashed. The eluent moves from left to right and the solid "moves" from right to left.

Universal Oil Products in the 1960s. SMB has been successful within the petroleum industry, sugar processing industry, and chiral separations. Its use in bioprocessing is not common partly because of the complexity of the mathematical models needed to design the process. Still, it has been used in research labs using ion exchange, affinity, and size exclusion columns. Four zones SMB is the most common device. Zone I is in between the injection of eluent and the collection of the extract. In zone I, the more retained compound has to be desorbed from the solid. Zone II is in between the collection of the extract and the injection of the feed. In zone II, the less retained compound has to be desorbed from the solid. Zone III is in between the injection of the feed and the collection of the raffinate. In this zone, the more retained compound has to be adsorbed on the solid. Zone IV is in between the collection of the raffinate and the injection of the eluent. In zone IV, the less retained compound has to be adsorbed on the solid. It is normally accepted that a SMB has to have anywhere from eight to twelve columns to equate the performance of a true moving bed (TMB). Three examples of some proteins purified by six columns SMB-affinity chromatography are 1) trypsin from extract of porcine pancreas (Huang et al. 1986), 2) human serum albumin and the separation between myoglobin and lysozyme (Nicoud 1996), and 3) monoclonal antibodies (Gottschlich and Kasche 1997).

The design and optimization of a SMB consists of the following three steps (Charton and Nicoud 1995):

1. Acquisition of the necessary parameters: equilibrium adsorption isotherms, number of theoretical plates per column and estimation of pressure drop.
2. Calculation of the TMB. It is necessary to calculate the flow rates of a TMB given the feed composition, eluent, stationary phase and adsorption isotherms.

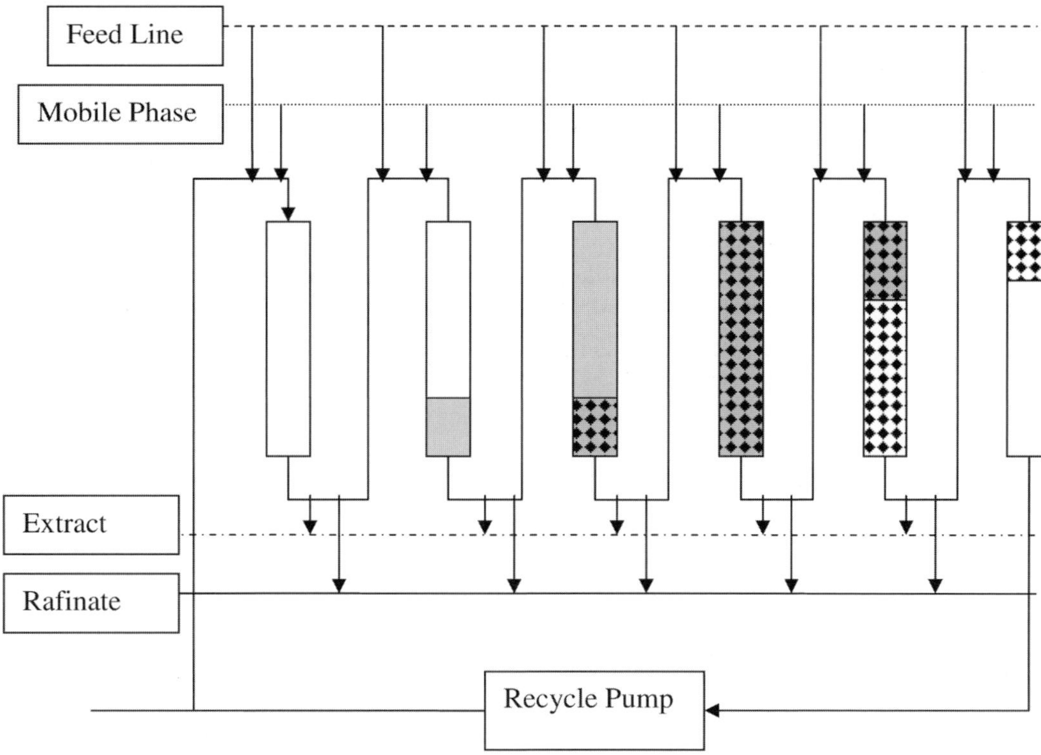

Figure 7.15. Schematic showing a simulated moving bed process at a particular time. Two species are being separated: gray and diamond.

3. Calculation of the SMB. Given the flow rates for the TMB, the flow rates of the SMB are calculated given the number of columns and number of theoretical plates per column. It is essential that all columns have the same retention time. Still, because of a dead volume at the recycle pump there is an asymmetry (the effluent for each column is not the same). Two approaches that have been followed to remedy this problem are: a) either the first or the last columns are made shorter. Therefore, the residence time in the shorter column plus the residence time in the dead volume is equal to the residence time in the other columns; b) the valve just before the recycle pump is switched with a delay equivalent to the time needed for the liquid to pass through the dead volume.

Andersson and Mattiasson (2006) have followed a different approach for the design of a SMB. Their design of the SMB was based on results of a single column that was used to determine the characteristics of the chromatography cycle: loading, washing, and bed volumes for elution, regeneration, and conditioning. The breakthrough of the column was used to choose loading. The final design consists of eight-column SMB that were used to purify lactoperoxidase and lactoferrin from whey. Three keys in the final design are:

1) To choose the number of columns in each zone (loading, washing, elution, regeneration, and conditioning)

2) To choose the switching times
3) To choose adequate flow rates

Because the breakthrough in a single column was almost instantaneous, they chose an eight-column loading zone, which was divided into two steps (step 1: columns 1-4; step 2: columns 5-8) in a way that the effluent of 1 to 4 is loaded into 5 to 8. Two columns were chosen for washing, regeneration, and conditioning. By this judicious choice of columns and flow rates, they were able to obtain a 338 purification fold for lactoperoxidase and 420 fold for lactoferrin from whey with 88% and 105% recovery, respectively. It is not clear how this complex scheme compares with a standard chromatographic separation because these results were compared with the equivalent separation of twenty columns in batch. Better benchmarks are needed to assess the benefits of SMB.

One of the problems with SMB is that all the columns must be identically packed (Abunasser and Wankat 2004). It has been proposed to replace a true SMB by a series of batch columns and holding (or recycle tanks); this is called an analog SMB. This analog had not been tested experimentally at the time this book was written. Still, computer simulations suggest that it may be an alternative to SMB.

Operating Tips

The first step in the design of a chromatographic separation is to select the most appropriate chromatography technique, the specific type of gel, and its operating mode (fixed bed, expanded bed, elution, adsorption). After that initial decision has been made, the elution protocol, type of buffer and flow rate must be chosen. The choice of gel determines the maximum flow rate. The optimum flow rate, however, must be determined for the desired resolution. The determination of the energy requirements will be decided by the pressure drop through the column for the chosen flow rate. Previous reports for similar separations could be used to choose the elution protocol. Otherwise, there are instruments in the market that would search for the optimum elution pattern and type of buffer automatically. If chromatography is a standard technique in a given laboratory it would save money and time to acquire such an instrument. Figure 7.16 shows pictures of two chromatography setups. The left panel shows and a process scale system, and the right panel shows a low pressure preparative scale one. The following guidelines should be useful to scale up a chromatography process. The bed height, eluent velocity, sample concentration, gradient slope/bed volume, and sample residence time should be kept constant. For an increase in capacity, one should increase the column diameter such that the required volumetric flow rate is increased while maintaining a constant eluent velocity, and should also increase the sample volume, volumetric flow rate, and gradient volume in proportion to the column's cross-sectional area. Sometimes it will not be possible to increase column diameter, because columns of the desired diameter may not be commercially available. Yamamoto, Nakamishi, and Matsumo (1988) claim that the resolution will not change if the column length increases in proportion to the eluent velocity times the gradient slope per unit bed volume. Therefore, the sample volume can be increased in proportion to bed volume, the eluent velocity remains constant, and the gradient slope/bed volume is increased in proportion to the column's length.

Although the scaling up arguments are quite straightforward, significant variations may be observed as the scale is changed. For example, it is common to observe a broadening of the peaks as the scale is increased because of additional mixing in larger tubes and valves. It is common practice to degas buffers in laboratory scale but not in process scale. Therefore, in large scale processes bubbles may form deteriorating the resolution. Gradients are also much more difficult to control in large scale. Mixing may be incomplete or the volume hold up in the mixer (necessarily larger than in lab scale) may lead to inaccuracies. Finally, a decrease in flow rate may be observed also because

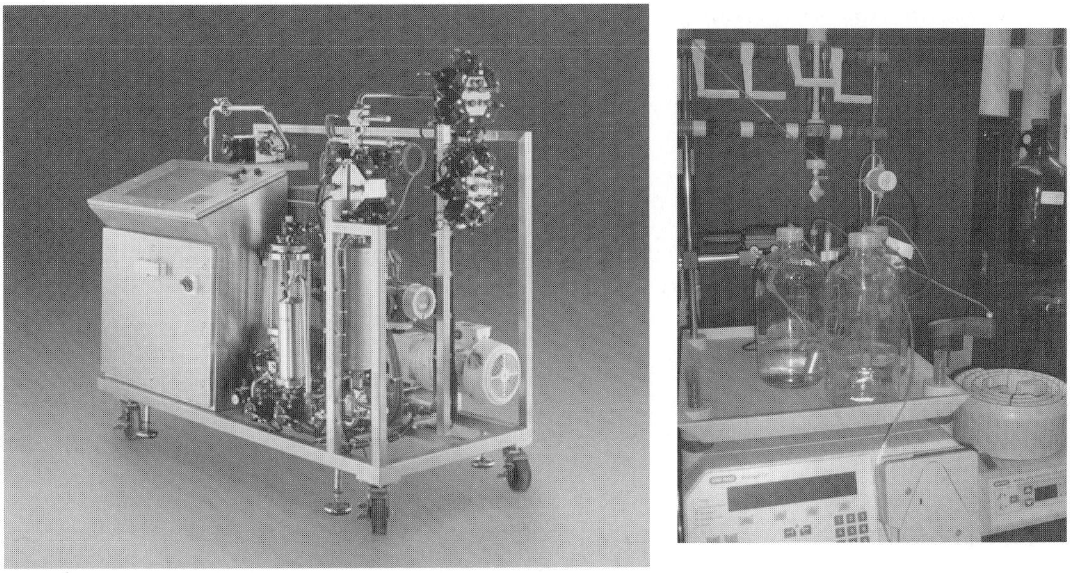

Figure 7.16. Picture of laboratory (right panel) and process (left panel) scale chromatography systems.

of a change in the compression of the gel as the walls of the column start playing a lesser role in keeping the gel together. Because of these considerations it is highly recommended to start the scaling up by not more than one hundred-fold.

After the process has been designed and built it needs to be validated; that is, one needs to prove that the process does what it is designed to do in a consistent manner. Documentation is an essential part of validation and must be a common practice even (or particularly) during the development stages. Because validation varies from country to country, it is best to become acquainted (i.e., see U.S. Food and Drug Administration [FDA] or the European Medicines Agency in countries that belong to the European Union, or an analogous organization for other countries of interest) with the validation process in the place where the plant is going to be built. There will be validation procedures for different parts of the process, including, of course the facilities. Of particular concern are ventilation systems (HVAC) and water used in different parts of the process (e.g., in the making of chromatography buffers).

Experiments

Three different experiments are described in this section. In the first one, a step-by-step guide on how to pack and run a size exclusion chromatography is described. The column is used to fractionate an extract from calf lenses prepared in previous chapters. In the second one, the packing and running of an ion exchange column is presented. The column is used to fractionate a mixture of γ-crystallins with a gradient developed for this particular mixture. Finally, in the third experiment, the separation of a human antibody from corn meal extract is presented using a commercial protein A column.

Size Exclusion Chromatography

Materials

100 mM sodium phosphate buffer at pH 7 and containing 3 mM Na_3N Blue Dextran and crystallins extract. Sephadex G-75

Equipment

A low-pressure chromatography system equipped with a UV detector, a fraction collector.

Swelling the Gel

1. Attach your column to a stand with a clamp.
2. Determine the volume of your column.
3. Using the instructions on the Sephadex G-75 bottle, mix the suggested amount of powder with enough phosphate buffer to get a mixture with the consistency of syrup.
4. Cover with Parafilm™ and allow the mixture to stand for 24 hours at room temperature.
5. To ensure maximum bead expansion heat mixture for 1 hour at 80°C, stirring constantly with a glass stirring rod. Stir in such a manner that the rod does not touch or rub the sides or bottom of the beaker. (**DO NOT USE A MAGNETIC STIR BAR. IT CAN BREAK THE BEADS.**)
6. Allow the gel to cool and remove any small beads floating on the surface using a pipette.

Packing a GPC Column

1. Place a waste beaker below the column and open the valve.
2. Slowly pour the gel into the column and fill to within an inch from the top. Use a funnel to minimize the formation of air bubbles.
3. Allow the gel to settle. This settling can be seen as it occurs. Buffer will drip out of the bottom and the gel will begin to settle. Both buffer and gel should be added as needed to ensure that the buffer level remains above the gel and to ensure that enough gel is added to pack the entire column.
4. Column is packed when the gel is 1 inch from the top and its level does not move upon addition of buffer.
5. To ensure packing, connect a buffer reservoir to the top of your column and run three column volumes of phosphate buffer through it.
6. Collect the buffer in a waste beaker.
7. Check periodically to make sure the column has not run dry or the column has not overflowed with buffer.
8. If gel packs considerably, more should be added.
9. Close valve at bottom of column and always make sure there is buffer on top of the gel.
10. Cover top of column with Parafilm™ and store in refrigerator.

Equilibrating a GPC Column

Equilibrating refers to a final settling of the gel in a column, which results in a constant flow of buffer through that column.

1. Connect the pump to the top of your column.
2. Start the pump and allow the phosphate buffer to be pumped through the column.
3. Collect the buffer in a waste beaker.

4. Using a stop watch and graduated cylinder, periodically check the flow rate through the column.
5. Equilibration is complete when a constant flow rate has been reached.

Channeling Test for a GPC Column

Channeling refers to cracks or channels through a gel, which allow solutes to bypass the pores, ligands, and so on, which the column uses for its basis of separation. Blue Dextran (2 mg/mL) is used to check for channeling, because it is too large to enter the pores of the gels used. Thus, it flows around the beads of the gel and forms a distinct front that is easily observed with the naked eye.

The front must be observed during the entire procedure. Record the length of time it takes for the Blue Dextran to run completely through the column and record the total eluent volume. This volume is equal to the void volume of the column.

1. Remove excess buffer from the top of the gel using a pipette.
2. Carefully place 1.5 to 3 mL of Blue Dextran on top of the gel.
3. Allow sample to move completely into the gel by gravity.
4. Replace the cover on the top of the column.
5. Turn on the pump.
6. Collect the eluent in a waste beaker.
7. Just before the Blue Dextran starts to exit the column, collect the eluent with the fraction collector instead of the waste beaker. Set the fraction collector to collect twenty to thirty drops per tube.
8. Collect until all the Blue Dextran has exited the column.

If cracking or channeling is observed in the column, the column must be repacked and the above procedures repeated.

Running a Sample through a GPC Column

1. Connect a UV detector to the outlet of the column.
2. Connect the UV detector to the recorder.
3. Connect fraction collector to the outlet of the column.
4. Set up fraction collector to collect by the drop for approximately the number of drops it takes to make 3 mL. This must be determined prior to use of the machine.
5. Open the column outlet and start the pump; run buffer through the column until you get a flat baseline. You may have to adjust the sensitivity of the recorder in order to achieve this.
6. Inject the sample using the injection valve and a loop of enough volume.
7. Begin collection of the eluent using the fraction collector.
8. Construct an absorbance versus eluent volume chromatogram.

Ion Exchange Chromatography

Preparation of Running Buffer

275 mM Acetate pH 5 containing 1 mM 2-Mercaptoethanol with and without 100 mM NaCl. Prepare 4 L of each one.

The CM Sephadex C-50 gel is swollen following the same procedure as for the gel permeation gel.

Table 7.7. Gradient Protocol for the Elution of an Antibody from a Protein A Column.

Time (min)	A (%)	B (%)
0	100	
5	100	
5		100
8		100
8		100
12	100	

Packing, Equilibrating, and Running a IEC Column

Follow the same procedure as for "Packing and Equilibrating your GPC Column." Seed the GPC fraction containing the γ-crystalllin fraction (approximatelly 20 mL). Elute the γ-crystallins using a 0 to 0.4 M NaCl gradient segmented according to the following protocol:

a. 0 to 0.1 M NaCl in three-column volumes
b. Remain at 0.1 M NaCl for one-column volume
c. 0.1 to 0.2M NaCl in three-column volumes
d. Remain at 0.2M NaCl for one-column volume
e. 0.2 to 0.3 M NaCl in three-column volumes
f. Remain at 0.3M NaCl for one-column volume
g. 0.3 to 0.4M NaCl in three-column volumes
h. Remain at 0.4M NaCl for three-column volumes

Affinity Chromatography

Materials

Human IgG, NaCl, Sodium phosphate monobasic and dibasic, acetic acid, Bradford reagent

Extraction of Antibody from Corn Meal

1. Extract 1 g of corn meal with 10 mL of 150 mM NaCl for 8 to 12 hours at 4^0C with stirring.
2. Remove the solid particles by centrifugation at 11,000 rpm for 1 hour at 4^0C.
3. Filter the supernatant through a filter paper.
4. Filter again through a 0.45 μm membrane.
5. Spike the solution with 3 mg/mL of human IgG.
6. Determine total protein using the Bradford test.

Preparation of Calibration Curve

Equilibrate the column with Buffer A. Seed your IgG sample using a 200 uL loop. Run the gradient summarized in Table 7.7. In Table 7.7 the mobile phase A is 50 mM phosphate/500 mM NaCl, pH 7, mobile phase B is 100 mM Acetic acid/500 mM NaCl, and the flow rate is 5 mL/min.

Separation of the Human Antibody

1. Equilibrate the column with Buffer A.
2. Seed your sample using a 200 μL injection loop.
3. Run the above gradient.

Summary

Although originating from a simple concept, chromatography has blossomed into a powerful technique used in both lab and industrial settings. Furthermore, column chromatography is the most common unit operation found in bioprocessing. It is a very powerful separation method and it is technologically matured. Finally, new technology in chromatography, specifically column chromatography, such as analog SMB or expanded beds points toward a bright future for this separation technique.

Nomenclature

A	cross-sectional area of the column (m^2)
B_{ii}	second virial coefficient for equal species (L/mol)
B_{ij}	second virial coefficient for dissimilar species (L/mol)
C_m	concentration of the eluent (Mol/L)
c_i	concentration of separand i in the mobile phase (Mol/L)
$C_i{*}$	liquid phase concentration that would exist at equilibrium with q_i (Mol/L)
$c_{i,o}$	concentration in the fluid phase (Mol/L)
c_i^e	concentration of the liquid phase (excluding the pores inside the beads) (Mol/L)
C_f	the concentrations of solute in the percolating (Mol/L)
C_p	the concentrations of solute in the stagnant liquid phase (Mol/L)
C_0	inlet concentration (Mol/L)
d_l	characteristic diffusional length
D	diameter of the column (m)
D_m	molecular diffusivity (m^2/sec)
D_p	Pore diffusivity (m^2/sec)
D_{eff}	effective dispersivity (m^2/sec)
D_L	axial dispersion coefficient (m^2/sec)
D_S	axial dispersion coefficient in the solid phase (m^2/sec)
d_p	average diameter of the beads (m)
E	Taylor dispersion coefficient (m^2/sec)
h	extent of column compression (m)
H	height of a theoretical plate (m)
H_e	expanded bed height (m)
I_{pi}'	incomplete _ function
k	retention factor
k_0	parameter describing the intrinsic characteristics of ion exchange, reverse phase, or hydrophobic interaction columns
k_{ext}	external mass transfer coefficient (m/sec)
K_e	equilibrium constant
K_D	partition coefficient
K_T	overall mass transfer coefficient (sec^{-1})
L_0	initial bead height (m)
L	actual height of the bed (m)
L_e	height of the expanded bed (m)
M	impulse width
N	number of theoretical plates
N_s	number of moles in stationary phase
N_M	number of moles in mobile phase

P	pressure (Pa)
p'_I	probability that species i is in the mobile phase
Pe	Peclet number
P_s	normalized stress profile
q_i	concentration of separand in the stationary phase averaged over an adsorbent particle (mol/L of solid phase)
$q_{i,sat}$	average adsorbent-phase concentration of solute i at saturation (mol/L of solid phase)
$q_{i,tb}$	average adsorbent-phase concentration of solute i at breakthrough (mol/L of solid phase)
Q	flow rate (m^3/sec)
r	radial distance (m)
r_f	radii of the fiber (m)
r_p	radius of the molecule (m)
R	ideal gas constant
$R(Z)$	particle radius in a expanded bed (m)
R_s	resolution
t	time (sec)
t^*	ideal adsorption time (sec)
t_b	breakthrough time (sec)
T	absolute temperature (K)
v_c	critical velocity (m/sec)
v	superficial velocity (m/sec)
V_0	void volume (m^3)
V_R	retention volume (m^3)
V_M	mobile phase volume (m^3)
V_S	volume of the stationary phase (m^3)
V_{feed}	volume of the feed (m^3)
V_c	volume of the column (m^3)
V_i	interstitial volume (m^3)
w	peak width (m^3)
Z	axial distance (m)

Greek Letters

α	selectivity
λ_p	tortuosity of a single particle
λ	compression of the bed
λ_T	tortuosity factor
τ	tortuosity
ε	bed porosity or void fraction of the bed
ε_p	particle porosity
μ	viscosity (Kg/m sec)
μi	chemical potential of species i
μ_2	second central volume (sec^2)
ϕ	angle of wall friction (rad)
ν	Poisson ratio for the material
σ_o	applied area-average axial stress
$\sigma(z)$	stress profile
ϕ_f	volume fraction of the fibers

$\mu_{iL}{}^0$ standard chemical potential of species i in the liquid phase
$\mu_{iS}{}^o$ standard chemical potentials of species i in the solid phase
τ injection time
σ^2 variance of the zone
γ obstructive factor
Γ incomplete Gamma function

References

Abunasser, N., and P. C. Wankat. 2004. One-column chromatograph with recycle analogous to simulated moving bed adsorbers: Analysis and applications. *Industrial and Engineering Chemistry Research* 43:5291-9.

Andersson, J., and B. Mattiasson. 2006. Simulated moving bed technology with a simplified approach for protein purification separation of lactoperoxidase and lactoferrin from whey protein concentrate. *Journal of Chromatography A* 1107:88-95.

Belew, M., and J. Porath. 1990. Immobilized metal ion affinity chromatography. *Journal of Chromatography* 516:333-54.

Brandt, S., R. A. Goffe, S. B. Kessler, J. L. O'Connor, and S. E. Zale. 1988. Membrane-based affinity technology for commercial scale purifications. *Bio/Technology* 6:779-82.

Bristow, P., and J. H. Knox. 1977. Standardization of test conditions for high performance liquid chromatography columns. *Chromatographia* 10:279-89.

Broughton, D. B. and C. G. Gerhold. 1961. Continuous sorption process employing fixed bed of sorbent and moving inlets and outlets. US patent 02985589.

Chao, S., C-H. Bu, and W. Yiu. 1990. Activation of troponin C by Cd^{2+} and Pb^{2+}. *Archives of Toxicology* 64:490-6.

Charton, F., and R. M. Nicoud. 1995. Complete design of a simulated moving bed. *Journal of Chromatography A* 702:97-112.

Fulton, S. P., A. J. Shaihidi, N. F. Gordon, and N. B. Afeyan. 1992. Large scale processing and high-throughput perfusion chromatography. *Bio/Technology* 10:635-9.

Furlong, A. M., D. R. Thomsen, K. R. Marotti, L. E. Post, and S. K. Sharma. 1988. Active human tissue plasminogen activator secreted from insect cells using a baculovirus vector. *Biotechnology and Applied Biochemistry* 10:454-64.

Geankoplis, C. J. 1993. *Transport Process and Unit Operations,* 3rd edition. Englewood Cliffs, NJ: Prentice Hall, p. 120.

Giddings, J. C. 1965. *Dynamics of Chromatography.* New York: Marcel Dekker.

Ghosh, R. 2002. Protein separation using membrane chromatography: opportunities and challenges. *Journal of Chromatography* 952:13-27.

Gopalakrishna, R., and W. B. Anderson. 1985. Monovalent cation-insensitive hydrophobic region on calmodulin facilitates the rapid isolation and quantitation of calmodulin free from other calcium-dependent hydrophobic proteins. *Journal of Applied Biochemistry* 7:311-22.

Gottschlich, N., and V. Kasche. 1997. Purification of monoclonal antibodies by simulated moving-bed chromatography. *Journal of Chromatography A* 765:201-6.

Guiochon, G. 2006. The limits of the separation power of unidimensional column liquid chromatography. *Journal of Chromatography A* 1126:6-49.

Hagel, L. 1989. Gel filtration. In *Protein Purification, Principles, High Resolution Methods and Applications,* edited by J.-C. Janson and L. Ryden, pp. 63-106. New York: VCH.

Hahn R. and A. Jungbauer. 2000. Peak broadening in protein chromatography with monoliths at very fast separations. *Analytical Chemistry* 72:4853-58.

Harrison, R. G., P. Todd, S. R. Rudge, and D. P. Petrides. 2003. *Bioseparations: Science and Engineering.* New York: Oxford University Press.

Hjerten, S., J. L. Liao, and R. Zhang. 1989. High-performance liquid chromatography on continuous polymer beds. *Journal of Chromatography* 473:273-5.

Houwing, J. 2003. Separation of proteins by simulated moving bed chromatography. PhD thesis, Technische Universiteit Delft, The Netherlands.

Huang, S. Y., C. K. Lin, W. H. Chang, and W. S. Lee. 1986. Enzyme purification and concentration by simulated moving bed chromatography: an experimental study. *Chemical Engineering Communications* 45(1-6):291-309.

Kaczmarski, K., and J-C. Bellot. 2004. Theoretical investigation of axial and local particle size distribution on expanded bed adsorption process. *Biotechnology Progress* 20:786-92.

Keener, R. N., J. E. Maneval, K. C. E. Ostergren, and E. J. Fernandez. 2002. Mechanical deformation of compressible chromatographic columns. *Biotechnology Progress* 18:587-96.

Katz, E., K. L. Ogan, and R. P. W. Scott. 1983. Peak dispersion and mobile phase velocity in liquid chromatography: the pertinent relationship for porous silica. *Journal of Chromatography* 270:51-75.

Lapidus, L., and N. R. Amundson. 1952. Mathematics of adsorption in beds. VI. The effect of longitudinal diffusion in ion exchange and chromatography columns. *Journal Physical Chemistry* 56:974.

Maisano, F., S. A. Testpro, and G. Grandi. 1989. Immobilized metal-ion affinity chromatography of human growth hormone. *Journal of Chromatography* 472:422-7.

Martin A. J. P., and R. I. M. Synge. 1941. A new form of chromatogram employing two liquid phases. I. Theory of chromatography. *Biochemical Journal* 35:1358.

Menkhaus, T. J. and C. E. Glatz. 2005. Antibody capture from corn endosperm extracts by packed bed and expanded bed adsorption. *Biotechnology Progress* 21:73-485.

Mondal, K., A. Sharma, and M. N. Gupta. 2003. Macroaffinity ligand facilated three phase partitioning (MLFTPP) for purification of glycoamylase and pullulanase using alginate. *Protein Expression and Purification* 28:190-5.

Nicoud, R. M. 1996. Large-scale fractionation of optical isomers by simulated moving bed chromatography. *Proceedings of Chiral Europe* '96, Strasbourg (France), pp. 117-9.

Ogston, A. C. 1958. The spaces in a uniform random suspension of fibers. *Transactions of the Faraday Society* 53:1754-57.

Pixley, R. A., and R. W. Coleman. 1986. Purification of human factor XII from plasma using zinc chelate affinity chromatography. *Thrombosis Research* 41:89-98.

Rivier, J., and R. McClintock. 1983. Reversed phase high performance liquid chromatography of insulins from different species. *Journal of Chromatography* 268:112-9.

Roy, I., and M. N. Gupta. 2002. Purification of a bacterial pullulanase on a fluidized bed of calcium alginate beads. *Journal of Chromatography A* 950:131-7.

Roy, I., S. Jain, S. Teotia, and M. N. Gupta. 2004. Evaluation of microbeads of calcium alginate as a fluidized bed medium for affinity chromatography of aspergillus niger pectinase. *Biotechnology Progress* 20:1490-5.

Sharma, S., I. Roy, and M. N. Gupta. 2001. Separation of phospholipase D from peanut on a fluidized bed of cross-linked alginate beads. *Biotechnology and Applied Biochemistry* 35:161-5.

Sharma, A., S. Sharma, and M. N. Gupta. 2000. Purification of wheat germ amylase by precipitation. *Protein Expression and Purification* 18:111-4.

Stickel, J. J., and A. Fotopoulus. 2001. Pressure-flow relationships for packed beds of compressible chromatography media at laboratory and production scale. *Biotechnology Progress* 17:744-51.

Sofer, G., and L. Hagel. 1997. *Handbook of Process Chromatography: A Guide to Optimization Scale-up and Validation.* New York: Academic Press.

Truei, Y.-H., T. Gu, G.-J. Tsai, and G. T. Tsao. 1992. Large-scale gradient elution chromatography. In *Advances in Biochemical Engineering/Biotechnology,* edited by A. Fiechter, pp. 1-44. New York: Springer-Verlag.

van Deemter, J. J., F. J. Zuiderweg, and A. Klinkenberg. 1956. Longitudinal diffusion and resistance to mass transfer as causes of nonideality in chromatography. *Chemical Engineering Science* 5:271-89.

Velayudhan, A., and M. R. Ladisch. 1993. Plate models in chromatography: analysis and implications for scale up. In *Advances in Biochemical Engineering/Biotechnology*, edited by A. Fiechter, pp. 147-60. New York: Springer-Verlag.

Wills, P. R., Y. Georgalis, J. Dijk, and D. J. Winzor. 1995. Measurement of thermodynamic nonideality arising from volume-exclusion interactions between proteins and polymers. *Biophysical Chemistry* 57:37-46.

Wong, J. W., R. L. Albright, and N. H. L. Wang. 1991. Immobilized metal ion affinity chromatography (IMAC) chemistry and bioseparation applications. *Separation and Purification Methods* 20:49-106.

Yamamoto, S., K. Nakamishi, and R. Matsumo. 1988. *Ion Exchange Chromatography of Proteins*. New York: Marcel Dekker.

Zhou, J. X., and T. Tressel. 2006. Basic concepts in Q membrane chromatography for large-scale antibody production. *Biotechnology Progress* 22:341-9.

Further Reading

Watler, P., O. Kaltenbrunner, D. Feng, and S. Yamamoto. 2003. Engineering aspects of ion-exchange chromatography. *Chromatographic Science Series* 88:123-71.

Freitag, R. 2003. Novel approaches to the chromatography of proteins. *Biotechnology and Bioprocessing* 27:455-502.

Chin, C.Y., and N-H. L. Wang. 2004. Simulated moving bed equipment designs. *Separation and Purification Reviews* 33:77-155.

Chapter 8
Precipitation

Introduction

Precipitation is one of the oldest methods used for protein purification. The objective in any protein fractionation procedure by precipitation is to find means to alter the solubility of a particular protein without affecting the solubility of the other proteins present in the mixture. Although the benchmark for protein precipitation is the fractionation of human plasma, the technique has evolved into new levels of sophistication. A few examples that illustrate novel uses of the method are the use of pH and NaCl for the precipitation of sesame protein concentrated (Inyang and Iduh 1996), the use of carbon dioxide (Thiering et al. 2001, Kirkland and Tanner 2005) and supercritical carbon dioxide (Marazban et al., 2003) to precipitate various proteins, the use of affinity ligands (Galaev and Mattiasson 1997) including metal chelates (Kumar, et al. 1998), the use of matrix ligands (compounds containing a strong anion sulfonate head and a hydrophobic tail that "trap" the protein) (Matulis, Lovrien, and Richardson 1996; Lovrien, Wu, and Matulis 2000), and so on.

As with many other protein purification techniques, protein precipitation has been developed in the laboratory and then scaled up to processes. The large volume of experimental work has facilitated the development of several heuristics that can be followed to design a precipitation-based protein purification protocol. Theoreticians have used that large experimental base to develop models of diverse complexity that can be used to correlate experimental data or to dig into the molecular bases of the precipitation process.

The solubility of a protein depends on a number of parameters that can be grouped into two categories: 1) those intrinsic to the protein and 2) solution properties. The first category includes the size, amino acid composition, amino acid sequence, number of ionizable residues, ratio, and distribution of polar to nonpolar residues, dissociation constants, conformation, charge, and ion-binding properties of the protein. The second category includes the availability of solvent, pH, ionic strength, dielectric constant, temperature, and specific ion (or solute) effects. Most of the protein-related parameters are redundant because the amino acid composition of the proteins, their number, and sequence are enough to determine the size and conformation of the proteins. For any given conformation the distribution and ratio of polar and non-polar and ionizable residues is fixed. Molecular simulation programs like Tripos (St. Louis, MO) can be used to quantify the hydrophobicity of a protein given its ternary structure. Finally, the pH and ionic strength of the solution fix the dissociation constant of the amino acids, and therefore, the overall charge and charge distribution. The ion-binding capacity of the protein is determined by the presence of the appropriate amino acids at the surface (e.g., histidine) and by the regulatory effect of pH and salts. Although each of these parameters affects the solubility of the protein, its size and surface properties seem to dominate its solution behavior. Non-prosthetic groups like heme, carbohydrates, and lipids will affect the solution properties of the protein by changing its charge, hydrophobicity, and ability to bind metal ions.

Although one can always find exceptions, the following rules of thumb can be used to predict the solubility behavior of a protein in a protein mixture. First, larger proteins are less soluble than smaller ones. Second, proteins whose surface is rich in hydrophobic amino acids are less soluble in water than those proteins whose surface is rich in hydrophilic amino acids. Third, denatured proteins are less soluble in water than native ones; this is a consequence of an increase in the hydrophobocity of the molecule as hydrophobic groups are exposed to the solvent as the protein unfolds. Solvent properties that affect protein solubility are pH (pHs close to the isoelectric point decrease solubility), dielectric constant (the higher the dielectric constant the lower the solubility), and ion strength (the higher the ionic strength the lower the solubility). It has been found that the integrity of the water of hydration (a thin layer of water closely associated with the surface of the protein) is important for the protein to remain in solution. If that layer is disturbed because the solvent is not available for hydration, the protein will precipitate. This is one accepted mechanism for the precipitation by some solvents such as ethanol. At the other end, the addition of cosolvents that favor hydration will stabilize the protein. An example of such a solvent is glycerol, which is known to increase protein solvation.

The heuristics enumerated above indicate that the solubility of a protein can be affected by variations in pH, temperature, ionic strength, and by the addition of cosolvents. Each of these environmental variables affects protein solubility in a distinct manner and although generalizations are possible, the quantitative effect of each of the factors named above are protein-dependent. Still, protein precipitation techniques have been historically classified according to the precipitation method used. It is a well-known fact that the solubility of a protein is at a minimum at its isoelectric point. Therefore, a common technique used to precipitate a protein is to adjust the pH at a value equal to the protein's isoelectric point. This is called isoelectric precipitation. Another way of weakening electrostatic repulsions is to add an electrolyte at high concentration (salting out). Protein solubility is also a function of temperature. The solubility of a native protein decreases as the temperature decreases, but an increase in temperature will denature the protein and, therefore, it will induce the formation of insoluble aggregates. Dehydration of the protein has been invoked as the main mechanism for protein precipitation caused by the addition of cosolvents. This argument holds only if the presence of the cosolvent (normally an organic compound) does not cause the protein to denature. The addition of ionic and nonionic polymers also decreases protein solubility and both have been used in protein precipitation studies. Although less common, affinity ligands attached, for example, to a non-ionic polymer have been used to improve the selectivity of precipitation; this is called affinity precipitation.

Most protein precipitates are not crystalline under process conditions unless precipitation is done with the specific purpose of obtaining crystals. Instead, protein precipitates consist of more or less disorganized structures containing a large amount of water (gels). This characteristic may be troublesome for inexperienced researchers, because, particularly when the protein concentration is low, the precipitate may go unnoticed. Often, protein precipitates look like a smear on the size of the centrifuge tube in laboratory experiments.

The main applications of protein precipitation are: a) volume reduction, b) product stabilization, c) aid to chromatography, and d) protein fractionation. Handling of large volumes is common in bioprocessing because most times the target protein is at a very low concentration. Large volumes increase processing cost because of the increase in capital cost and associated operating cost.

Precipitation offers a simple way of concentrating the product. There are no simple rules to be followed to keep a particular protein stable. Some proteins are more stable in solution (with the right buffer composition and storage temperature), but most are better preserved in solid form. Precipitation can be used to eliminate most of the liquid of a protein solution just before freeze drying. Precipitation has also been used as an aid to chromatography. Concentration of the sample is particularly important in modern biotechnology in which minute amounts of proteins may be available. This is typically the case in proteomics. Therefore, precipitation has been revived as a protein concentration means without which the limit of detection of many analytical protocols could not be reached. Finally, protein precipitation may be used to isolate a protein from a mixture. This is called protein fractionation and the best known example, as stated earlier, is Cohn's method for the fractionation of plasma proteins (Cohn 1950).

Theoretical Foundations

Theoretical treatments of protein precipitation range from semi-empirical approaches (Agena, Pusey, and Bogle 1999) to molecular thermodynamic one (Jenkins 1998). Molecular thermodynamic approaches are attractive because they make a connection between macroscopic phenomena and molecular features. The interested reader may want to look at the works of Chang and Bae (2003) who used perturbation theory in a generalized van deer Walls partition function to explain the effect of salts and shape of the protein on its solubility (Kim, Bae, and Ryu 2000; Kim and Bae 2003) who used molecular thermodynamics to explain the effect of ionic strength, protein and salt sizes, and the degree of pre-aggregation on protein solubility; and Anderson et al. (2000) who used fluorescent anisotropy to determine the potential of mean force between peptide molecules in the presence of salt, and thus offered a molecular inside into the precipitation mechanism. Because the theoretical treatment depends on the precipitation method, in the next sections we describe some theoretical approaches specific for precipitation by non-ionic polymers and by the addition of salts.

The understanding of the precipitation process is useful for the conceptual design of new precipitation methods and for the optimization of existing ones. However, after a concept has been proved in laboratory scale it is necessary to scale it up. Understanding and quantifying the steps leading to precipitation formation helps in the design of "precipitation vessels." Because the mechanisms are similar to all precipitation methods, they are described in this section.

The formation of a precipitate consists of the following steps (Petenate and Glatz 1983): a) initial mixing, b) nucleation, c) growth governed by diffusion, d) growth governed by fluid motion, and e) precipitate aging.

Nucleation is the formation of miniature "crystals" that will eventually grow to sedimentation size. If the amount of nuclei is too large the precipitate will be too fine because the growth of the precipitate will be spread over too many sites. This kind of precipitate will be difficult to sediment. Therefore, the control of nucleation will determine the quality of the precipitate. The strategy to follow is to keep the degree of supersaturation (difference between actual concentration and equilibrium solubility) low. This strategy works because the rate of nucleation grows exponentially with the degree of supersaturation.

It is important to achieve good mixing between the solution and the precipitant solution at the beginning of precipitation. Otherwise, supersaturation and undersaturation regions will be present in the solution. The time required for good mixing is,

$$t = \frac{l_e^2}{8D_{AB}} \tag{8.1}$$

in which D_{AB} is the molecular diffusivity of the molecules being mixed and l_e is the diameter of eddies (Bell et al. 1983). Here (eddies are pieces of the fluid where mixing is diffusion limited and the mixing between eddies [in a turbulent regime] is instantaneous). According to Equation 8.1, the mixing time is of the order of a few seconds.

After the nuclei have been formed, new molecules moving by Brownian motion collide with them and some remain attached. As a consequence the nuclei grow in size. This diffusion limited growth continues until the particles are 0.1 to 10 μm in size. A consequence of this random process is that not all the nuclei will grow at the same rate. Therefore, a population of sizes is obtained. By assuming a single size, the molecular mass of the particles at a time t (M) can be estimated from (Harrison et al. 2003),

$$M = M_0(1 + K_A N_0 t) \qquad 8.2$$

in which M_0 is the molecular weight of the solute, K_A is a second order rate constant (second order), and N_0 the solute number concentration.

After a critical size of approximately 1 μm has been reached, particles grow by colliding and sticking to each other. In this case, the motion of fluid will define the growth. Of course, growth by diffusion continues, but it is negligible compared to growth by fluid motion (mixing). This growth can be estimated by (Harrison et al. 2003),

$$\frac{N}{N_0} = \exp\left(\frac{4\alpha\phi\gamma t}{\pi}\right) \qquad 8.3$$

in which N is the number concentration of particles (it decreases as time goes by) at time t, N_0 is the number concentration of particles at time zero, α is the "collision effectiveness factor", ϕ is the volume fraction of the particles (assumed constant), and γ is the shear rate. For turbulent flow the shear rate can be estimated from (Camp and Stein 1943),

$$\gamma = \left(\frac{P/V}{\mu}\right)^{1/2} \qquad 8.4$$

in which P/V is the power dissipated per unit volume, and μ is the viscosity of the solvent. Constant sheer (constant P/V) is a proposed scaling-up criterion together with geometrical similarity; that is, the volume of the reactor is increased by changing its diameter.

After the particles reach a critical size they begin to break (remember that the solution is subjected to shear because of the mixing) and re-associate. After a while, the size of the precipitate stabilizes. The time that the precipitate needs to reach a stable size is called "aging" time. Sometimes, stirring may be increased during aging to accelerate the disruption of the precipitate and allow the formation of better particles. The maximum equilibrium size of the aggregates, d_{max}, is proportional to the impeller speed (n) (Hoare 1982),

$$d_{max} \propto n^{-\delta} \qquad 8.5$$

in which δ varies from 0.1 to 0.8. It is illustrative to consider the different times involved in the various steps of the precipitation process. Initial mixing happens in a matter of seconds. Therefore, the time required for nucleation is directly proportional to the rate at which the precipitant is added. The time required for growth because of fluid flow is of the order of several minutes. For example, Harrison et al. (2003) calculated a time of approximately eighteen minutes for a precipitate of α2-macroblobuline at a shear rate of 866 s^{-1} to reach 20 μm. Finally, the time that a precipitate needs to reach a steady size when exposed to shear is of the order of one hour. Therefore, the total time for precipitation can be estimated from

$$t = t_{supersaturation} + t_{mixing} + t_{aging} \qquad 8.6$$

in which the time needed to reach supersaturation is equal to the time used to add the precipitant. If the precipitation reaction is done in batch, this time will be equal to the processing time. If the precipitation is done in a continuous reactor, this time will be equal to the reactors' residence time, which is equal to the volume of the reactor divided by the feed flow rate.

Precipitation Modes

As stated before, there are several means by which the solubility of a protein can be affected. In the following sections, we describe the most common protein precipitation modes. Some of them are used in both laboratory and process scale. Some others have not been able to overcome laboratory stage mostly because of cost considerations. Although some of the cases discussed below include fractionation of protein mixtures, one must keep in mind that the main use of protein precipitation in process scale is in volume reduction.

Precipitation by Ionic Polymers

Ionic polymers or poly(electrolytes) have been used to precipitate proteins (Simpson 2004; Yu, Liu, and Chen 2002). Although the most common polymer used has been polyacrylic acid (Kim, Hirasawa, and Kim 2002), several other polymers have been used; for example, hexametaphosphate, CM-cellulose of different degrees of polymerization and degrees of substitution values, polygalacturonic acid, Na alginate, and carrageen (Gillberg and Tornell 1976; Clark and Glatz 1990). The applications have been broad, including, for example, the use of polyacrylic acid to precipitate lysozyme from tobacco extract (Zhang et al. 2005). Research has been focused on elucidating the effect of polymer molecular weight (Shieh and Glatz 1994) and dosage (Chen and Berg 1993) on the formation of the precipitate. Phase diagrams for polyelectrolyte/protein mixtures have been determined by using light scattering (Mattison, Brittain, and Dubin 1995). Figure 8.1 shows one of those phase diagrams for albumin and poly(dimethyldiallylammonium chloride (PDMDAAC). The net charge of the protein is plotted versus the square root of the ionic strength. The straight line corresponds to the pH at which soluble complexes between the electrolyte and the protein are formed. The curved line corresponds to the pH at which a precipitate is observed. The two lines divide the space into three regions. Above the straight line the polymer and the protein do not associate; thus, they form a true solution. In between the lines two liquid phases are at equilibrium. One phase is rich in water and the other is rich in a polyelectrolyte/protein mixture. Finally, below the curved line a precipitate is formed. It is conceivable that different proteins will have different phase diagrams and that, therefore, conditions may be found to preferentially precipitate a particular protein from a mixture.

It is worth mentioning the use of "intelligent" polymers whose solubility behavior is affected by other polymer. For example, mixtures of poly(methacrylic acid) (PMA) and poly(*N*-ethyl-4-vinylpyridium bromide) (PEVPB) reversibly precipitate at some pHs (Dainiak et al. 1998-1999). If an antigen is attached to PMA, antibodies can be precipitated by adding PEVPB and adjusting the pH. Similarly, Nath, Patrickios, and Hatton (1995) found that the addition of a protein induced the precipitation of the polyampholytes, which were soluble otherwise.

It has been argued that polymer bridging is the dominant mechanism for protein precipitation by (at least) polyacrylic acid (Kim, Hirasawa, and Kim 2001). However, others have argued that high

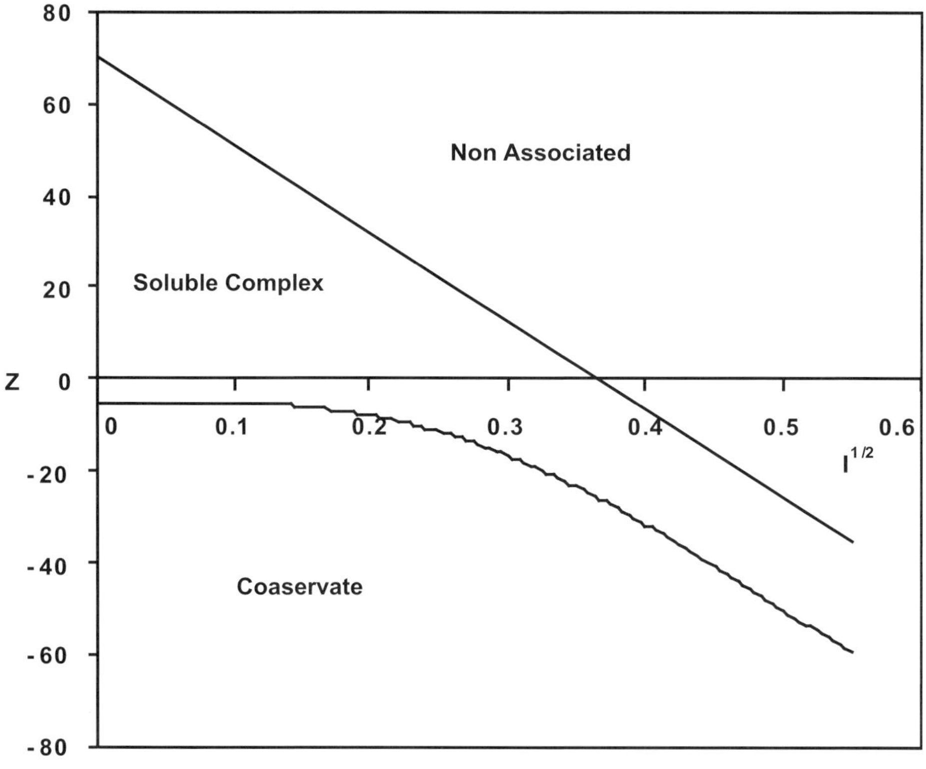

Figure 8.1. Phase diagram for a protein/water solution

molecular weight polyelectrolytes act by polymer-bridging whereas low molecular weight ones act by charge neutralization (Chen, Walker, and Berg 1992). Those interested in studies aimed at answering fundamental questions may want to see the works of Niederauer and Glatz (1994), Clark and Glatz (1992), or Jiang and Prausnitz (1999). Jiang and Prausnitz's work is particularly quite comprehensive and is based on molecular thermodynamics principles. One of their main conclusions is that polyelectrolytes have both salting in and salting out effects, depending on the concentration used.

Precipitation by Salting Out

Salting out consists of the use of multivalence salts to increase the ionic strength of the solution, and thus precipitate a protein. It is by far the most common protein precipitation method. The effect of ionic strength can be understood from three viewpoints; although each probably contributes to the observed decrease of protein solubility upon the addition of salts. The simplest way of explaining salting out of proteins is to apply mass action law. Suppose a protein is dissolved in water. The species that will be present in solution are: 1) the protein at a concentration P and with Z- negative charges and Z+ positive charges such as the net charge is |Z+|-|Z-| different from zero (the protein is not isoelectric); 2) cations at a concentration C=P|Z-|/Z+| and anions at a concentration A=P|Z+|/|Z-|. Therefore, the solubility product, K, of the protein plus counterions is,

$$K = [protein][C]^{|Z-|}[A]^{|Z+|} \qquad 8.7$$

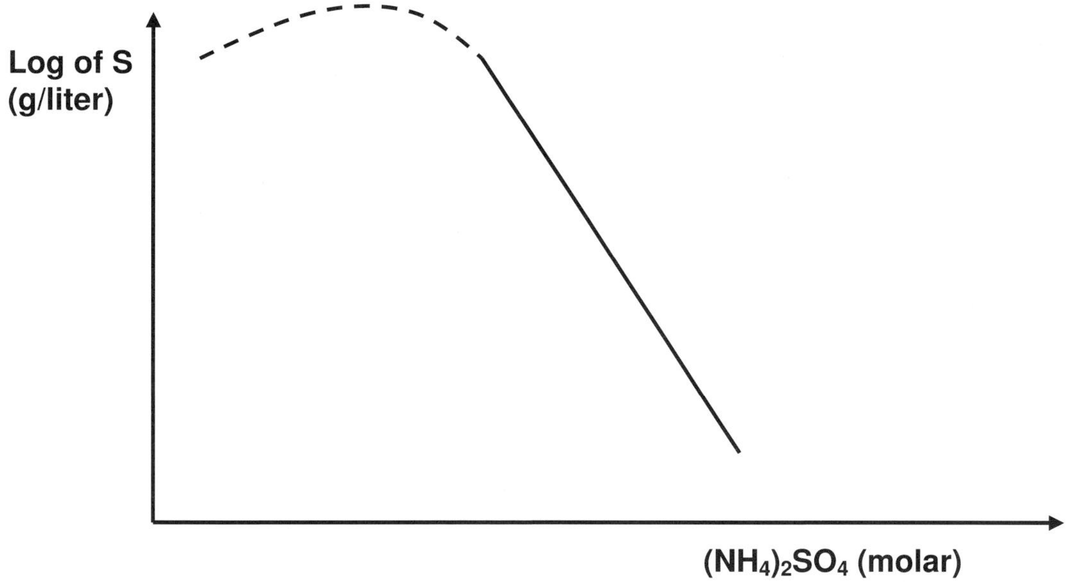

Figure 8.2. Typical solubility vs. salt concentration graph for a protein including salting-in (dashed portion of the curve).

Therefore, if a salt containing either the cation or the anion is added the concentration of the protein in solution must decrease to keep K constant. Thus, the protein will precipitate. It is observed, however, that the solubility of a protein at its isoelectric point is also affected by the presence of salts. Therefore, a simple mass action mechanism is not sufficient to explain protein precipitation by salting out.

Because proteins behave roughly like colloids, arguments used to explain the stability of colloids in the presence of strong electrolytes can be used to explain the effect of ionic strength on protein solubility. This topic was discussed in detail in Chapter 4 and it will not be repeated here.

A third argument that has been used to explain the effect of salts on protein solubility is the effect that salts may have on protein hydration. The idea here (loosely) is that the layer of water molecules surrounding the polypeptide chain makes the protein stable in solution (hydration layers around the protein molecules will repeal each other and therefore, the molecules will not aggregate). Salts will "suck" this water changing the hydration of the protein and, therefore diminishing its solubility.

Proteins are such complex molecules that a detailed description of the precipitation phenomenon may escape science for a while. From a bioseparation point of view, we are better off by taking a pragmatic approach. The experimental evidence overwhelmingly supports the following correlation of protein solubility with ionic strength (Dixon and Webb 1961),

$$\ln S = \ln S_o - K_s I \qquad 8.8$$

I is the ionic strength, K_S is the salting-out constant and it is a function of the particular protein, and β is a function of the type of protein, pH and temperature. β represents the intrinsic protein solubility (the solubility at zero ion strength). Equation 8.8 is known as the Cohn equation (Cohn 1925). However, the solubility of many proteins slightly increases at low salt concentrations and then decreases (see Figure 8.2). This increase in solubility at low salt concentrations is called "salting in" and explains why proteins are more soluble in dilute buffers than in pure water.

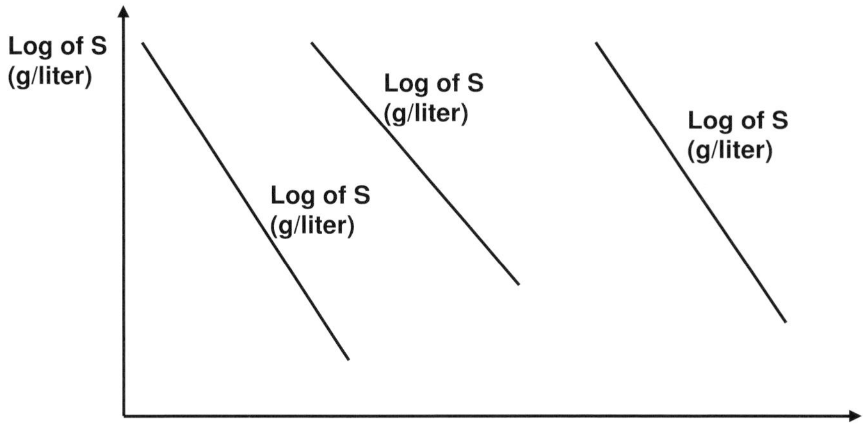

Figure 8.3. Solubility graphs for three proteins vs. the concentration of ammonium sulfate. The protein molecular weight decreases from left to right.

In the absence of any salting in phenomenon a plot of the ln of S versus salt concentration looks like the one presented in Figure 8.3. The extrapolation at zero salt concentration is the intrinsic protein solubility.

Equation 8.8 cannot predict the behavior depicted in Figure 8.2. Kirkwood developed a theory (Kirkwood 1943), based on statistical mechanics of electrolytes, that can be reduced to the form of Cohn's equation (Equation 8.7); that is,

$$\ln\frac{S}{S_0} = K_i I - K_s I \qquad 8.9$$

in which I is the ionic strength and K_i and K_s are given by,

$$K_i \propto \left(\frac{u}{\varepsilon T}\right)^2 \qquad 8.10$$

$$K_s \propto \frac{V_e}{\varepsilon T} \qquad 8.11$$

in which u is the dipolar moment of the protein, V_e is the volume excluded by the protein, and ε is the dielectric constant of the solvent, which decreases as the polarity of the solvent decreases. This explains the experimental observation that salting in is predominant in non-polar solvents.

The most common salt used for salting out is ammonium sulfate, which is highly soluble in water (4.1 M at 25°C). The common practice is to add the salt at a concentration expressed as percent saturation (541.8 g/L at 25°C). Because of the large amount used and because of non-idealities if you prepare 1 L of solution containing 541.8 g/L you will not get a saturated solution but a 77% one. Nomograms and tables have been published (Scopes 1982) to facilitate the preparation of solutions of this salt.

Precipitation by Addition of Solvents

Dehydration of the protein has been invoked as the main mechanism for protein destabilization caused by the addition of cosolvents. This argument holds only if the presence of the cosolvent (normally an organic compound) does not cause the protein to denature. Although precipitation by organic solvents can be quite effective, it has been limited to laboratory scale work and to the fractionation of plasma proteins.

The best-known solvent precipitation process is Cohn's method for blood fractionation. The overall strategy consists of the addition of various amounts of ethanol (up to 40%) combined with a judicious choice of temperature (from 0 to -5°C) and ionic strength. The process consists of a series of precipitation steps in which the precipitate is redissolved and then re-precipitated under slightly different conditions. The raw material is human or bovine plasma and the main products are albumin and globulins. In the first step, plasma is treated with 8% ethanol at pH 7.2, and an ionic strength of 0.14 at -5°C. Under these conditions, a precipitate is formed (P1) and a supernatant is removed (S1). The supernatant is precipitated with 25% ethanol at pH 6.9, and an ionic strength of 0.09 at -5°C. Again, a precipitate (PII and III) and a supernatant are formed (SII and III). The supernatant II and III are treated with 18% ethanol at pH 5.2, an ionic strength of 0.09 at -5°C. Precipitate IV-1 and supernatant IV-1 are formed. Supernatant IV-1 is treated with 40% ethanol at pH 5.8, an ionic strength of 0.09 at 5°C. Precipitate IV-4 and Supernatant IV-4 are formed. Supernatant IV-4 is treated with 40% ethanol at pH 4.8, an ionic strength of 0.11 at –5°C. Precipitate V and supernatant V are formed. The last supernatant is discharged and the remaining work is done with all the precipitates. Precipitate V is treated with 10% ethanol at pH 4.5, at an ionic strength 0.1 and -3°C. The precipitate consists of impurities and the supernatant is treated with 40% ethanol, at pH 5.2, an ionic strength of 0.01 at -5°C. Under these conditions albumin is precipitated. The supernatant is discarded. Precipitates II and III are treated with 20% ethanol at pH 7.2, an ionic strength of 0.005 at -5°C. Precipitate II+III W and supernatant II+III W are formed. This supernatant is treated with 25% ethanol at pH 5.7, an ionic strength of 0.005 at -5°C. Supernatant III-O and precipitate III-O are formed. Precipitate III-O contains lipoproteins. Precipitates II+III W are treated with 17%, pH 5.2, at an ionic strength of 0.015 at -6°C to obtain two fractions: supernatant III and precipitate III. Precipitate III is treated with 1% ethanol, pH 5.4, an ionic strength of 0.08 at 0°C. Supernatant III-2.3 and precipitate III-2.3 are obtained. Precipitate III-2.3 is dissolved in buffer of ionic strength 0.1 at 0°C and pH 7. The dissolved fraction is supernatant III-2 that contains prothrombin and the insolubles are precipitate III-3 that contains plasminogen.

Supernatant III-2.3 is treated with 15% ethanol, an ionic strength of 0.06, pH 6, and -5°C to obtain supernatant III-1 and precipitate III-1 (isoagglutinins). Supernatant III is treated with 17% ethanol, pH 5.2, an ionic strength of 0.05 and -6°C to obtain precipitate II-3 (γ-globulin) and supernatant II-3. Supernatant II-3 is treated with 25% ethanol, pH 7.4, an ionic strength of 0.05, and -5°C to obtain supernatant II-1.2 and precipitate II-1.2 (that also contains γ-globulin). There is a total of fifteen steps to obtain six protein fractions: albumin, lipoproteins, prothrombin, isoagglutinins, plasminogen, and γ-globulin.

Isoelectric Precipitation

The solubility of proteins is at its minimum at their isoelectric point, but they do not necessarily precipitate. Therefore, it is common to adjust the pH at the isoelectric point of the protein that needs to be precipitated and then the actual precipitation is done by the addition of salts or non-ionic polymers. An exception to this is the isoelectric precipitation of casein.

A classical example of isoelectric precipitation is the precipitation of casein from skim milk. The amount of casein in milk ranges from 24 to 29 g/L. Casein products are used in a variety of food products including bakery, cheese, ice cream, foods, pasta, and pharmaceuticals. Casein exists in milk in the form of micelles that are stable at the pH of the milk, 6.6. As the pH is lowered to pH 4.6 (the isoelectric point of casein) the micelles coagulate and a precipitate is formed. The acidification is done by the addition of strong acid, such as sulfuric acid or by acid fermentation of lactic acid producing bacteria.

Precipitation by Non-Ionic Polymers

It is well known that the addition of non-ionic polymers to a protein solution decreases protein solubility (Thrash, Otto and Deits 1991; Ingham 1990). Precipitation by non-ionic polymers has been used, for example, for the crystallization of glycosylated and non-glycosylated variants of agglutinin (Dao-Thi et al. 1996) for affinity precipitation by attaching a tag to temperature sensitive polymers (Mori, Nakata, and Endo 1994; Kumar, Galaev, and Mattiasson 1998; Mattiasson, Dainyak, and Galaev 1998), for the precipitation and co-precipitation of plasma proteins (Robinson et al. 1989), and for the isolation of milk proteins (Ortin, Cebrian, and Johansson 1992). One of the most commonly used non-ionic polymers in protein precipitation studies is poly(ethylene)glycol (PEG). The polymer is readily soluble in water, various molecular weights are commercially available, and it has been approved by the U.S. Food and Drug Administration (FDA) for human consumption. PEG is slightly hydrophobic and quite rigid. Scattering experiments have shown that PEG has a cylindrical shape in solution. Two arguments have been used to explain the mechanism of precipitation by non-ionic polymers. The simplest one consists of considering that the presence of the polymer decreases the volume of the solution available to the protein (Atha and Ingham 1981). This would force the proteins to be closer to each other and aggregates would form. A more sophisticated argument is considering that because of stearic constraints the polymers are excluded from the space separating two protein molecules (Mahadevan and Hall 1990, 1992a,b). This particular distribution of the polymer causes a "pressure imbalance" that pushes the proteins against each other. Purely stearic arguments fail to explain the effect of protein charge on protein precipitation by non-ionic polymers. On the contrary, the second argument can nicely (and rigorously) incorporate the effect of pH on protein precipitation by non-ionic polymers. Still these models are approximate because they assume that the solvent is a continuous fluid; therefore, they ignore the contribution of hydration forces to protein stability.

The solubility of a protein in the presence of a non-ionic polymer can be correlated by,

$$\ln S = \ln S_o' - K_p c_p \tag{8.12}$$

c_p is the concentration of polymer, and K_p and β' are constants. For a rod-shape polymer (such as PEG) (Juckes 1971):

$$K_p = \frac{\overline{V}}{2.303}\left(\frac{r_r + r_p}{r_r}\right)^3 \tag{8.13}$$

in which r_p is the radius of the protein, r_r is the radius of the rod, and is the partial specific volume of the polymer. In the next paragraphs we illustrate the use of PEG as a precipitant of a series of transgenic human antibodies.

Two human antibodies expressed in corn were used: 1) a fully non-glycosylated antibody and 2) a mixture of 25% non-glycosylated, 50% single GlcNac, and 25% glycosylated MAB. The isoelectric point of the non-glycosylated protein is 9.7 to 9.1, whereas the isoelectric point of the glycosylated mixture expands the range to 9.4 to 8.7. Because of the presence of carbohydrates the glycosylated protein is slightly larger (MALS yields a molecular weight of the glycosylated and non-glycosylated protein of 135,600±900 and 121,600±5000 respectively. Both preparations have

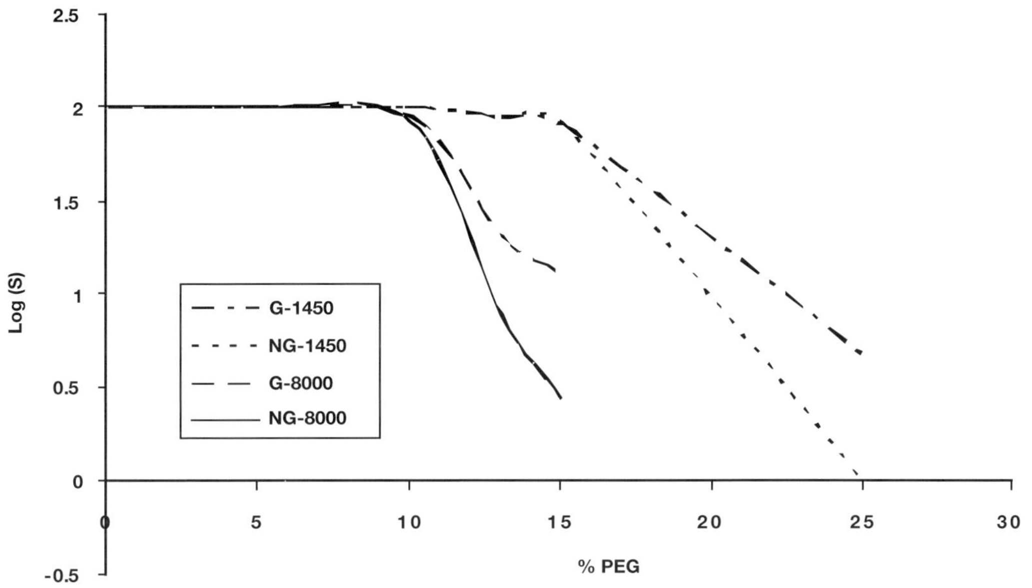

Figure 8.4. Solubility of a glycosylated and non-glycosylated antibody vs. PEG concentration. NG-1450 and G-1450 are the solubility curves for the non-glycosylated and glycosylated proteins in PEG 1450. NG-8000 and G-8000 are the solubility curves for both proteins in PEG 8000.

also similar electrophoretic mobilities. The proteins were exposed to various concentrations of PEG at different pHs and temperatures.

Upon addition of various amounts of PEG, the decrease in the solubility of the glycosylated sample was between 0.1 to 7%. On the contrary, the decrease in the solubility of the non-glycosylated antibody ranged from 0.1 to 40% (pH 10). Near its isoelectric point the protein becomes quite unstable after 24 hours. The effect of polymer molecular weight and pH on the solubility of both preparations is shown in Figure 8.4. The figure shows that after a threshold concentration of PEG is reached, the solubility of the antibody decreases as the concentration of PEG increases as observed with many other proteins. The preparation that contains the glycosylated protein is more soluble than the non-glycosylated protein under all conditions and the differences in solubility are enhanced at a higher molecular weight PEG. This confirms that one of the roles played by the carbohydrates chains is to increase protein stability. It is also obvious that Equation 8.12 is valid only in a narrow range of PEG concentrations. Figure 8.4 shows that both proteins are more soluble in PEG 1450 than in PEG 8000 (this is a well-known trend of protein solubility with polymer molecular weight). Both proteins are more soluble at pH 9 than at 6 (Figure 8.5), which is unexpected, because proteins are less soluble at or near their isoelectric point. At very acidic (pH 4) or very basic (pH 10.7) pH levels, the solubility of both preparations increases; however, the solubility of the preparation containing the glycosylated antibody increases even further. The results obtained at pH 10.7 are particularly striking because the glycosylated preparation does not precipitate in the PEG concentration range covered by these experiments. This indicates some synergism between the different variants of the protein because the glycosylated preparation contains 25% non-glycosylated protein. Miekka and Ingham (1980, 1978) argued that the effect of hetero-association or self-association on the composition of precipitates obtained by adding PEG are important only when the proteins are already associated before the addition of the polymer. That does not seem to

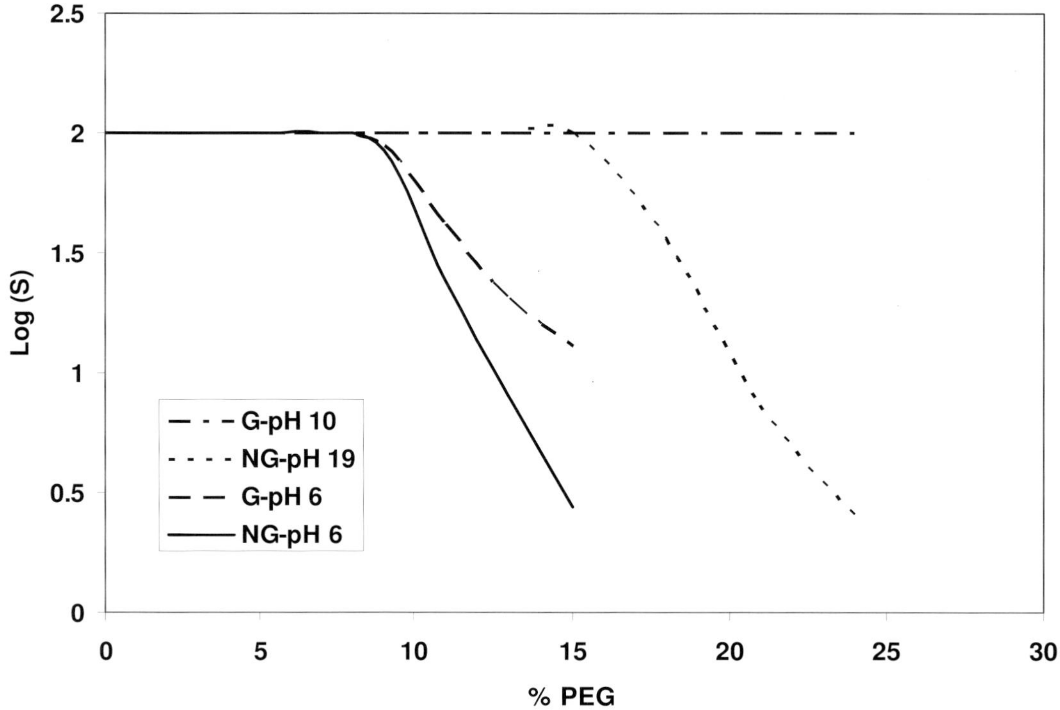

Figure 8.5. Solubility of a glycosylated and non-glycosylated antibody vs. PEG concentration. The notation is the same as in Figure 8.4.

be the case here because MALS shows that the preparations consist of 95% monomers and 5% dimers.

One common argument (Atha and Ingham 1981) used to explain the precipitation of proteins by non-ionic polymers is that the presence of the polymers excludes the proteins from regions that they would otherwise occupy. Obviously, this argument does not hold here. Combinations of exclusion volume arguments and electrostatic forces cannot be used to explain the observed trends either (Guo, Guo, and Narsimhan 1996; Guo and Narsimhan 1996; Mahadevan and Hall 1992b). A common feature of current models is that they treat the solvent as a continuous fluid whose only role is to modulate the interactions between the species. The phenomenon observed in this work suggests that hydration forces (that can be accounted for only by recognizing the molecularity of the solvent) play an important role in determining protein solubility and that they have to be included to properly explain the precipitation of proteins by non-ionic polymers and probably by any other precipitant.

Temperature is not a variable commonly explored in the precipitation of proteins by non-ionic polymers. Considering that temperature will affect the equilibrium between phases, the overlook of temperature effects is surprising. The solubility of the glycosylated and non-glycosylated antibodies in the presence of PEG was studied at 5, 25, and 40°C. The solubility of both preparations decreases as the temperature decreases as expected. Still, the non-glycosylated antibody is more sensitive to changes in temperature than the glycosylated preparation. For example, at 4°C and at a PEG concentration of 12% only 1% of the non-glycosylated antibody remains in solution, whereas about 8% of the glycosylated preparation remains in solution. The differences in solubility between 4 and 40°C are quite dramatic for the non-glycosylated protein. For example, at

12% PEG nearly 70% of the protein remains in solution at 40°C, whereas only 1% remains in solution at 5°C. These results indicate that the intrinsic solubility of the antibody is strongly affected by the presence of the carbohydrate chains. This must be taken into account when the decision of shutting down the glycosylation machinery is made in the production of human antibodies in plants.

Affinity Precipitation Including Precipitation Using Lectins

Affinity precipitation is feasible as long as there is some specificity of the ligand used and that the ligand-protein complex can be made insoluble. Affinity precipitation is very easy to implement but it is very costly. The actual selectivity of the technique is difficult to assess, because most of the experimental base has been obtained with pure proteins. Still, the developing of new compounds may allow a better development (and more use) of this potentially interesting technique.

Affinity precipitation may be divided into two categories according to the type of macroligands: homogeneous or heterogeneous (Niederauer and Glatz 1994). Homogeneous macroligands must have the affinity "function" in more than one place. The existence of this polyfunctionality is necessary because they function by bridging more than one protein molecule together (the enzyme must have more than one affinity site too). The two functional ligands are connected by a spacer whose length must be tuned up to maximize the selectivity of the ligand. One of the earliest examples of the use of homogeneous macroligands in affinity precipitation is the use of bis-NAD to precipitate dehydrogenases (Larsson and Mosbach 1979; Mansson and Larsson 1979). It has been pointed out that the conditions for affinity precipitation using homogeneous ligands are not very stringent. This assessment is questionable because the conditions for binding can be very specific and narrow regarding ionic strength, pH and presence of impurities.

Two other examples of homogeneous macroligands are Cu containing polymers and tryazine dyes. Binding of histidine groups to Cu(II) ions has been used to selectively precipitate proteins that differ in the amount of histidines by using bis-chelating agents; for example, two Cu(II) ions joined by a spacer, such as PEG or ethylene glycol bis (β-aminoethyl ether)N_1N'-tetraacetic acid. Triazine dyes are quite popular mostly because of their low cost. They precipitate proteins by forming large insoluble complexes because of heavy cross-linking with the enzyme. One problem with triazine dyes is that their specificity is broad. A typical example of the "promiscuity" of tryazine dyes is the binding of all NAD-dependent dehydrogenases to Cibacron blue.

Heterogeneous macroligands are relatively new. The idea is to use a ligand whose intrinsic solubility is limited; for example, the use of polymers whose solubility is sensitive to pH, ionic strength, or temperature. The complex between the ligand and the protein is first formed and then the conditions of the solution are changed to decrease the solubility of the polymer. The advantage of this approach is that the target protein does not need to have more than one binding site. A potential disadvantage is that the conditions needed for precipitation of the polymer may also weaken the ligand-protein complex. Several heterogeneous macroligands have been used consisting of a variety of polymers. The two conditions that the polymer must satisfy are: 1) it must have reactive groups such as the affinity ligand that can be added and 2) it may change its solubility as the pH, temperature, or ionic strength of the solution are changed. For example, Schneider et al. (1981) attached p-aminobenzamidine to a soluble polyacrylamide and used the new compound to selectively precipitate trypsin from bovine pancreas extract. Although the recovery and yield were high, the low pH used to precipitate the complexes (pH 3) makes this approach non-practical for most proteins. Another polymer that has been used is poly-N-isopropylacrylamide (whose solubility depends on the ionic strength and temperature) conjugated with monoclonal antibodies. Copolymers of N-isopropylacrylamide with N-acryloxysuccinimide or glycidyl methacrylate have also been explored. Ligands such as p-aminobenzamidine were added to the copolymer to precipitate trypsin and monoclonal antibodies.

Lectins are an interesting group of natural affinity precipitation molecules. Lectins are a group of proteins that bind to various polysaccharides present in glycosylated proteins. Some lectins are very specific, but others are promiscuous as they recognize a group of different sugars. The use of lectins as an affinity precipitation ligand is potentially important to separate mixtures of glycoforms of the same protein. This heterogeneity may be found, for example, in human antibodies produced in plants whose glycosylation machinery is not turned off.

We have explored the use of several lectins, including Griffonia (Bandeiraea) Simplicifolia Lectin II and jacalin for the precipitation of a mixture of non-glycosylated and glycosylated antibodies by PEG. A complex interplay between the proteins and PEG was observed. After the precipitation was completed, the precipitate was resuspended in buffer and analyzed by high-pressure liquid chromatography (HPLC) in tandem with a multiple-angle light-scattering instrument. For systems containing jacalin, the precipitate consists of equal amounts of the glycosylated mixture and the lectin at low PEG concentrations. At higher PEG concentrations, the antibody completely precipitates out. On the contrary, the precipitate is leaner in the non-glycosylated antibody up to a PEG concentration of 7%. At higher PEG concentrations, the antibody quickly precipitates. There are some obvious differences in the precipitation pattern of the non-glycosylated protein and the glycosylated mixture and between the composition of the precipitate formed at low and high PEG concentrations (Figure 8.6). At the onset of precipitation (Figure 8.6 upper panels), there are not major differences in the composition of the non-glycosylated and glycosylated mixture precipitates. However, in the middle of the precipitation curve (Figure 8.6 lower panels) the precipitate of the non-glycosylated protein is richer in the lectin (peak at approximately 9 mL) than the corresponding precipitate of the glycosylated mixture. For both systems shown on the lower panels of Figure 8.6 there is a portion of the precipitate that cannot be resuspended. We speculate that this precipitate consists of aggregates of the lectin with the GlcNAc glycosylated antibody.

New Breakthroughs

In recent years, there has been an explosion of publications in which precipitation is used as a preparation step for liquid chromatography/mass spectroscopy analysis (Xu, Zhou, and Korfmacher 2005; Hou, Watters, and McLeod 2004; Yang et al. 2004) LC-ESI-MS (Souverain, Rudaz, and Veuthey 2004), for the preparation of samples and analysis of drug levels in plasma (Dams et al. 2003; Chen et al. 2005), and as a concentration step previous to two-dimensional electrophoresis (Zellner et al. 2005) in proteome work. The use of 96-well form has been common (Kitchen et al. 2003). The list of precipitants include acetonitrile (Polson et al. 2003), NaOH-$ZnSO_4$ (Rouan et al. 2001), trichloroacetic acid (Zellner et al. 2005), perchloric acid (Souverain, Rudaz, and Veuthey 2004), ethanol (Zellner et al. 2005), ammonium sulfate (Jiang, He, and Fountoulakis 2004), acetone (Jiang, He, and Fountoulakis 2004), acetone/trichloroacetic (Chen et al. 2005), and chloroform/methanol (Jiang, He, and Fountoulakis 2004). It is somehow surprising that the methods used are the very traditional methods that have been used in the last 100 years. Obviously, advances in protein purification will benefit not only the processing sector of the industry but also the exploratory work motivated by the genomic project.

Precipitation has also been used for the isolation of pharmaceutical-grade plasmid DNA. For example, Polytag Technology SA (Mannedorf, Switzerland) is commercializing a thermo-responsive affinity macroligand based on soluble poly N-alkylacrylamides containing a single stranded oligonucleotide whose sequence is complementary to a motif in the plasmid DNA. This approach seems to be promising, but, unfortunately, it does not provide any information about the cost of the macroligands. Still, by a smart selection of pH and temperature, the polymer may be recoverable, which is a must in affinity precipitation.

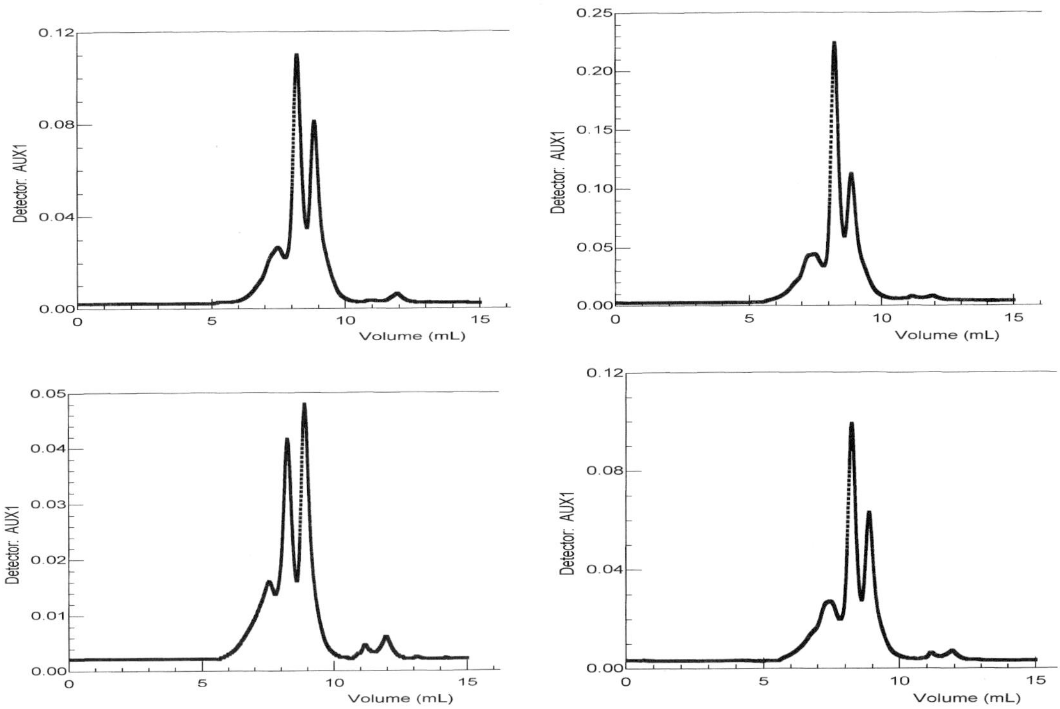

Figure 8.6. Size exclusion chromatograms for the resuspended precipitate of a glycosylated and non-glycosylated antibody in the presence of a lectin. Upper left panel: non-glycosylated protein at 16.6% precipitation; upper right panel: glycosylated protein at 29.6% precipitation; lower left panel: non-glycosylated protein at 46.3% precipitation; lower right panel: glycosylated protein at 65.3% precipitation. Note the differences in precipitate composition at different PEG concentrations for both the glycosylated and non-glycosylated proteins.

Experiments

In this section, we describe the precipitation of human serum albumin and its mixtures with hemoglobin upon the addition of PEG. These two proteins are chosen because the concentration of each can be monitored at different wavelengths, which facilitates the study of precipitation of the mixture.

Materials

Phosphate (pH 6), acetate (4), tris (pH 9), and carbonate (pH 10.7) 0.040M buffers. PEG of several molecular weights, human serum albumin, and human hemoglobin.

No specialized equipment is needed. Only a spectrophotometer, an orbital mixer and a centrifuge.

Methods

Preparation of Stock Solutions

A) PEG (50% *w/w)*

1. Weigh 25 g of PEG into a bottle.
2. Dissolve the PEG in 25 g of de-ionized water.

B) Protein stock solution.

1. Weigh 250 mg of HSA or human hemoglobin in two volumetric flasks. Weight 125 mg of HSA and 125 of human hemoglobin in a third volumetric flask.
2. Add buffer at a selected pH to complete 25 mL.

Experimental Procedures

Prepare eight tubes from 2% to 25% PEG.

1. Shake (stir) the stock solutions.
2. Place a 15 mL conical tube on a balance.
3. Add the appropriate amount (by weight) of PEG stock solution.
4. Add 1 mL of the protein stock solutions.
5. Add enough buffer of the desired pH to complete 5 g.
6. Measure and note the pH in each tube using a pH electrode.
7. Mix the contents of the test tube thoroughly. Mix them by hand first, then use a vortex mixer (be sure that the stock solution of PEG is actually mixed with the rest of the buffer) and finally a rotary shaker (20 minutes).
8. Centrifuge the tubes for 30 minutes at 10,000 x g to facilitate the separation between the supernatant and the sediment.
9. Remove 2.5 mL of the supernatant (without disturbing the precipitate) and measure absorbance at 280 nm or 450 nm in quartz cuvettes.
10. Blanks are prepared in the same fashion but the 1 mL of protein stock solution is replaced by 1 mL of buffer. Measure absorbance at 280 and 450.
11. Calculate protein concentration from $Ad^{sample} - Ad^{blank} = \varepsilon^{*}L^{*}C$.
12. Prepare a log S versus PEG concentration for both proteins and their mixtures. Evidence of co-precipitation should be found.

Summary

Protein precipitation is a powerful method for volume reduction in the early bioprocessing stages. Although its selectivity can be increased by using affinity ligands, its use as a fractionation method has been limited to a few proteins. Recent interest in the technique has been fostered by the development of the field of proteomics because of the need to concentrate protein samples before mass spectroscopy analysis.

Nomenclature

- c_p concentration of polymer (g/L)
- d_{max} maximum aggregate size (μm)
- D diffusion coefficient (cm^2/sec)
- I ionic strength (mol/L)
- K solubility product constant
- K_A rate constant (cm^3/sec molecule)
- K_S salting out constant (L/mol)
- l_e diameter of eddies (cm)
- M molecular weight of the particles (g/mol)
- M_0 molecular weight (g/mol)

N	number concentration (molecules/cm^3)
N_o	solute number concentration (molecules/cm^3)
n	Impeller speed
P/V	power dissipated per unit volume (g/cm sec^3)
r_p	radius of the protein (Å)
r_r	radius of the rod (Å)
S	solubility (mol/L)
S_o	intrinsic solubility (mol/L)
T	absolute temperature (K)
t	time (sec)
u	dipolar moment (C cm)
V_e	excluded by the protein (nm^3)
$\underline{V}$	partial molar volume of the polymer (cm^3/mol)

Greek Letters

α	collision effectiveness factor
ϕ	volume fraction
γ	shear rate (sec^{-1})
ρ	density (g/cm^3)
μ	viscosity (g/cm sec)
ε	dialectric constant of the solvent (dimensionless)

References

Agena, S. M., M. L. Pusey, and I. D. L. Bogle. 1999. Protein solubility modeling. *Biotechnology and Bioengineering* 64(2):144-50.

Anderson, C. O., J. F. Niesen, H. W. Blanch, and J. M. Prausnitz. 2000. Interactions of proteins in aqueous electrolyte solutions from fluorescence anisotropy and circular-dichroism measurements. *Biophysical Chemistry* 84(2):177-88.

Atha, D. H., and K. C. Ingham. 1981. Mechanism of precipitation of proteins by polyethylene glycols: analysis in terms of excluded volume. *Journal of Biological Chemistry* 256(23):12108-17.

Bell, D. J., M. Hoare, and P. Dunnill. 1983. The formation of protein precipitates and their centrifugal recovery. In *Advances in Biochemical Engineering/Biotechnology,* edited by A. Feichter, pp. 1-72. Berlin: Springer-Verlag.

Camp, T. R., and P. C. Stein. 1943. Velocity gradients and internal work in fluid motion. *Boston Society of Civil Engineers* 30:219.

Chang, B. H., and Y. C. Bae. 2003. Salting-out in the aqueous single-protein solution: the effect of shape factor. *Biophysical Chemistry* 104(2):523-33.

Chen, W., and J. C. Berg. 1993. The effect of polyelectrolyte dosage on floc formation in protein precipitation by polyelectrolytes. *Chemical Engineering Science* 48(10):1775-84.

Chen, W., S. Walker, and J. C. Berg. 1992. The mechanism of floc formation in protein precipitation by polyelectrolytes. *Chemical Engineering Science* 47(5):1039-45.

Chen, Y. Y., S. Y. Lin, Y. Y. Yeh, H-H. Hsiao, C-Y. Wu, S-T. Chen, and A. H. J. Wang. 2005. A modified protein precipitation procedure for efficient removal of albumin from serum. *Electrophoresis* 26(11):2117-27.

Clark, K. M., and C. E. Glatz. 1992. A binding model for the precipitation of proteins by carboxymethyl cellulose. *Chemical Engineering Science* 47(1):215-24.

Clark, K. M., and C. E. Glatz. 1990. Protein fractionation by precipitation with carboxymethyl cellulose. *ACS Symposium Series* 490:170-87.

Cohn, E. J. 1925. The physical chemistry of proteins. *Physiological Reviews* 5:349-437.

Cohn, E. J. 1950. Separation and preservation of the formed elements and of the plasma proteins of human blood. *Proceedings of the University Laboratory of Physiological Chemistry* 1:1-15.

Dao-Thi M. H., T. W. Hamelryck, F. Poortmans, T. A. Voelker, and M.J. Chrispeels. 1996. Crystallization of glycosylated and nonglycosylated phytohemagglutinin-L. *Proteins* 24(1):134-7.

Dainiak, M. B., V. A. Izumrudov, V. I. Muronetz, I. Y. Galaev, and B. Mattiasson. 1998-1999. Affinity precipitation of monoclonal antibodies by nonstoichiometric polyelectrolyte complexes. *Bioseparation* 7(4-5):231-40.

Dams, R., C. M. Murphy, R. E. Choo, W. E. Lambert, A. P. De Leenheer, and M. A. Huestis. 2003. LC-atmospheric pressure chemical ionization-MS/MS analysis of multiple illicit drugs, methadone, and their metabolites in oral fluid following protein precipitation. *Analytical Chemistry* 75(4):798-804.

Dixon, M., and E. C. Webb. 1961. Enzyme fractionation by salting out; a theoretical note. *Advances in Protein Chemistry* 16:197-219.

Galaev, I. Y., and B. Mattiasson. 1997. New methods for affinity purification of proteins. Affinity precipitation: a review. *Biochemistry (Moscow)* (Translation of *Biokhimiya (Moscow)*) 62(6):571-7.

Gillberg, L., and B. Tornell. 1976. Preparation of rapeseed protein isolates. Precipitation of rapeseed proteins in the presence of polyacids. *Journal of Food Science* 41(5):1070-5.

Guo, R., M. Guo, and G. Narsimhan. 1996. Thermodynamics of precipitation of globular proteins by nonionic polymers. *Industrial and Engineering Chemistry Research* 35(9):3015-26.

Guo, M., and G. Narsimhan. 1996. A model for the prediction of precipitation curves for globular proteins with nonionic polymers as the precipitant agent. *Separation Science and Technology* 31(13):1777-1804.

Harrison, R. G., P. Todd, S. R. Rudge, and D. P. Petrides. 2003. *Bioseparations Science and Engineering.* New York: Oxford University Press.

Hoare, M. 1982. Protein precipitation and precipitate ageing. Part II: Growth of protein precipitates during hindered settling of exposure to shear. *Transactions of the Institution of Chemical Engineering* 60:157-63.

Hou, W., J. W. Watters, and H. McLeod. 2004. Simple and rapid docetaxel assay in plasma by protein precipitation and high-performance liquid chromatography-tandem mass spectroscopy. *Journal of Chromatography B* 804(2):263-7.

Ingham, K. C. 1990. Precipitation of proteins with polyethylene glycol. *Methods in Enzymology* 182:301-6.

Inyang, U. E., and A. O. Iduh. 1996. Influence of pH and salt concentration on protein solubility, emulsifying and foaming properties of sesame protein concentrate. *Journal of the American Oil Chemists Society* 73(12):1663-7.

Jiang, J., and J. M. Prausnitz. 1999. Molecular thermodynamics for protein precipitation with a polyelectrolyte. *Journal of Physical Chemistry B* 103(26):5560-9.

Jiang, L., L. He, and M. Fountoulakis. 2004. Comparison of protein precipitation methods for sample preparation prior to proteomic analysis. *Journal of Chromatography A* 1023(2):317-20.

Jenkins, W. T. 1998. Three solutions of the protein solubility problem. *Protein Science* 7(2):376-82.

Juckes, I. R. M. 1971. Fractionation of proteins and viruses with polyethylene glycol. *Biochimica et Biophysica Acta* 229:535-46.

Kim, W-S., I. Hirasawa, and W-S. Kim. 2002. Aging characteristics of protein precipitates produced by polyelectrolyte precipitation in turbulently agitated reactor. *Chemical Engineering Science* 57(19):4077-85.

Kim, S. G., and Y. C. Bae. 2003. Salt-induced protein precipitation in aqueous solution: Single and binary protein systems. *Macromolecular Research* 11(1):53-61.

Kim, S. G., Y. C. Bae, and S. O. Ryu. 2000. Salt-induced protein precipitation in aqueous solution: the effect of pre-aggregation. *Korean Journal of Chemical Engineering* 17(6):638-42.

Kim, W.-S., I. Hirasawa, and W.-S. Kim. 2001. Effects of experimental conditions on the mechanism of particle aggregation in protein precipitation by polyelectrolytes with a high molecular weight. *Chemical Engineering Science* 56(23):6525-34.

Kirkwood, J. G. 1943. The theoretical interpretation of the properties of dipolar ions. In *Proteins, Amino Acids, and Peptides,* edited by E. J. Cohn and J. T. Edsall, pp. 276-303. New York: Reinhold.

Kitchen, C. J., A. Q. Wang, D. G. Musson, A. Y. Yang, and A. L. Fisher. 2003. A semi-automated 96-well protein precipitation method for the determination of montelukast in human plasma using high performance liquid chromatography/fluorescence detection. *Journal of Pharmaceutical and Biomedical Analysis* 31(4):647-54.

Kumar, A., I. Y. Galaev, and B. Mattiasson. 1998. Metal chelate affinity precipitation: a new approach to protein purification. *Bioseparation* 7(4/5):185-94.

Kumar, A., P-O. Wahlund, C. Kepka, I. Y. Galaev, and B. Mattiasson. 2003. Purification of histidine-tagged single-chain Fv-antibody fragments by metal chelate affinity precipitation using thermoresponsive copolymers. *Biotechnology and Bioengineering* 84(4):494-503.

Kirkland, R. A., and R. D. Tanner. 2005. Removing proteins from an aerated yeast fermentation by pulsing carbon dioxide: replacing salting-out as a method of protein precipitation. *Applied Biochemistry and Biotechnology* 121-124:685-93.

Larsson, P. O., and K. Mosbach. 1979. Affinity precipitation of enzymes. *FEBS Letters* 98:333-8.

Lovrien, R., C. Wu, and D. Matulis. 2000. Ligand-protein coprecipitative isolation by matrix stacking and entanglement. *Separation Science and Technology* 35(11):1795-1811.

Mahadevan, H., and C. K. Hall. 1992a. Experimental analysis of protein precipitation by polyethylene glycol and comparison with theory. *Fluid Phase Equilibria* 78:297-321.

Mahadevan, H., and C. K. Hall. 1992b. Theory of precipitation of protein mixtures by nonionic polymer. *AIChE Journal* 38(4):573-91.

Mahadevan, H., and C. K. Hall. 1990. Statistical-mechanical model of protein precipitation by nonionic polymer. *AIChE Journal* 36(10):1517-28.

Mansson, M. O., and P. O. Larsson. 1979. Recycling by a second enzyme of NAD covalently bound to alcohol dehydrogenase. *FEBS Letters* 98:309-13.

Marazban, S., I. Darrat, and B. L. Knutson. 2003. CO2 and fluorinated solvent-based technologies for protein microparticle precipitation from aqueous solutions. *Biotechnology Progress* 19(2): 448-54.

Mattison, K. W., I. J. Brittain, and P. L. Dubin. 1995. Protein-polyelectrolyte phase boundaries. *Biotechnology Progress* 11(6):632-7.

Mattiasson, B., M. B. Dainyak, and I. Y. Galaev. 1998. Smart polymers and protein purification. *Polymer-Plastics Technology and Engineering* 37(3):303-308.

Matulis, D., R. Lovrien, and T. Richardson. 1996. Coprecipitation of proteins with matrix ligands: scaleable protein isolation. *Journal of Molecular Recognition* 9(5-6):433-43.

Miekka, S. I., and K. C. Ingham. 1978. Influence of self-association of proteins on their precipitation by poly(ethylene) glycol. *Archives of Biochemistry and Biophysics* 191(2):525-36.

Miekka, S. I., and K. C. Ingham. 1980. Influence of hetero-association on the precipitation of proteins by poly(ethylene glycol). *Archives of Biochemistry and Biophysics* 203(2):630-41.

Mori, S., Y. Nakata, and H. Endo. 1994. Purification of rabbit C-reactive protein by affinity precipitation with thermosensitive polymer. *Protein Expression and Purification* 5(2):153-6.

Nath, S., C. S. Patrickios, and T. A. Hatton. 1995. Turbidimetric Titration Study of the Interaction of Proteins with Acrylic Polyampholytes. *Biotechnology Progress* 11(1):99-103.

Niederauer, M. Q., and C. E. Glatz. 1994. Model of the polyelectrolyte precipitation of genetically engineered enzymes possessing charged polypeptide tails. *Journal of Macromolecular Science, Pure and Applied Chemistry A* 31(1):127-53.

Ortin, A., J. A. Cebrian, and G. Johansson. 1992. Large scale extraction of alpha-lactalbumin and beta-lactoglobulin from bovine whey by precipitation with polyethylene glycol and partitioning in aqueous two-phase systems. *Preparative Biochemistry* 22(1):53-66.

Petenate, A. M., and C. E. Glatz. 1983. Isoelectric precipitation of soy protein. II. Kinetics of protein aggregate growth and breakage. *Biotechnology and Bioengineering* 26:3059-78.

Polson, C., P. Sarkar, B. Incledon, V. Raguvaran, and R. Grant. 2003. Optimization of protein precipitation based upon effectiveness of protein removal and ionization effect in liquid chromatography-tandem mass spectrometry. *Journal of Chromatography B* 785(2):263-75.

Robinson, M. W., D. G. I. Scott, P. A. Bacon, K. W. Walton, J. S. Coppock, and D. L. Scott. 1989. What proteins are present in polyethylene glycol precipitates from rheumatic sera? *Annals of the Rheumatic Diseases* 48(6):496-501.

Rouan, M. C., C. Buffet, L. Masson, F. Marfil, H. Humbert, and G. Maurer. 2001. Practice of solid-phase extraction and protein precipitation in the 96-well format combined with high-performance liquid chromatography-ultraviolet detection for the analysis of drugs in plasma and brain. *Journal of Chromatography B* 754(1):45-55.

Schneider, M., C. Guillot, and B. Lamy. 1981. Application to the isolation and purification of trypsin from bovine pancreas. *Annals of the New York Academy of Sciences* 369:257-63.

Scopes, R. K. 1982. *Protein Purification: Principles and Practice.* New York, NY: Springer-Verlag.

Simpson, R. J. 2004. Concentrating solutions of proteins. In *Purifying Proteins for Proteomics*, edited by R. J. Simpson, pp 701-32. Woodbury, NY: Cold Spring Harbor Laboratory Press.

Shieh, J-Y., and C. E. Glatz. 1994. Precipitation of proteins with polyelectrolytes: role of polymer molecular weight. In *Macromolecular Complexes in Chemistry and Biology,* edited by P. Dubin, pp. 273-84. Berlin, Germany: Springer.

Souverain, S., S. Rudaz, and J. L. Veuthey. 2004. Protein precipitation for the analysis of a drug cocktail in plasma by LC-ESI-MS. *Journal of Pharmaceutical and Biomedical Analysis* 35(4):913-20.

Thiering R., G. Hofland, N. Foster, G. J. Witkamp, and L. van De Wielen. 2001. Carbon dioxide induced soybean protein precipitation: protein fractionation, particle aggregation, and continuous operation. *Biotechnology Progress* 17(3):513-21.

Thrash, S. L., J. C. Otto, Jr., and T. L. Deits. 1991. Effect of divalent ions on protein precipitation with polyethylene glycol: mechanism of action and applications. *Protein Expression and Purification* 2(1):83-9.

Xu, X., Q. Zhou, and W. A. Korfmacher. 2005. Development of low volume plasma sample precipitation procedure for liquid chromatography/tandem mass spectrometry assays used for drug discovery applications. *Rapid Communications in Mass Spectroscopy* 19(15):2131-6.

Yang, L., N. Wu, R. P. Clement, and P. J. Rudewicz. 2004. Validation and application of a liquid chromatography-tandem mass spectrometric method for the determination of SCH 211803 in rat and monkey plasma using automated 96-well protein precipitation. *Journal of Chromatography B* 799(2):271-80.

Yu, J., H. Z. Liu, and J. Y. Chen. 2002. Selective precipitation of water-soluble proteins using designed polyelectrolyte. *Separation Science and Technology* 37(1):217-28.

Zhang, C., R. Lillie, J. Cotter, and D. Vaughn. 2005. Lysozyme purification from tobacco extract by polyelectrolyte precipitation. *Journal of Chromatography* 1069(1):107-12.

Zellner, M., W. Winkler, H. Hayden, M. Diestinger, M. Eliasen, B. Gesslbauer, I. Miller, M. Chang, A. Kungl, E. Roth, and R. Oehler. 2005. Quantitative validation of different protein precipitation methods in proteome analysis of blood platelets. *Electrophoresis* 26(12):2481-9.

Chapter 9
Polishing

Introduction

Industrial enzymes are normally commercialized in concentrated solutions. Therefore, the extract from the last separation unit does not need further processing. The situation is quite different for proteins for therapeutic use, because the potency of the drug must be preserved during storage and the drug must be made ready for delivery.

The most common preservation process is freeze drying (Carpenter et al. 1997; Sarciaux et al. 1999). In freeze drying, or lyophilization, a sample that has been quickly frozen is kept at approximately -70°C and a vacuum (<100 μm Hg) is applied. The water in the frozen sample sublimes and it is removed by the vacuum. It is accepted that the drying occurs in two phases: primary and secondary drying. During the primary drying phase loose water (the water that we associate with a solution) is removed. During the secondary phase, bound water or hydration water is removed. The powder is stored at low temperatures and must be reconstituted for use. Because most protein-based drugs are administered intravenously, the protein crystals are dissolved with water or saline in the ampoule used for drying and injected into the body. The reconstituted material must have the same properties as the original drug. To stabilize the product, additives are incorporated into the protein solution before drying. Common additives are sugars, surfactants, and biocompatible polymers.

An alternative to the freeze drying/reconstitution approach is to encapsulate the protein into a matrix so that it may be delivered orally. The approach in this case is quite different and is treated separately in this chapter.

Besides preserving and stabilizing the drug, the manufacturer must guarantee that the preparation is virus free. There are well-established techniques to implement virus removal. Still, the particular technique used must be in the "approved list" of regulatory agencies. Otherwise, the product will not be accepted. Virus removal has a large economic impact because it represents up to 40% of the total downstream processing cost.

The rest of this chapter presents the principles of freeze drying, followed by a discussion on standard procedures and current technology. Then, some other polishing methods, including spray drying and crystallization, are introduced. Third, the chapter covers various additives normally used in formulations (for a recent review on protein formulation, see Sellers and Maa 2005). Fourth, a discussion of formulations aimed at the oral administration of protein-based products ensues. This is followed by a detailed discussion of the acceptable practices for virus removal. Finally, we present a few laboratory experiments that can be used for the determination of process conditions.

Freeze Drying

A typical freeze dryer consists of a chamber with shelves where the filled vials are placed, a condenser, a refrigeration system, and a vacuum pump. In addition to cooling, heating, and vacuum controls, modern freeze dryers are equipped for clean in place and sterilize in place cycles, and have computerized cycle control and cycle monitoring functions.

Freeze drying consists of three different phases: freezing, primary drying phase, and secondary drying phase. The protocol followed to freeze the sample determines the quality of the solid obtained. During the primary drying phase, all the frozen water is removed. In the secondary drying phase, the water closely associated with the protein is removed. Still, it is not possible to remove all of the water of proteins and peptides during the secondary phase. In fact, there is a minimum amount of water that must remain with the product to improve its stability and reconstitution. This amount is product-dependent and needs to be determined experimentally. Two level factorial design (see Appendix) can be used for this purpose, one level would be the protein in solution and the other would be the preparation resuspended in buffer after freeze drying and additional drying by placing the sample in a P_2O_5 desiccator for 24 hours.

A standard freezing procedure is as follows. The freezing stage is started by placing the vials (that contain the solution to be dried) on the freeze drying shelves and the temperature is set at 5°C. Then, the shelf temperature is reduced to -5 to -10°C and kept at that value for 30 minutes. After that initial cooling, the freezing stage starts by suddenly lowering the temperature to its target. Sometimes annealing is included to help crystallize the bulking agent by increasing the temperature to near the glass transition temperature.

The freezing rate must be controlled because it affects the structure of the frozen matrix. Fast rates of freezing produce small crystals, which may form a fine pore structure and increase the resistance of water to sublimate. Slow rates of freezing make larger crystals, increasing sublimation rate and therefore shortening the primary drying time. Key parameters during the freezing stage are the minimum freezing temperature, the freezing rate, and the hold time at the freezing temperature. Normally, laboratory-scale freeze dryers can achieve lower temperatures and higher freezing rates. Therefore, scaling up may not be possible at the lab conditions.

The vials can be frozen from the bottom of the flask or by immersion. The first method is common in industry, whereas the second is found in laboratory freeze dryers. The cooling method affects the structure and morphology of the matrix and of the product and may cause or prevent the formation of a "skin" on the surface of the frozen product.

The product needs to be cooled below the lowest eutectic point and this temperature must be maintained during the primary drying phase. If the product does not crystallize, the drying needs to be done at temperatures below the glass transition temperature. The presence of amorphous phases may be caused by product formulation or process.

The additional freezing time to allow the crystallization of some excipients (mannitol or glycine) or to increase the size of the crystals is called annealing (equivalent to precipitate aging). When annealing is used to increase the size of crystals, the temperature of the shelves is increased slightly to allow small crystals to melt. The two variables to control are the annealing temperature and the annealing time. The temperature should be 10°C higher than the devitrification temperature. Of course, if the temperature is increased too much, eutectic melting will occur. After annealing, the temperature is decreased again to the primary drying temperature.

During the primary drying phase vacuum and heat are applied to the product to sublime the frozen water. Heating is necessary because sublimation of water will lower the temperature of the vial. It is important to have enough condenser capacity so that all the water is removed. Otherwise, if the temperature of the ice in the condenser is warmer than the product, water vapor will move toward the product and drying will stop. The control of the drying and heating rate is important. If drying is too fast, the water vapor may carry the product with it. If the product is heated too fast, it will melt or collapse, degrading the product and creating undesirable morphologies. During the primary stage of drying the temperature must be maintained below Tg′ (the glass transition temperature) to prevent collapse of the cake, which decreases sublimation rates and challenges the reconstitution of the product. Tg′ must be as high as possible (this is regulated by formulation). For example (Pikal 1996) if the sample needs to be maintained at -50°C, the pressure must be an unpractical 29.6 microns of Hg.

The modeling of freeze drying is quite complex and the reader is referred to the work of Liapis and Bruttini (2006) for a comprehensive treatment of freeze drying. Simple mass and energy balances allow gaining insight into the drying process and into processing strategies. The amount of water removed, m, per unit area, per unit time in the primary phase can be calculated from (Chang and Patro 2004; Pikal 1996),

$$\frac{dm}{dt} \propto \frac{(P_0 - P_c)}{(R_s + R_p)} \tag{9.1}$$

P_0 is the vapor pressure of ice at the product temperature, P_c is the pressure in the chamber, R_p is the product resistance (cm^2 mmHg hr g^{-1}), and R_p is the resistance of the stopper. The goal is to operate at the highest Δp. This is achieved by working under the highest possible temperature (large P_0) and lower vacuum (lowest P_c). A heat balance for the sublimation process (amount of heat removed = rate of heat for sublimation + rate of heat for temperature increase + losses) is (Chang and Patro 2004; Pikal 1996),

$$AK(P_0)(T_s - T_p) = \frac{(P_0(T) - P_c)}{(R_s + R_p)} \Delta H_s \tag{9.2}$$

In Equation 9.2, A is the heat transfer area of the vial, K is the heat transfer coefficient of the vial, T_s is the shelf temperature, T_p is the product temperature, and ΔH_s is the enthalpy of sublimation. This equation allows you to calculate the value of $P_0(T)$ needed for a particular temperature protocol. The chamber pressure is set at 10 to 30% of the vapor pressure of ice at the target product temperature. Chamber pressures of 100 to 200 x 10^{-3} mm Hg are preferred.

The end of primary drying is determined by monitoring either temperature or pressure. Thermodynamics "predicts" that as water sublimes the temperature will decrease. Therefore, the product's temperature remains lower than the shelf temperature during the primary drying phase. At the end of the primary phase, the temperature of the product will increase until reaching the shelf temperature. This temperature increase can be used to determine the end of the primary phase and the beginning of the secondary phase. Although scientifically sound, the determination of the primary drying phase by controlling the temperature of the sample is technically difficult, because the temperature of the sample needs to be monitored. Two ways of doing this are by increasing the temperature of the shelf (this is obvious) and by setting the temperature during primary drying equal to the temperature of secondary drying. Lower pressures, below the ice vapor pressure of the product, that allow sublimation compensate for the higher temperature. When sublimation ends, the product automatically heats up at the temperature of the shelf. A better option is to monitor the slowing down of pressure fluctuations in the chamber.

As soon as the primary drying has finished, the secondary drying phase is started by increasing the temperature without causing melting because there is no free water. The secondary phase is done

Table 9.1. Boiling Temperature at Several Pressures of Solvents Found in Freeze Drying (adapted from Freeze Drying 101, SP Industries Inc., Warminster, PA) with permission. Copyright by SP Industries, Inc.

Boiling temperature in °C at various pressures							
Compound	1 mm	10 mm	40 mm	100 mm	400 mm	760 mm	MP
Acetonitrile	–47	–16.3	7.7	27	62	81	–41
Methanol	–44	–16.2	5	21.2	49.9	64.7	–97.8
2 propanol	–26.1	2.4	23.8	39.5	67.8	82.5	–85.8
Water	–16	11.2	24	51	83	100	0
Acetic acid	–17.2	17.5	43	63	99	118	16.7
Benzyl alcohol	58	92.6	119.8	141.7	183	204.7	–15.3
Ethanol	–31.3	–2.3	19	34.9	63.5	78.4	–112
Tert-butyl alcohol	–20.4	5.5	24.5	39.8	68	82.9	25.3

MP = melting point

at the highest possible vacuum and continues until the desired amount of water is removed. The upper limit in temperature rise is the glass transition point of the product, because products containing 10% or less of water can still collapse if the glass transition temperature is exceeded. At very low pressures, heat is transferred by conduction from the shelf through the bottom of the container. Because this process is not very efficient (glass is an insulator) an inert gas such as N_2 may be introduced into the drying chamber to improve heat transfer. Following this approach reduces energy cost because it increases the drying rate. Secondary drying is controlled by knowing the residual moisture level as a function of shelf temperature and holding time. The shelf temperature is set from 25 to 30°C for proteins.

Solvents may be incorporated into the solution to be dried to improve the solubility of product. For example, pancreatin is dried in 50% isopropanol and blood is fractionated using cold ethanol. Some common solvents that can be used are acetonitrile, methanol, ethanol, propanol, tert-butyl-alcohol, acetone, cyclohexane, and dimethyl sulfoxide (DMSO).

The use of solvents always increases the risk of accidents. Therefore, it is extremely important to know the flammability range of the chosen solvent (Table 9.1). For example, acetonitrile boils at 81°C, melts at -48°C, auto-ignites at 524°C, has a flash point of 5.6°C and a vapor pressure of 72.8 mm Hg at 20°C, a vapor density of 1.41 times heavier than air, and a density of 785 (so it is lighter than water). Its upper explosive level is 16% and its lower explosive level 4.4%. Therefore, if the material is above 5.6°C and between 4.4 and 16% by weight mixed with air, the material can catch fire if exposed to spark or flame. Acetonitrile/water solutions are considered flammable down to the 15 to 18% level. The trick in freeze drying materials with a solvent such as isopropanol, which creates a viscous fluid at the bottom of the condenser, is to collect the solvent in a low temperature chilled catch pot connected to the condenser drain and drain the catch pot from the system at the right time. In this way, one can keep the material from entering the vacuum pump by trapping it first in the condenser catch pot arrangement. Vacuum dryers utilize water-cooled condensers for this purpose. At the end of the liquid collection cycle, the isopropanol will be almost gone, leaving mostly water. The condenser has a low enough temperature to condense and solidify the remaining subliming vapors, which are now trapped at the appropriate temperature.

Crystallization

A key in developing a protein crystallization protocol is the ability to predict how the protein phase diagram will look and how it is affected by pH, ionic strength, additives, and concentration and

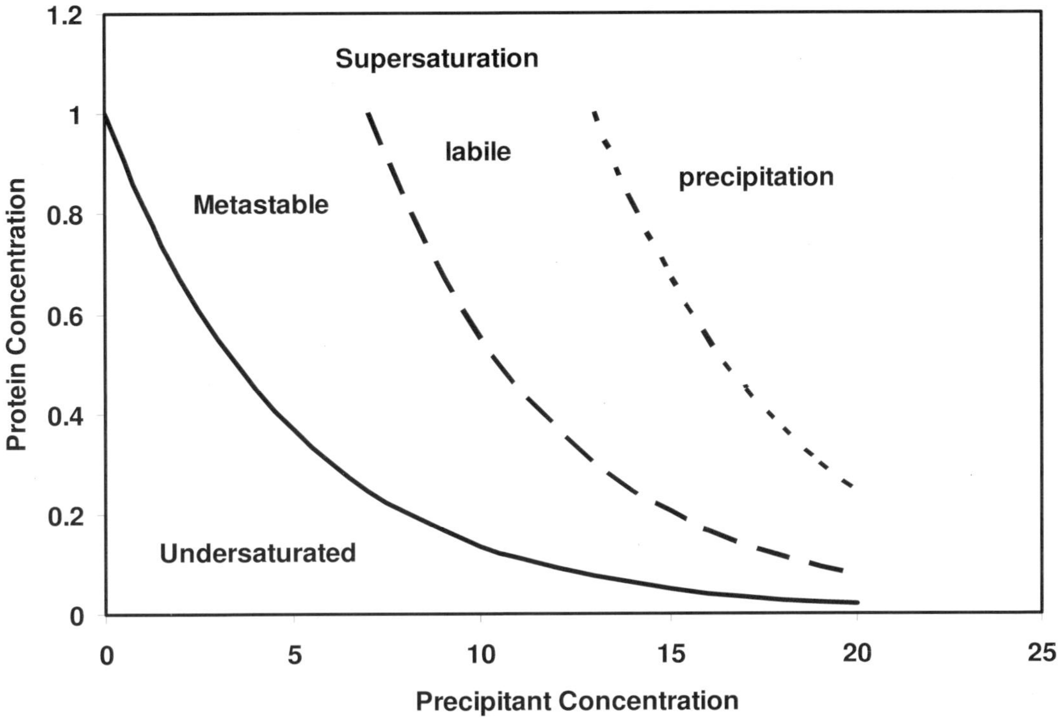

Figure 9.1. Schematic of a generic phase diagram for a protein/precipitant mixture (adapted from Asherie, N. 2004. Protein crystallization and phase diagrams. *Methods* 34:266-72. Copyright Elsevier Inc.

chemical characteristics of the buffer. The first stage in the formation of a crystal is nucleation. In theory, nucleation will happen spontaneously from a supersaturated solution. In practice, this will occur only when the solubility of the protein is exceeded by a factor of three or more (Chernov 1997). If supersaturation is too high, disordered structures such as aggregates or precipitates will form. It is important to stay away from the precipitation zone because aggregates and precipitates have a faster kinetics than crystals. Figure 9.1 shows a phase diagram for an arbitrary protein. The solid line separates the diagram into two regions: supersaturated and undersaturated. In between the saturation curve and the metastable line, the number of nuclei is so small that crystallization will not occur. In between the metastable and the labile line, crystallization will occur spontaneously. To the right of the labile line precipitates will form. Depending on the conditions, two liquid phases may form, a phase rich in water and the other rich in protein. This can be good, because sometimes crystals are formed in the protein-rich phase. Asherie (2004) argues that the liquid/liquid phase transition is inherently metastable (it will move to crystals), but this conclusion is arguable, because the liquid/liquid phase transition is thermodynamically well-characterized and stable. After normalizing the data it has been shown that the solubility and coexistence (LL) curves for two proteins, γD crystallins and hen lysozyme, are nearly identical (Figure 9.2). It has been argued that this may be a universal phenomenon for all globular proteins (Asherie 2004). However, it seems to be risky to make such a generalization with only two data points.

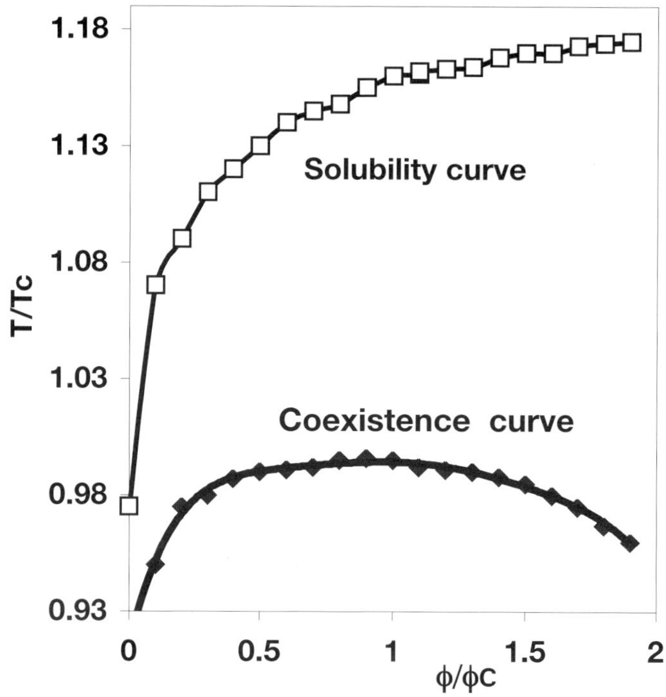

Figure 9.2 The phase diagram of lysozyme and bovine γD crystallin. T is the temperature and ϕ is the protein's volume fraction. Both axis are normalized by the critical temperature and volume fraction,ϕ_C (adapted from Asherie, N. 2004. Protein crystallization and phase diagrams. *Methods* 34:266-72. Copyright Elsevier Inc.

Spray Drying

In addition to freeze drying, spray drying (Prinn, Costantino, and Tracy 2002; Maa, Nguyen, and Andya 1998; Elkordy, Forbes, and Barry 2002), supercritical drying (Moshashaee et al. 2003), vacuum drying (Sharma and Kalonia 2004; Mattern et al. 1997), and a Bubble Drier® (Sievers et al. 2001) have been used to produce powders. Only spray drying is discussed here. In spray drying, a liquid is atomized (small droplets) at the top of a chamber. Dry hot air is pumped from the bottom of the chamber. Moisture is removed as the droplets fall from top to bottom. The dry solids are separated from the remaining liquids by cyclones. Figure 9.3 shows a schematic of a spray drying system. Spray drying is commonly used in the milk industry, which has contributed greatly to the development of the technique. The problem with the use of spray drying to dry protein preparations is that it requires high temperatures to function. Besides, although air is normally used to dry food products, nitrogen is needed to dry proteins because, otherwise, they oxidize. Still there are some reports of spray drying of proteins in which the integrity of the proteins has been maintained. The use of spray drying is attractive because it produces very fine particles, which can be used in atomizers for the delivery of therapeutic proteins directly into the lungs.

Although spray drying is a well-established technique in other industries, the protein pharmaceuticals sector opens new challenges. As stated above, the two main concerns are the thermal labiality of the proteins and the risk of oxidation if N_2 is not used. A third problem is the variety of mor-

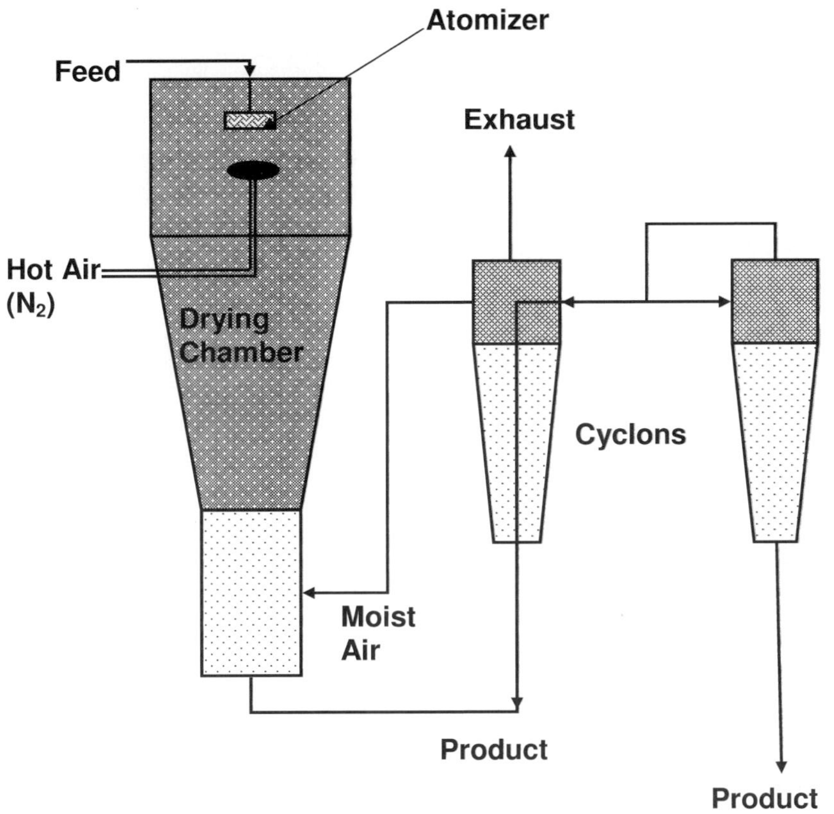

Figure 9.3. Schematic of a typical spray dryer.

phologies that can be obtained depending on operational conditions. Figure 9.4 (Prinn, Costantino, and Tracy 2002) shows four different types of morphologies of bovine serum albumin (BSA) dried in a laboratory scale spray dryer. High protein concentrations and high liquid feed produces the best morphology (spherical shapes). The other morphologies were obtained at low protein concentrations and/or low flow-rates.

Protein Formulation

Protein pharmaceuticals are commercialized in liquid or solid forms. The obvious advantage of liquid formulations over solid ones is the ease of application. On the other hand, solids are more stable. Formulation is the conditioning of the preparation, either liquid or solid, to increase the stability of the protein. For many years, liquid formulations have been preferred, but lately, pharmaceutical companies have been choosing solid formulations as default because solid formulations are more stable and they can be stored for a longer time (more than two years) (Jameel and Patro 2005). The U.S. Food and Drug Administration (FDA) requires less than 10% deterioration after storage for 2 years under the specified storage conditions. Kelner, Wang, and Wang (2000) recommend following a three-stage process to develop the right formulation: preformulation, formulation, and sta-

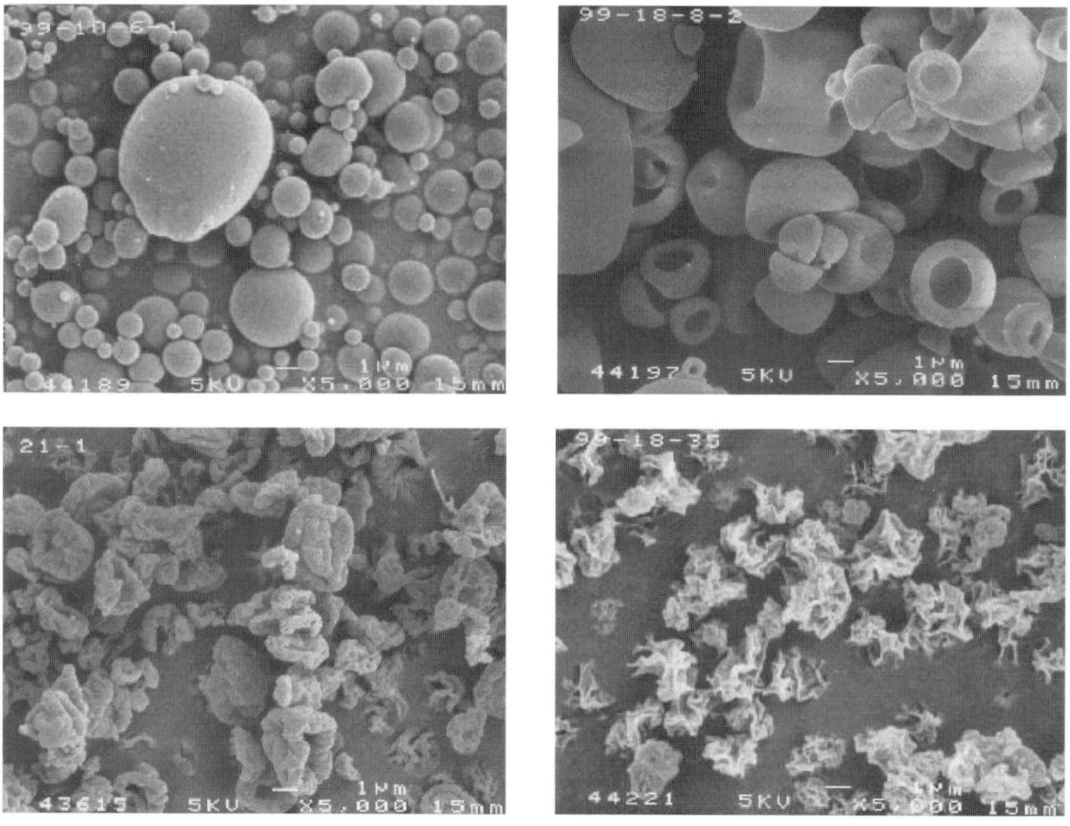

Figure 9.4. Morphologies of four spray drying preparations of BSA (reproduced from Prinn, K. B., H. R. Costantino, and M. Tracy. 2002. Statistical modeling of protein spray drying at the lab scale. *AAPS PharmaSciTech* 3(1):1-8. Copyright by the American Association of Pharmaceutical Scientists.

bility studies. Protein preformulation consists of determining the stability of the protein and the most important decomposition pathways under different solution conditions (pH, ionic strength). The most common way to determine the stability of a protein preparation is by measuring T_m (see Chapter 4), which is not necessarily good, because T_m may not change much and the protein still loses biological activity. A better way would be to measure its biological activity. The most common stability problems are deamidation, oxidation, and aggregation (Table 9.2) (Chang and Hershenson 2002). Because of market requirements, the stability studies are done under accelerated conditions and then extrapolated, which is always risky. Table 9.3 (Chang and Hershenson 2002) shows typical conditions for accelerated testing of most common protein degradation pathways.

Freeze drying is, by far, the most common process to obtain solid preparations. During the lyophilization process, the protein is subjected to several stresses including low temperature, high concentrations, exposure to interfaces (ice-water), pH changes, annealing, and dehydration. Freezing of a protein solution increases the concentration of all the components of the solution. Both an increase in salts and protein concentrations may induce the formation of irreversible aggregates. Denaturation of proteins at ice-water interface is well documented (Chang, Kendrick, and Carpenter 1996). This problem is unique to freeze drying because to obtain a high-quality product the freezing process is done so that many nucleation sites are generated and thus, the interfacial area

Table 9.2. Main Protein Degradation Pathways (adapted from Chang, B. S., and Hershenson, S. 2002. Practical approaches to protein formulation development. *Pharmaceutical Biotechnology* 13:1-25. Copyright by Wiley-VCH Verlag GmbH & Co.; with permission.

Problems	Potential Causes	Possible Solutions
Non-covalent aggregation	Solubility, structural changes, heat, shear, surface, denaturants, impurities	pH, ionic additives, amino acids, surfactants, protein concentration, raw material purity
Covalent aggregation	Disulfide scrambling, other unknown mechanisms	pH, inhibit non-covalent aggregation
Deamidation	pH < 5.0 or pH > 6.0	pH optimization
Cyclic imide	pH around 5	pH optimization
Cleavages	Protease impurity, other unknown mechanisms	pH, product purity, inhibitors
Oxidation	Active oxygen species, free radicals, metals, light, impurity	Excipient purity, free-radical scavenger, active oxygen scavengers
Surface denaturation, adsorption	Low protein concentration, specific affinity, protein hydrophobicity	Surfactants, protein concentration, pH

Table 9.3. Potential Stresses Affecting Proteins During Freeze Drying. Accelerated Stress Conditions are Shown in Column Two (adapted from Chang, B. S., and Hershenson, S. 2002. Practical approaches to protein formulation development. *Pharmaceutical Biotechnology* 13:1-25. Copyright by Wiley-VCH Verlag GmbH & Co.; with permission.

Stresses	Routine Ranges	Practical Applications	Problems to Monitor
Temperature	0-50°C	Storage, shipping, handling, delivery	Precipitation, aggregation, recovery loss, solubility, increased reaction rates for all degradations
Light	>1.2 million lux hrs illumination	Light exposure, container, package	Oxidation, cleavage
Freezing	Multiple freeze thaw, liquid Oxidation, cleavageN_2 freeze	Frozen storage, accidental freezing, lyophilization	Precipitation, aggregation, pH change, crystallization of excipients
Oxidation	Oxygen purge, peroxide spike	Storage, excipient stability, impurity	Oxidations, inactivation
Humidity	0-100% relative humidity	Storage, container integrity, powder	Moisture content, moisture related degradations
Mechanical	Agitation, shear stress ($3000 s^{-1}$)	Manufacturing, filling, shipping, handling, delivery	Precipitation, aggregation, recovery loss
Other denaturants		Impurities, pH denaturing excipients	Precipitation, aggregation, recovery loss, structural changes.

is very large. High protein concentrations reduce losses, because if the concentration is high, only a small fraction of the protein is exposed to the water/ice interface. Buffers control the pH of a solution by maintaining chemical equilibrium between the acid and the base of the buffer. If one of the two species freezes first, the pH of the buffer may change drastically. Phosphate is a buffer commonly used in protein formulations that has this problem because Na_2PO_4 crystallizes more quickly than NaH_2PO_4. Dehydration is an obvious problem in freeze drying. Some recommend stopping the drying before the water of hydration is eliminated. However, most freeze drying processes leave less than 10% water, which means that some of the water of hydration has been removed.

Table 9.4. Freeze Drying Additives and Their Collapse Temperature (Freeze Drying 101, SP Industries Inc., Warminster PA; adapted with permission from SP Industries Inc.). Copyright by SP Industries.

Compound	Collapse Temperature (°C)	
Dextran	−9	Mackenzie (1976)
Sucrose (15%)	−22.5	Bellows and King (1972)
PVP	−23	Mackenzie (1976)
Sucrose (45%)	−29	Bellows and King (1972)
Lactose	−32	Bellows and King (1972)
Inositol	−27	Pikal et al. (1983)
Sorbitol	−57	Mackenzie (1966)
10% NaCl-KCl-H2O	−24	Mackenzie (1966)
5% NaCl-Manitol-H2O	−50	Ito (1971)
5% Mannitol, Sorbitol, H2O	−34	Ito (1971)
Lactose (1.3% residue moisture)	60°C glass transition of very dry lactose	Ito (1971)
Sucrose (8% residual moisture)	60°C glass transition of very dry sucrose	Bonelli et al. (1997)
Fickle	−19	Bonelli et al. (1997)
Ovalbumin	−10	DHEW Publication
PVP (k-30)	−23	DHEW Publication
Sephadex G-200	−10	DHEW Publication
Sucrose	−32	DHEW Publication
1.5 Sucrose 3.5% Mannitol	−21	DHEW Publication

Excipients are used to stabilize the protein against stresses caused by freezing and drying and as bulking agents. They are also important for long-term (storage) stability. The excipient may be stabilizers, other proteins, amino acids, polymers, sodium chloride and other salts, surfactants, covalent modifications of the target protein, and bulking additives. Most times a combination of the latter is found to be the optimal excipient. Table 9.4 shows a collection of excipients that have been used for a series of proteins. Excipients have been traditionally chosen by trial and error (Liu, Wang, and Nail 2005). The most common stabilizers are non-reducing sugars (reducing sugars must be avoided because of the risk of them participating in Millar reactions) at a minimum concentration of 5% (Kelner, Wang, and Wang 2000). Sugars are thought to protect the protein by either the formation of an amorphous glass or by the formation of hydrogen bonds with the protein as the water of hydration is removed. Another common stabilizer is serum albumin. However, there is a trend to stop using it because of the risk of introducing pathogens. Other stabilizers are amino acids, polymers, salts, and surfactants. Some amino acids may be good in stabilizing one protein but then destabilize others. Therefore, the addition of amino acids must be carefully tested. The most common polymer used is hydroxypropyl-β-cyclodextrin. The idea is that the inner hydrophobic core of the molecule interacts with hydrophobic patches on the protein surface inhibiting aggregation. Salts weaken protein-protein electrostatic interactions and surfactants minimize adsorption of the proteins at the glass surface (Chang, Kendrick, and Carpenter 1996). Buffers and salts are chosen based on solubility and stability considerations. The most common buffers used are acetate, citrate, glycine, histidine, and phosphate. For pH-sensitive proteins, phosphate must be avoided because of its change in pH upon freezing as discussed previously. Another consideration in buffer selection is that the optimum pH for protein stability may be far from physiologic values; a trade-off is then needed. Finally, proteins can also be stabilized by structural modifications. The most common modification of proteins to increase their stability is to attach PEG covalently to lysine residues; this is known as pegylation.

The bulking agent is a crystalline chemical to improve the physical characteristics of the solid. The most common bulking agent is glycine. Glycine forms crystals, and therefore, increases the water/freeze interface; thus when used alone, it has either a negative or no effect in recovery. On the opposite side, the presence of a crystalline substance produces a stronger matrix (speeding drying) and yields a product with better morphologic properties.

Two compounds found frequently together as excipients are sucrose (amorphous protectant) and glycine (a crystallizing agent). Then, it is important to understand the effect of these mixtures on the recovery of enzymatic activity after freeze drying to develop formulations that maximize that recovery. Liu, Wang, and Nail (2005) found that different combinations of both excipients offer different protection in the freeze drying of various enzymes. The total concentration of Gly and sucrose in this study was 20 mg/mL, the protein concentration was 25 µg/mL, and the NaCl concentration of 30 mM/L. The highest activity was recovered when only 10 mg/mL of sucrose was used and the lowest recovery was found when 10 mg/mL sucrose was combined with NaCl and 290 mM of Gly. The activity recovery of G-6-PDH and ADH deteriorated quickly as the concentration of glycine was increased. This is expected because of the crystalline structure of glycine upon freezing. The ratio sucrose to Gly seems to have no effect in freeze thawing. They also observed that the addition of a surfactant like Tween 90 increases recovery independently of the ratio between Gly and sucrose. Finally, they found that the faster the freezing rate, the higher the number of aggregates formed. This supports the idea that denaturation is happening at the water/ice interface, because when the freeze drying rate is fast, small crystals are formed that offer more interfacial area. The idea is that surfactants may compete with the protein for the interface. The addition of a surfactant eliminates the deleterious effect of the addition of glycine.

Knowledge of the phase diagram of the combination of excipients is important to make an informed decision in selecting optimum concentrations and ratios. The phase diagrams can be very complex and the solid phases need to be closely examined to determine their morphologies. Shalaev and Kanev (1994) determined the phase diagram for mixtures of glycine and sucrose in water-rich solutions and in low moisture compositions. Water and glycine form a system with a simple eutectic at -3.6°C with a concentration of glycine of 11.34%. Water solutions of sucrose form a eutectic involving incongruently melting hydrates (Kanev et al. 1989). They observed the following transitions in aqueous mixtures of glycine and sucrose as the temperature was varied: a glass transition at low-moisture characterized by T_g=f(R,W), in which R is the ratio between glycine and sucrose and W is the water content, a glass transition at high water content characterized by T_g'=f(W), a mechanical glass transition (shrinkage) characterized by T_{am}=f(R,W), a mechanical glass transition at high water content characterized by T_{am}', a devitrification phase transition characterized by T_{D2}=f(R,W,α) in the low moisture solutions (α is the heating rate in °C/minute), and a devitrification phase transition in the low moisture regime characterized by T_{D1}=f(R,α). Their experimental data was fitted with the following polynomials,

$$T_g = 58.2 - 9.14608W + 0.256997W^2 \\ -3.15358 * 10^{-3} W^3 - 0.363185R(W) - 17.6481R \quad 9.3$$

For R=0 to 1.8 and W from 0 to 20%

$$T_{AM} = 73.4 - 9.14608W + 0.256997W^2 \\ - 3.15358 * 10^{-3} W^3 - 0.363185R * W - 17.6481R \quad 9.4$$

For R=0 to 1.8 and W from 0 to 20%

$$T_g' = -49.2 - 27/3205 * R + 6.11668R^2 \quad 9.5$$

For R from 0 to 1.8

$$T_{AM}' = -35.4 - 28.7326R + 20.0681R^2 - 5.25873R^3 \quad 9.6$$

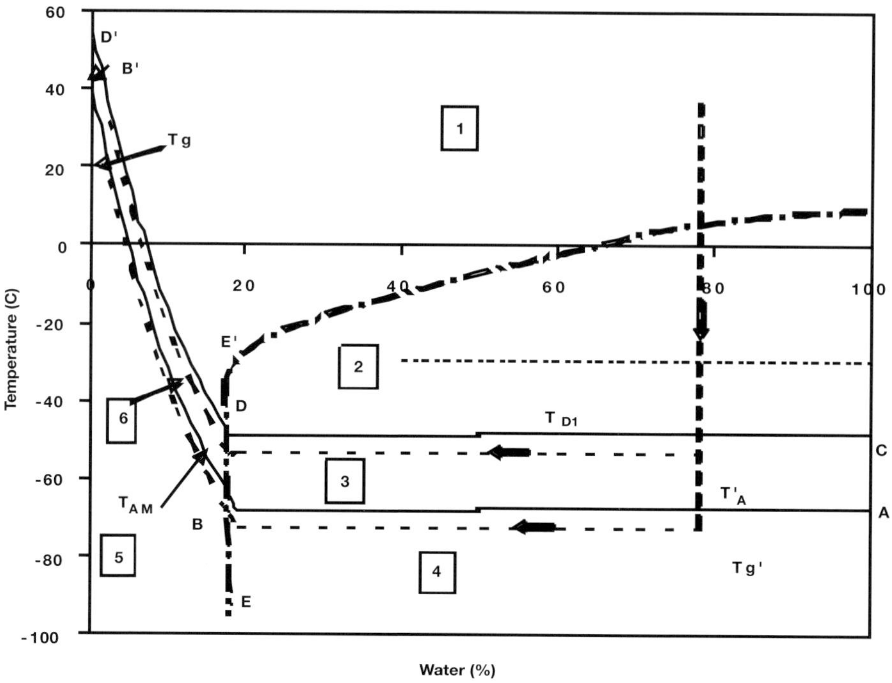

Figure 9.5 Phase diagram for glycine/sucrose/water mixtures. The glycine/sucrose composition is constant. The block arrows show two freeze drying paths, below and above the glass transition temperature (redrawn from Shalaev, E. Y., and A. N. Kanev. 1994. Study of the solid-liquid state diagram of the water-glycine-sucrose system. *Cryobiology* 31:347-82. Copyright by Elsevier Inc.

For R from 0 to 1.8

$$T_{D1} = -19.6 - 11.3278R + 0.0582\alpha - 0.87965\alpha^2 \qquad 9.7$$

For R=0.77 to 3 and α from 0.28 to 3.3.

These equations can be used to generate phase diagrams for an arbitrary value of R. Figure 9.5 shows a phase diagram for R=0.8 and a cooling rate of 0.5°C per minute. The solid lines divide the plane into six regions. The line EE′ divides the plane into two regions: an amorphous phase and an ice + amorphous phase. Region 1 corresponds to a solution. Region 2 corresponds to a solution at equilibrium with ice. Region 3 corresponds to an amorphous phase plus ice, which has the mechanical properties of a solid; region 4 corresponds to ice plus glass, region 5 is glass, and region 6 is an amorphous phase with the mechanical properties of a solid. The dashed lines correspond to possible freeze drying pathways. The horizontal line corresponds to sublimation and the curved line corresponds to secondary drying.

Other Delivery Systems

Most protein therapeutics are injectable because they cannot be orally administrated; they do not survive the extreme acidity of the stomach or macrophagocytosis in the lungs. Considering the problems associated with injectable drugs, there is a considerable interest in developing alternative delivery systems. One protein that is of particular concern is insulin because of the frequency at which it needs to be injected. One route that has been explored is to encapsulate the protein into a biodegradable polymer so that the protein can reach the intestine. For example, Hoffart et al. (2002) found that polymeric nanoparticles made by a mixture of biodegradable polyester and a polycationic polymer were able to deliver heparin to rabbits. Following this idea, Attivi et al. (2005) developed a delivery system based on water in oil in water emulsions using poly-ε-caprolatone (biodegradable) and a positively-charged non-biodegradable polymer. They developed an experimental design (see Appendix) to study the effect of the ratio of polymers, volume, and pH of the aqueous solution of polyvinyl alcohol. The experimental design yielded a few attractive formulations. They found that when the amount of insulin entrapped was 25 IU per 100 mg of polymer and the particle size was 350 nm, the amount of insulin released was 4.8 IU per 100 mg of polymer after 7 hours. The release consists of an initial burst (5 to 15% of the total release) followed by slow release during 7 hours. The system was not checked *in vivo*.

Another viable route is to deliver a protein via inhalers. Two technologies have been explored: liquid and solid aerosols. In spite of abundant research on this topic (see Krishnamurthy, 1999), the problem of maintaining appropriate protein bio-availability persists. One intrinsic problem of liquid aerosols is the denaturation of the protein by the propellant. A possible solution to this is the use of tetrafluorethane. Other problems are water evaporation during delivery, adsorption of the protein to the container, changes in the pH upon storage, and short shelve life. Dry powder (generated by spray drying) aerosols resolved some of these problems. Because the stability of the protein depends on the residual water, secondary drying steps have been proposed to increase protein stability. The drawback is the fabrication of a drug that is more difficult to be absorbed by the lungs. It is important that the preparation survives macrophages (they naturally eliminate particles in the lungs, proteases, and free radicals) when it is inhaled. Several remedies have been developed to protect the protein from destruction; for example, pegylation of the protein, encapsulation of the drug, and addition of lung-tissue mimics to the formulation.

Pegylated protein formulations have shown better survival rates than unmodified proteins. This is an interesting alternative because pegylation is a well-developed technique. Another alternative that has shown some success is the delivery of insulin as a dry aerosol that has been encapsulated in biodegradable polymers. Particles larger than 6 μm seem to be able to survive macrophages because of their low mass densities. Finally, hypothesizing that macrophages would not destroy anything that looks like lung tissue, dipalmitoylphosphatidylcholine has been added to the preparation because it is a natural surfactant of the lungs.

Virus Removal and Inactivation

Viruses may be present in a final protein preparation because the raw material was already contaminated (such as when human plasma is used as raw material) or because contamination took place during the process. It has been claimed that virus removal may represent up to 40% of the total separations cost. This is somewhat misleading because some operations used for separations/purification also play the role of virus removal steps. Still, virus removal is an intrinsic part of any bioseparation strategy.

Here we focus on virus removal in human plasma, because the problem may be particularly acute and because it is treated in great detail by the FDA. However, most of the guidelines will apply to any process. The viral control in the process of human plasma (raw material) starts with a careful control of the source. The FDA recommends careful screening of donors, and maintains donors' registries to screen blood from unsuitable donors, tests blood obtained from donors for disease, quarantines blood until the units are declared safe, and fully investigates incidents to correct deficiencies.

The viruses of most concern in human plasma are hepatitis B and C, human immunodeficiency viruses types 1 and 2 and human parvovirus B19. Hepatitis A and B19 are small (15-30 nm), non-enveloped viruses that are very resistant to heat and chemical treatment and they are difficult to eliminate by ultrafiltration. Although the diseases caused by these two viruses are less serious they may be dangerous. Other viruses like human T-lymphotropic virus and cytomegalovirus are not considered risky because they are not present in plasma and because they do not tolerate refrigeration and freezing and thawing. The most clinically significant viruses, HIV, HBV, and HCV, are susceptible to inactivation methods based on chemical treatments or heat because they are enveloped viruses. These viruses are large and thus, they can be efficiently removed by ultrafiltration using the proper membrane. Viruses responsible for prion diseases, such as transmissible spongiform encephalopathies (mad cow disease) are not easily detectable (the FDA does not recognize an effective method) and no effective method for their removal is known. This, of course, is the subject of current investigation.

Viruses can be removed or inactivated but whatever route is chosen must be validated according to FDA guidelines. Therefore, different aspects of validation are discussed in the last section of this chapter. Common removal processes are filtration, chromatography (affinity and ion exchange), and polyethylene glycol fractionation, whereas common inactivation processes are heating, addition of solvent, addition of detergents, irradiation, photoinactivation, and the addition of other chemicals. Viral inactivation techniques must destroy at least one viral component essential to replicate. For example, photosensitizing techniques use light-activated dyes that are irradiated and cause the dyes to convert to molecules that alter DNA or membrane lipoproteins, exposing the preparation at high temperatures denatures viral proteins and nucleic acids, rendering viruses incapable of replication; radiation destroys viral nucleic acids, and detergents destroy the viral envelope in lipid-enveloped viruses.

Removal Methods

Thc following rcmoval methods are accepted by the FDA:

a) Disposable ultrafiltration membranes operated in cross-flow mode. The company must prove that the pressure and flow rates used during validation are being followed. Filtration of macromolecules (such as dextran) must be used to test the integrity of the membrane post-production because forward flow and bubble point methods are not applicable.
b) Purification methods such as affinity and ion exchange chromatography that are routinely used during the purification process may be validated for viral removal. The company must prove that flow rates, column height, and protein load parameters used during validation are followed consistently. The ability of a column to remove viruses may deteriorate upon repeated uses. Therefore, if the manufacturer has established a number of uses for a particular resin, the manufacturer must not exceed that number during production. Consistency must be established for the regeneration procedures used in between batches.

Inactivation Methods

Heat and solvent/detergent treatments are two inactivation methods recommended by the FDA. Heat treatment is relatively simple but there are a few guidelines that must be followed. First, the heat treatment process must be validated by the manufacturer if a standard procedure is not already recognized and established by the regulatory agency. Second, the manufacturer must prove that the temperature and time of inactivation are met consistently. This is extremely important, because the efficiency of the heat inactivation process depends on both the temperature reached and the time of exposure. Third, the heat treatment process must be done with adequate equipment. Fourth, if the sterilization process does not reach completion because of some process failure, the material must not be re-treated unless the re-treatment has been validated and the manufacturer has proved that re-heating does not affect the quality of the product. A good example of the use of heat treatment in the process of human plasma is the viral inactivation in the production of human albumin and plasma protein fraction, which must be heated in a bulk tank, according to FDA guidelines, at 60°C ± 0.5°C for 10 to 11 continuous hours. Some other products that are heat treated are antihemophilic factor (human) and intravenous immune globulin (human).

The use of solvents and/or detergents to inactivate viruses is more cumbersome than heat inactivation because the solvent or the detergent must be removed later on. In this case, it is required that the same concentration of solvent or detergent and the time of exposure are consistently used. After the viruses have been inactivated, the removal of the detergent or solvent must be validated too. The manufacturer must prove (and document) that the removal has been done during the process to the extent claimed during validation. Examples of solvent and detergent treatment are found in the production of intravenous immune globulin (human).

Validation

Validation is the demonstration that a particular viral removal or inactivation process is as effective as claimed. In some cases, the removal or inactivation process is stated in the regulations; thus the manufacturer does not need to prove its efficiency but rather to prove that the operating parameters are consistently met. Otherwise, it is the responsibility of the manufacturer to validate the effectiveness of the method, as well as to demonstrate that the validated operating parameters are met consistently during production. Validation studies are always done when a new product is developed but also if a process is modified. In the former case, specific steps for viral removal are combined with the ability of process equipment to remove viruses. In the later case, the manufacturer must prove that the ability of the process to remove viruses has not been affected by the modification. On some occasions, a manufacturer may discover that an existent process unit has the ability to remove or inactivate viruses. In this case, it is important to limit the validation process to prove that the step is efficient without any attempt to optimize it, as its optimization may change the product and thus, the manufacturer will have to prove that the product quality has not been compromised.

When a process has been developed from scratch, the viral removal methods may have been tested first in laboratory and then in pilot scale. The scaling up to process scale may change the efficacy of the method and this needs to be carefully evaluated. Besides, if during the scaling up process changes are done, the viral clearance processes must be revalidated. Validation is done using small-scale laboratory replicas of the production process. This is necessary because of the hazards involved in introducing viruses in the production scale process to validate the viral clearance step and because it is not possible to produce sufficient amounts of a virus to validate its removal at process scale. Validation studies are normally done with model viruses that mimic the characteristics of the human viruses to minimize risks. A careful selection of the model viruses is necessary to demonstrate the viral clearance protocol. Unfortunately, only HIV and HAV have appropriate in vitro systems with which their titers can be measured. The FDA recognizes that working with scale-down models involves various challenges and that, therefore, the manufacturer must carefully prove the validation process.

The first challenge that the manufacturer faces is to set up a laboratory scale process that is an exact replica of the process scale one. Some operating parameters such as temperature and pH must be kept the same in both scales whereas some other parameters such as flow rates and column loads must be properly scaled down. Of course, the starting material must be the same in both scales. It is accepted that if the laboratory scale and large-scale processes have the same performance, as measured by capacity, yield, and purity, then the manufacturer has successfully proven its scale-down process. The demonstration must be statistically significant, which means that several runs must be done using identical virus loads. Replication of the demonstration must yield consistent results and they must survive acceptable statistical tests (see Appendix).

It is important to notice that the manufacturer must validate the virus removal protocol at the most unfavorable conditions that one may find in process scale production. For example, process conditions under which the removal or inactivation of viruses would be expected to be the least effective.

Manufacturers must normally demonstrate that process modifications do not alter the quality of their product as established by regulatory agencies. These changes in the process (e.g., the change from one resin to another one) may affect the virus clearance protocol too. Thus, the manufacturer must demonstrate either that the virus clearance protocol is not affected by the change in the process or revalidate the protocol. Considering the impact of virus clearance protocol on the total process cost, the impact of a new virus clearance protocol must be carefully considered when a change in the production process is proposed.

The viral clearance validation study consists of "spiking" experiments, in which "large" amounts of a virus are added to the process. Preliminary studies must be aimed at finding what would be the level of a contaminated raw material pool in the worst-case scenario. Then, the spike is chosen such as exceeds that viral load. This is done because: 1) it challenges the viral clearance process at very astringent limits and 2) simplifies the quantification of the residual titer. The reduction in the concentration of the virus is measured using accepted protocols and compared with appropriate controls. Obviously, the infectivity of the virus must be demonstrated too. It is important to notice that total removal of the virus is normally not necessary because of the large excess used in the spiking process.

The infectivity of the virus may be measured *in vivo* or *in vitro*. In vivo testing is becoming rare because of the cost of the animals and the increased public opinion pressure against animal testing. Therefore, *in vitro* testing is common practice. The infectivity of the virus may be measured *in vitro* using tissue-based or biochemical assays. Tissue-based tests may be "plaque assays" or assays that measure "cytopathic effect." Antigen-based or nucleic acid assays are examples of biochemical assays used in the quantification of viral clearance protocols.

Conventional human plasma fractionation steps reduce some of the viral load. However, most plasma derivatives are subjected to at least one specific viral clearance step. There are substantial variations from product to product and manufacturer to manufacturer regarding the number of steps validated and the number of different viruses studied. The reader is referred to the detail guidelines established by the appropriate regulatory agency.

Often, more than one manufacturing step is validated as a viral clearance step. Only under very special conditions may the removal efficiencies be considered additive. If both processes target the same virus and they remove the virus by two different mechanisms (e.g., sieving by an ultrafiltration membrane and heat inactivation in a heating step) then the overall clearance of the process is the product of the two steps (often calculated by adding log reduction factors). Therefore, efficiencies of the same process in series (e.g., two ultrafiltration steps) cannot be added.

It is not uncommon to change the process parameters according to the results of validation studies. However, to have the process approved both validation studies and process scale production must have the same operating parameters. This complex and somehow convoluted strategy to obtain approval by regulatory agencies may be seen as excessive. However, the risk in social health associated with the presence of different viruses in biopharmaceuticals justifies this astringent control.

Experiments

Two experiments are described here. In the first one, γ-crystallin fractions obtained in the experimental section of Chapter 7 are dried. In the second, human immunoglobulin G (IgG) is freeze dried under different process conditions; that is, various protein concentrations, two different cooling rates, and in the presence and absence of a surfactant.

Materials

The protein fractions obtained in Chapter 7. Human IgG, potassium phosphate buffer, and Tween 80. Liquid nitrogen.

Equipment

A bench top freeze dryer. A dynamic light scattering instrument.

Method 1 (γ-Crystallins)

a. Place all of the fractions in a single tare freeze drying jar and record the weight of your sample. Select a jar that is large enough for the liquid to form a thin film that will coat the sides of the jar after it is frozen.
b. Place the jar in a cold ethanol (-60°C) freezing chamber. Commercial chambers rotate the jar and allow for the formation of a thin film on its walls. This is always better than the standard laboratory practice of immersing the sample in liquid nitrogen. Leave the sample in the freezing bath for 1 hour.
c. Remove the jar from the cold ethanol chamber and place it on one of the nozzles of a freeze drying instrument previously chilled down to -40°C. Open the vacuum (it should reach less than 10^{-3} mm Hg in about 10 minutes).
d. Dry the sample for 24 hours, removing the jar at regular intervals to check the weight.
e. Make a plot of water removed versus time and calculate the drying time.

Method 2 (α- and β-Crystallin Fractions)

a. Place known amounts (from 3 mL to 30 mL) of the protein solution in five tare freeze drying jars.
b. Freeze the five samples by immersing them in liquid nitrogen.
c. Note the thickness (h) of the frozen solids and place the jars in five nozzles of a freeze drying instrument previously chilled down to -40°C. Open the vacuum (it should reach less than 10^{-3} mm Hg in about 10 minutes).
d. Dry the sample for 24 hours removing the jar at regular intervals and determine the weight.
e. Make a plot of the time needed for the samples to dry versus the initial height of the frozen solids to the power of 1.5 and determine the slope, m_d. This slope can be used to scale up the freeze drying unit from (Snowman 1996),

$$t = m_d h^{1.5} \tag{9.8}$$

Method 3 (IgG)

a. Prepare IgG solutions in phosphate buffer at concentrations ranging from 0.5 to 10 mg/mL with and without the addition of 0.1% w/w Tween 80.
b. Freeze the samples in either liquid nitrogen or in a cold ethanol freezing chamber and proceed as in b. in previous section.
c. Dry the samples for 24 hours.
d. Reconstitute the samples in ultra pure water and determine the presence and quantity of aggregates by dynamic light scattering as described in Chapter 4.

Summary

The most important finishing step in the purification of a protein for human use is freeze drying. The process is mature but its application to protein pharmaceuticals needs to be done with extreme care to guarantee the viability of the preparation after processing. Spray-drying is an interesting alternative to produce powders that can be administered to an individual via inhalers. The technique needs to reach maturity. Finally, virus removal has a tremendous economic impact and represents a challenge for process validation by regulatory agencies. Although virus removal is an intrinsic part of any bioprocess it is frequently ignored in bioseparation books.

Nomenclature

A	area of the vial (cm^2)
Δh_s	enthalpy of sublimation (J/g)
h	height (mm)
K	heat transfer coefficient of the vial (J/°C cm^2 hr)
m	water removed per unit area per unit time (g/cm^2 hr)
P_c	the pressure in the chamber (mmHg)
P_0	the vapor pressure of ice at the product temperature (mmHg)
R	ratio between glycine and sucrose
R_p	the product resistance (cm^2 mmHg hr g-1)
R_s	resistance of the stopper (cm^2 mmHg hr g^{-1})
t	time (hrs)
T_{AM}	mechanical glass transition (°C)
T_{AM}'	mechanical glass transition temperature (°C)
T_{D2}	devitrification transition temperature (°C)
T_{D1}	devitrification transition temperature at high water content (°C)
T_g	glass transition temperature (°C)
T_g'	glass transition temperature at high water content (°C)
T_p	product temperature (°C)
T_s	shelf temperature (°C)
W	water content

Greek Letters

α	heating rate (°C/minute)

References

Attivi, D. W. P., N. Ubrich, C. Damge, M. Hoffman, and P. Maincent. Formulation of insulin-loaded polymeric nanoparticles using response surface methodology. 2005. *Drug Development and Industrial Pharmacy* 31:179-89.

Asherie, N. 2004. Protein crystallization and phase diagrams. *Methods* 34:266-272.

Bellows, R. J., and C. J. King. 1972. Freeze drying of aqueous solutions: maximum allowable operating temperature. *Cryobiology* 9:559-61.

Bonelli, P., C. Schebor, A. L. Cukierman, M. P. Buera, and J. Chirife. 1997. Residual moisture content as related to collapse of freeze-dried sugar matrices. *Journal of Food Science* 62(4): 693-5.

Carpenter, J. F., M. J. Pikal, B. S. Chang, and T. W. Randolph. 1997. Rational design of stable lyophilized protein formulations: some practical advice. *Pharmaceutical Research* 14:969-75.

Chang, B. S., B. S. Kendrick, and J. F. Carpenter. 1996. Surface-induced denaturation of proteins during freezing and its inhibition by surfactants. *Journal of Pharmaceutical Sciences* 85: 1325-30.

Chang, B. S., and S. Y. Patro. 2004. Freeze-drying process development for protein pharmaceutical. In: *Lyophilization of Biopharmaceuticals*, Volume 2, edited by H. R. Constantino, and M. J. Pikal, pp. 113-38. Arlington, VA: American Association of Pharmaceutical Scientists.

Chernov, A. A. 1997. Crystals built of biological macromolecules. *Physics Reports* 288:61-75.

DHEW Publication (NIH-78-1422).

Elkordy, A. A., R. T. Forbes, and B. W. Barry. 2002. Integrity of crystalline lysozyme exceeds that of a spray-dried form. *International Journal of Pharmaceutics* 247:79-90.

Hoffart, V., N. Ubrich, C. Simonin, V. Babak, C. Vigneron, and M. Hoffman. 2002. Low molecular weight heparin loaded polymeric nanoparticles: formulation, characterization and release characteristics. *Drug Development and Industrial Pharmacy* 28:1091-9.

Ito, K. 1971. Freeze drying of pharmaceuticals. Eutectic temperature and collapse temperature of solute matrix upon freeze drying of three-component systems. *Chemical and Pharmaceutical Bulletin* 19(6):1095-102.

Jameel, F., and S. Y. Patro. 2005. Impact of formulation for optimizing lyophilization process development. *American Pharmaceutical Review* 8:46-51.

Kanev, A. N., V. I. Kosyakov, D. V. Malakhov, and E. Y. Shalaev. 1989. An equilibrium phase diagram for the system water-sucrose. *Izvestiya Sibirskogo Otdeleniya Akademii Nauk SSSR, Seriya Khimicheskikh Nauk* 2:11-5.

Kelner, D. N., W. Wang, and D. Q. Wang. 2000. Protein formulation. *Biotechnology and Bioprocessing* 27:549-604.

Krishnamurthy, R. 1999. Protein stability in pulmonary delivery formulations: a review. *Pharmaceutical Technology* 23:48-60.

Liapis, A. I., and R. Bruttini. 2006. Freeze drying. In *Handbook for Industrial Drying,* Volume 1, 2nd edition, edited by A. S. Mujumdar, pp. 309-44. New York: Marcel Dekker.

Liu, W., D. Q. Wang, and S. L. Nail. 2005. Freeze-drying of proteins from a sucrose-glycine excipient system: effect of formulation composition on the initial recovery of protein activity. *AAPS PharmSciTech* 6(2):E150-7.

Maa, Y. F., P. A. Nguyen, and J. D. Andya. 1998. Effect of spray drying and subsequent processing conditions on residual moisture content and physical/biochemical stability of protein inhalation powders. *Pharmaceutical Research* 15:768-75.

Mackenzie, A. P. 1976. Principles of freeze-drying. *Transplantation Proceedings* 8(2 Suppl 1):181-8.

Mackenzie, A. P. 1966. Basic principles of freeze-drying for pharmaceuticals. *Bulletin of the Parenteral Drug Association* 20:101-30.

Mattern, M., G. Winter, R. Rudolph, and G. Lee. 1997. Formulation of proteins in vacuum-dried glasses. Part 1. Improved vacuum-drying of sugars using crystallizing amino acids. *European Journal of Pharmaceutics and Biopharmaceutics* 44:177-95.

Moshashaee, S. B. M., R. T. Forbes, E. A. Quinn, H. Nyqvist, and P. York. 2003. Supercritical fluid processing of proteins: lysozyme precipitation from aqueous solution. *Journal of Pharmacy and Pharmacology* 55:185-92.

Pikal, M. J., S. Shah, D. Senior, and J. E. Lang. 1983. Physical chemistry of freeze-drying: measurement of sublimation rates for frozen aqueous solutions by a microbalance technique. *Journal of Pharmaceutical Sciences* 72:635-50.

Pikal, M. J. 1996. Freeze dried proteins: the effect of formulation on process design and product stability. In *Proceedings of the 10th International Drying Symposium*, edited by C. Strumillo and Z. Pakowski, pp. 1245-54. Lodz, Poland: Lodz Technical University Press.

Prinn, K. B., H. R. Costantino, and M. Tracy. 2002. Statistical modeling of protein spray drying at the lab scale. *AAPS PharmaSciTech* 3(1):1-8.

Sarciaux, J.-M., S. Mansour, M. J. Hageman, and S. L. Nail. 1999. Effects of buffer composition and processing conditions on aggragation of bovine IgG during freeze drying. *Journal of Pharmaceutical Sciences* 88:1354-61.

Sellers S. P., and Y-F. Maa. 2005. Principles of biopharmaceutical protein formulation. *Methods in Molecular Biology* 308:243-63.

Shalaev, E. Y., and A. N. Kanev. 1994. Study of the solid-liquid state diagram of the water-glycine-sucrose system. *Cryobiology* 31:347-82.

Sharma, V. K., and D. S. Kalonia. 2004. Polyethylene glycol-induced precipitation of interferon α-2a followed by vacuum drying: development of a novel process for obtaining a dry, stable power. *AAPS PharmSci* 6:1-13.

Sievers, R. E., E. T. S. Huang, J. A. Villa, J. K. Kawamoto, M. M. Evans, and P. R. Brauer. 2001. Low-temperature manufacturing of fine pharmaceutical powders with supercritical fluid aerosolization in a Bubble Dryer. *Pure and Applied Chemistry* 73:1299-303.

Snowman, J. W. 1996. Lyophilization. In *Downstream Processing of Natural Products*, edited by M. S. Verral, p. 275-99. New York: John Wiley and Sons.

Further Reading

Carpenter, J. F., M. J. Pikal, B. S. Chang, and T. W. Randolph. 1997. Rational design of stable lyophilized protein formulations some practical advice. *Pharmaceutical Research* 14(8):969-76.

Chang, B. S., and S. Hershenson. 2002. Practical Approaches to Protein Formulation Development. *Journal of Pharmaceutical Technology* 13:1-25.

Patro, S. Y., E. Freund, and B. S. Chang. 2002. Protein formulation and fill-finish operations. *Biotechnology Annual Review* 8:55-84.

Chapter 10
Separations in a Chip

Introduction

The development of miniature devices has been fueled by the needs of the National Aeronautics and Space Administration (NASA) and other space agencies worldwide to increase the scientific load of aircraft without increasing their load overall. One of the first attempts in space to downsize a protein separation process was the performance of aqueous two-phase systems experiments in microwells about 20 years ago (Todd et al. 1989). In recent years, advances in miniaturization have been fostered by the genomic revolution, because of the need of performing a large number of separations of small (of the order of a few microliters) and complex mixtures in a short period of time.

The smallest devices developed so far are of the order of a few micrometers in size. Borrowing techniques from the microelectronics industry, a diverse kind of "unit operations" have been imprinted in a variety of substrates such as silicon, glass, and polymeric materials. That is the same approach followed to etch a wafer or to fabricate the chip that is "powering" your computer.

Microdevices have paramount importance in proteomics. Often, samples contain up to 20,000 proteins in six-fold concentration differences need to be analyzed. Proteomic samples have been traditionally separated by two-dimensional gel electrophoresis. After the separation, the proteins are removed from the gel and they are chemically processed (e.g., enzymatic digestion). The resulting sample is purified and concentrated and then analyzed by electrospray ionization or matrix-assisted laser desorption mass spectroscopy. The peptide fingerprint obtained needs to be identified using a database searching software. The entire process takes several days. An alternative to using two-dimensional gel electrophoresis is a sequence of strong cation exchange and reverse-phase chromatography steps after the protein samples are chemically processed, purified, and concentrated. This is called shotgun proteomics. The problem with this approach is the added complexity to interpreting mass spectroscopy data.

Microdevices have the potential to transform the field of proteomics by speeding up the process. In addition to the obvious advantage of being able to handle minute amounts of samples, microdevices have better mass and heat transfer rates than conventional units. This is particularly important for microreactors, because highly exothermic reactions can be done isothermically. Still, the fact that the flow is laminar implies that any mixing between streams will happen only by diffusion, which is a very slow process. An additional advantage of microdevices is that they are simple to

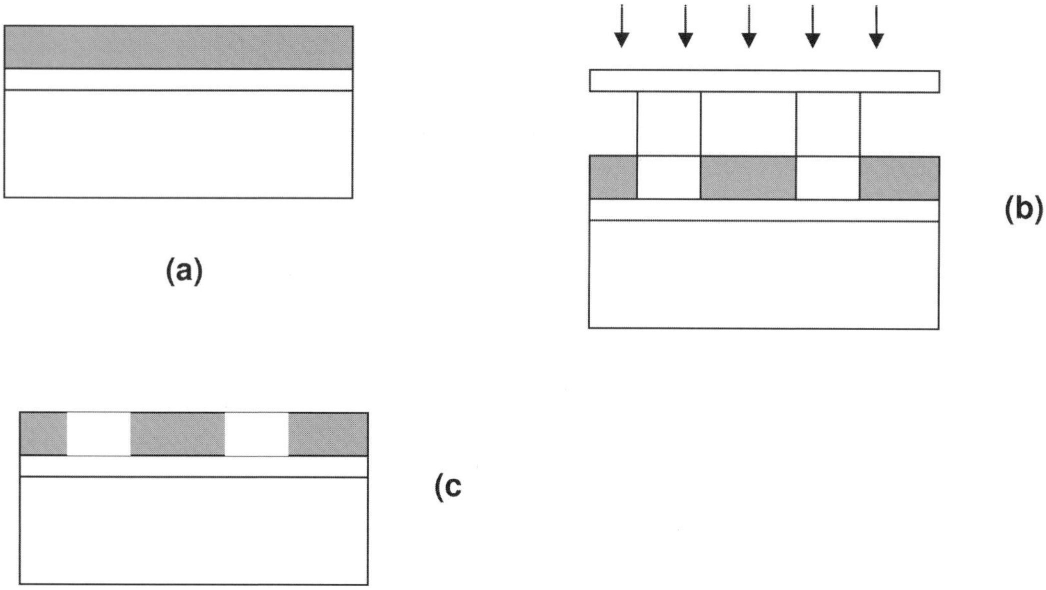

Figure 10.1. Schematic showing the fabrication of a pattern by photo etching. (a) A photoresist (gray block) of controlled thickness is deposited on a substrate. (b) Portions of the photoresist are protected such as the desired pattern is blocked and the resist is irradiated. (c) The photoresist is developed and the pattern is revealed.

scale up by simple parallelization. Moreover, the behavior of fluids in microdevices is easy to model because they work in laminar flow regime, and therefore, the problems associated with the characterization of turbulent flow do not exist.

The first microdevices were fabricated with silicon or glass wafers following well-established approaches of the microelectronics industry. However, the use of silicon or glass is not convenient in the handling of proteins because they have a tendency to adsorb to these hydrophobic materials. This practice evolved into the use of a variety of polymeric materials including polydimethylsiloxene, polymethyl-mathacrylate, polycarbonate, and Teflon. The fabrication methods include microlithography (soft and photolithography), molding, embossing and micromachining with lasers, electrochemical and ultrasonic technologies. A fabrication technique that combines lithography, electroplating, and molding has also been used (Koch, Evans, and Brunnschweiler 2000). Still, the most common fabrication method is photolithography. In this method, a thin layer of a photoresist is spun-coated on top of a silicon wafer (in this case a thin layer of silicon oxide will be present) or on top of a glass slide. The photoresist is then cured. After that, ultraviolet (UV) light is shone on top of the photoresist through a glass having the desired pattern (opaque regions will not be "carved" in a negative photoresist, whereas the opposite will happen in a positive photoresist) (Figure 10.1). After the photoresist is exposed to UV light the resist is developed (the degraded regions are washed out) to reveal the desired pattern.

Miniaturization does have a few disadvantages. One of the most severe ones is the incorporation in the microdevice of the appropriate instrumentation to monitor the process. An approach that has been followed is to fabricate the microdevice with optical materials (quartz); thus, the visualization by UV spectroscopy or infrared (IR) spectroscopy is facilitated. The problem with this approach is that the portability of the device is jeopardized. Some have used confocal microscopy with laser-induced fluorescence detection to monitor micro-electrophoretic devices. Although the design of the device is such that the separations are faster and with sharper peaks than in regular capillary electrophoresis; the size and complexity of a confocal microscope bring questions to this approach. Another alternative is to interface the microdevice directly with a mass spectrophotometer. These examples show that the current trend seems to be to sacrifice the portability of the device in order to incorporate the necessary analytical techniques. It is reasonable to think that in the future a complete miniaturization will be achieved. A second clear disadvantage of microdevices is that they cannot compete with techniques whose resolution depends on the distance that the fluid travels from the seeding to the detection point.

A visit to the microfluidics Web page of Sandia National Laboratory will give the reader a clear idea of how much this new field has advanced. For example, they have been able to "fold" a 0.9-m chromatographic column in 100 mm^2, an advance that decreases the differences between standard chromatography and chromatography in a chip. They have also developed electrokinetic pumps able to develop pressures up to 68.95 MPa and flow rates over 100 μL per minute (Paul et al., 1998, 2000). These pressures are at the upper end of pressures used in high performance liquid chromatography. Therefore, the differences in performance of microdevices and standard separations lab equipment are disappearing.

In the remaining part of this chapter, I discuss the behavior of fluids in microscopic channels. Then, the chapter presents examples of microreactors, micromixers, valves, pumps, membrane-based separation devices, electrophoresis, and chromatography in a chip. We need to understand that most of the material presented has not passed the experimental phase. However, it is only a matter of time and these microdevices will become a common sight in research laboratories and processing plants.

Fluidics

Fluidics refers to the movement of a fluid in microchannels, channels that are less than 1 mm and greater than 1 μm in width. The Reynolds number in these tiny microchannels is normally below 200, which implies that the flow in microchannels is laminar. The small size of the channel changes the behavior that a fluid would have in a regular pipe because boundary layers represent an important fraction of the entire flow. Flow in microchannels is unique because molecular effects are more important and because some continuum effects are enhanced; for example, the stress and the acceleration of a fluid in a microchannel are different than a macroscopic tube. Moreover, if the walls of the channel bear an electric charge, electrokinetic effects will become important too. In spite of this, it is generally accepted that continuum models (traditional hydrodynamics) still apply.

The pressure drop in a cylindrical capillary, or a cylindrical microchannel, is given by,

$$\Delta p = \frac{8\mu LQ}{\pi a^4} \qquad 10.1$$

in which μ is the viscosity of the fluid, Q is the flow rate, L is the length of the capillary and its radius. The friction factor (viscous and inertial effects that affect the movement of the fluid) for laminar flow in a round capillary is given by,

$$f = \frac{64}{\mathrm{Re}} \qquad 10.2$$

in which the Reynolds number is given by,

$$\mathrm{Re} = \frac{\rho \bar{v} L}{\mu} \qquad 10.3$$

in which $\bar{v}$ is a characteristic velocity and ρ is the density of the fluid. This can be extended to ducts of arbitrary form by using the Poiseulle number

$$Po = \frac{f\,\mathrm{Re}}{4} = -\frac{1}{\mu}\frac{dp}{dx}\frac{D_h^2}{2\bar{v}} \qquad 10.4$$

in which D_h is the hydraulic diameter. These relationships predict that as a fluid moves through, there is not dependence of the pressure drop with surface roughness. As the Reynolds number increases (>2000), there is a transition from laminar to turbulent flow in which the friction factor increases dramatically and surface roughness is known to affect the flow. For smooth pipes the friction factor in the turbulent regime is given by,

$$f = 0.3164\,\mathrm{Re}^{-0.25} \qquad 10.5$$

For rough pipes, the friction factor deviates from this relation. Theoretical calculations indicate that the flow in microchannels is laminar and that, therefore, the friction factor should be a function of the inverse of the Reynolds number. Some experimental evidence supports this prediction of theory but other evidence contradicts it. In some cases, the experimental friction factor is lower than expected, but in some others, the friction factor is larger. Several explanations have been offered to rationalize these observations. They cover a gamut of phenomena including the effect of surface roughness, the presence of electrical charges at the surface, variations in the viscosity of the fluid caused by the microscopic size of the channel, microrotational effects of individual fluid molecules, unpredicted transition to turbulence, entrance effects, and experimental inaccuracies. It seems that surface roughness and variations in viscosity close to the wall of the capillary as well as early transition to turbulence are not the reason. It is possible that entrance effects and errors in the measurement of the diameter of the channel are the reason for the discrepancies. Whatever is the reason, the designer must be aware of this because the predicted operational conditions of the device may be substantially different from the experimental values.

Mixing

As stated before, mixing (when needed) is of great concern in microdevices because of the dominant laminar flow regime, which means mixing happens only by diffusion, an inherently slow process. To improve mixing in microchannels one may induce turbulence. Under turbulent flow, each fluid element (an artificial conception) is well mixed by diffusion (they are very small so diffusion is enough). This fluid's elements, or eddies, collide and exchange mass accelerating mixing.

Two types of mixing devices have been developed so far: active and passive. Active mixers use a mechanical mixing aid such as moving parts or varying pressure gradients. Although they are very efficient, they are also very difficult to build. Passive mixers use as a sole energy source the flow field created by the pump. Chaotic advection is a mixing phenomenon that is caused by changes in the direction of flow. It has a large potential in microfluidic devices because it is relatively easy to implement. Recently, Liu et al. (2000) developed a device that seems to be a quite efficient mixer. Figure 10.2 shows a drawing of it. It consists of a series of C-sections where the fluid changes directions by 180 degrees; the C-sections are perpendicular to each other creating a three-dimensional pattern in the movement of the fluid. It is not clear if the device has been used in microfluidic applications such as enzymatic reactors or refolding reactors. It is important to realize that this concept is related to the presence of turns that produce mixing, which is a problem in the

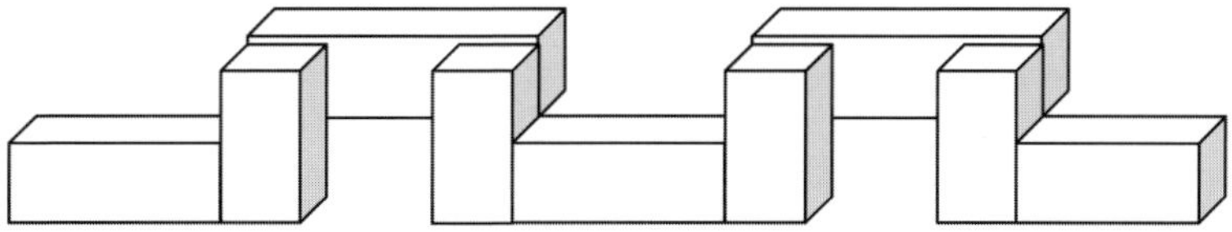

Figure 10.2 Mixer that uses changes in the direction of the flow (adapted from Liu, R. H., M. A. Stremler, K. V. Sharp, M. G. Olsen, J. G. Santiago, R. J. Adrian, H. Sref, and D. J. Beebe. 2000. Passive mixing in a three-dimensional serpentine microchannel. *Journal of Microelectromechanical Systems* 9(2):190-8). Copyright by the Institute of Electrical and Electronic Engineers and the American Society of Mechanical Engineers.

design of capillary electrophoresis or chromatography in a chip when the length of the channel is such that the channel must be folded over itself several times. Mohammadi and Santiago (2001) used computational fluid mechanics to optimize the geometry of the channel. The results of this work highlight the importance of mathematical modeling for the design of microdevices.

The mixing of fluid eddies was cleverly use by Lin, Fu, and Chien (2004) to effect mixing in microchannels. The mixing device has a T-shape. The fluids to be mixed flow through each arm of the T toward the junction and then leave the T. By alternating the flow in both channels they were able to segregate the flow and thereby improve mixing. The idea is to "intercalate" slices of fluid from both streams. They achieved this by using a switching DC field that generates a switching electro-osmotic flow. By making the slices very small, mixing in between slices takes place by diffusion very quickly. The device reaches 97% mixing within 1 mm downstream from the T-junction using 60 V/cm driving voltage and 2 Hz switching frequency.

Good and fast mixing is a key feature for the good design of microreactors. Static mixers that contact the fluid streams (reactants) and then divide them into two flows perpendicular to the interface area can be used to improve mixing. The merging of these two streams increases the mixing interface for optimal mixing. The process is repeated several times (Ehrfeld et al. 1999).

Pumps

There are several types of micropumps that have been used in microdevices. Micropumps may be classified into two groups: membrane-based and field-induced flow. Membrane pumps use, for example, a microfabricated silicon diaphragm, to pump a liquid. The membrane may be actuated by different means, including piezoelectric (Koch et al. 1998; Richter, Linnemann, and Woisas 1998), electrostatic (Nguyen et al. 1998), and electromagnetic actuators (Gong et al. 2000). The momentum of piezoelectric, thermal, pneumatic, or electrostatic actuations is transferred to the solution and thus, they generate flow. One type of pump is called the diffuser pump. This pump has an actuated pump chamber and a pair of nozzle/diffuser elements to rectify the flow. The flow is originated by the change in pressure when a fluid passes different channel geometries. There is no mechanical parts to realize the valve function; therefore, this pump is considered a valveless pump. Bubble-powered pumps use the repeated growth and collapse of a single bubble to push the liquid through the channel. Electro-osmotic, electrohydrodynamic, and magnetohydrodynamic pumps are examples of field-induced flow pumps. One of the advantages of field-induced flow pumps is that they do not require moving parts. For example, electrodynamic pumping is done by applying large electric fields between two electrodes orthogonal to the flow direction. Field-induced flow pumps can reach much higher pressures than membrane based ones. Of this kind, electro-osmotic pumps are able to reach pressures higher than any other type. It is reasonable to think that this kind of pump represents the future of pumping in microdevices. Therefore, the rest of this section is dedicated to electro-osmotic pumps.

Electro-osmotic flow was explained in the first chapter of this book in the context of capillary electrophoresis. This type of voltage-induced flow is an annoyance in capillary electrophoresis (CE) and, for many years, different ways of eliminating it (or controling it) have been proposed. For a fixed buffer composition and voltage drop, electro-osmotic flow should be constant. However, anyone familiar with CE knows that variations in this flow are the cause of the lack of reproducibility of capillary electrophoresis. This must be taken into account when using electro-osmotic pumps, which use electro-osmotic flow to generate flow rates and pressure gradients. One of the problems with electro-osmotic flow in a single channel is that it is very slow and produces low pressure drops. It is easy to conceive that an array of a large number of short tubes will produce a much higher flow rate. Therefore, an alternative that has been explored is to use a number of microchannels (up to 100) to increase the pressure drop and flow rate. For example, Lazar and Karger (2002) used up to 100 microchannels to produce flow rates from 10 to 400 nL and pressures up to 0.55 MPa.

Yao et al. (2003) developed an electro-osmotic flow pump with porous pumping structures. The maximum flow rate for this type of device is,

$$Q_{\max} = \left(\frac{\varepsilon \xi \ A}{\mu} \right) \frac{\psi}{\tau} f \frac{V_{eff}}{L_M} \qquad 10.6$$

in which ξ is the zeta potential, A is the cross-sectional area, Ψ is the porosity, V_{eff} is the effective pump voltage, τ is the tortuosity, ε is the electrical permitivity, and L_M is the membrane thickness. The maximum pressure gradient is equal to,

$$\Delta P_{\max} = \left(\frac{8 \varepsilon \xi}{a_p^2} \right) f V_{eff} \qquad 10.7$$

in which,

$$f = \int \left(1 - \frac{\varphi}{\xi} \right) \frac{2r}{a_p^2} dr \qquad 10.8$$

where a_p is the pore's radius and φ is the electrical potential. Instead of arranging a large number of tubes to increase flow, a porous material will mimic that geometry except that the tubes will not be straight but tortuous, and only a fraction of the porous material, the void volume, is available to flow. Figure 10.3 shows a schematic of such a device. A potential difference is applied at opposite sides of a porous frit, which consists of a glass material composite by 80.60% silica, 12.60% boric oxide, and 4.2% sodium oxide. Its pore diameter varies from 1.2 to 0.55 μm and its thickness from 1 to 4.8 mm with a diameter of 40 mm. The electro-osmotic flow will move the solvent in the direction of the arrows; the magnitude of the flow will depend on the following: the voltage difference applied, the zeta potential of the porous material, the pH, and the type and concentration of the carrier electrolyte. One problem with any electro-osmotic flow-driven pump is the formation of molecular oxygen in the anode and of molecular hydrogen in the cathode as a result of electrolysis of water. The generation of these gases negatively affects the functioning of the pump creating fluctuations in the flow rate. Yao et al. (2003) proposed to recombine hydrogen and oxygen using a platinum catalyst. Oxygen diffuses through the downstream chamber toward the vapor head. Hydrogen diffuses through the upstream chamber and then through a polytetrafluoethylene (PTFE) mem-

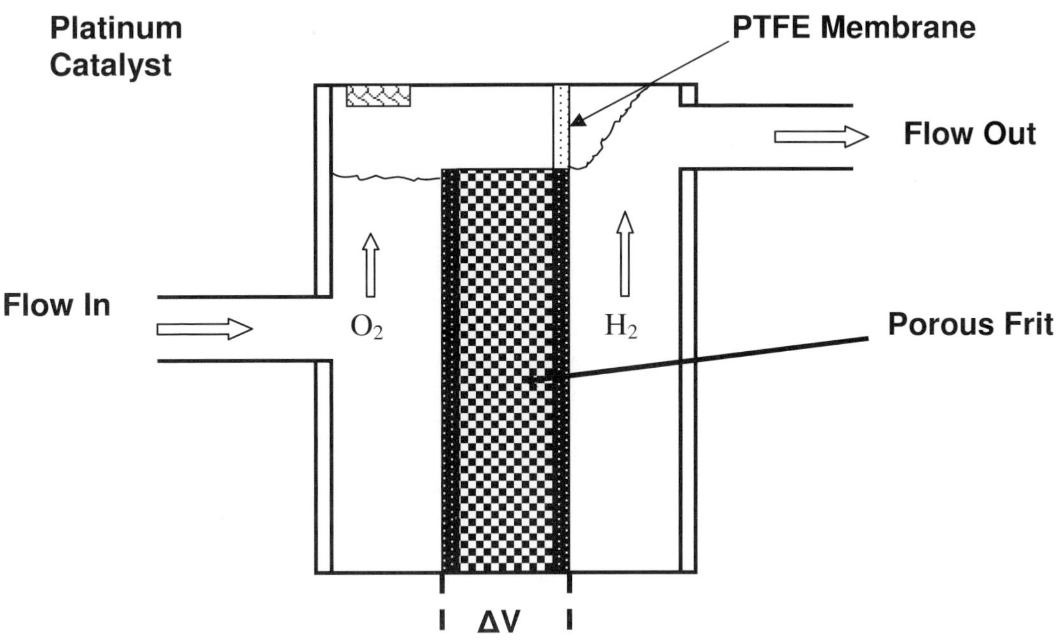

Figure 10.3. Electro-osmotic pump. The porous frit acts as a large collection of capillary tubes (pores). An electrostatic potential difference is applied at opposite phases of the frit to generate electro-osmotic flow.

brane. Of course, part of the gases remain soluble in the liquid and are pumped out. The pump can be thermodynamically characterized. The efficiency, η, of the pump is,

$$\eta = \frac{Q\Delta P}{V_{app} I} \qquad 10.9$$

in which ΔP is the pressure capacity (in Pa), Q is the flow rate in m^3/sec, V_{app} is the applied voltage, and I is the current. The efficiency of the pump is at its maximum when the pump operates at half its maximum ΔP, it is highly dependent on pore radius (scaling as $1/a^2$) and has a weak dependence on the thickness of the frit. The efficiency increases steadily with decreasing ion strength of the buffer (less than 0.1 mM is of no use because of large fluctuations in pH). Therefore, there is a trade-off between pH stability, thermodynamic efficiency and absolute pressure and flow rate performance of this kind of pump. Although the optimum performance from a thermodynamic viewpoint would be to use pore sizes of 0.1 μm, they argue that the pump will have poor flow rate performance because of the effects of finite electrical double layers.

The electro-osmotic flow depends on the type and concentration of the buffer or buffer additives. The effect of additives was explored by Reichmuth, Chirica, and Kirby (2003a) and Reichmuth and Kirby (2003b) who studied the effect of zwitterions to improve the performance of electrokinetic micropumps. In the presence of additives, the thermodynamic efficiency of the pump becomes,

$$\frac{\eta}{\eta_0} = \frac{\left(1+\gamma_e C_z\right)^2}{\mu^* \sigma^*} \qquad 10.10$$

in which σ^* is the normalized conductivity, $=\sigma/\sigma_0$, μ^* is a normalized viscosity, μ/μ_0, σ_0 and μ_0 are the conductivity, and the viscosity of the solution at zero zwitterionic solute concentration, C_z is the concentration of zwitterion, and γ_ε is defined by,

$$\gamma_\varepsilon = \left(\frac{d\varepsilon}{dC_z} \right) \frac{1}{\varepsilon_o} \qquad 10.11$$

in which ε is the permittivity and ε_0 is the permittivity of the solution in the absence of the additive. Equation 10.10 suggests that an increase in zwitterionic concentration, which increases the permitivity as its concentration is increased, will significantly increase the efficiency of the pump. They tested trimethylammoniopropane sulfonate (TMAPS), cyclohexylamino alkyl sulfonates (CHES, CAPS, and CABS), and tris(hydroxymethyl)methylamino alkyl sulfonates (TAPS). The addition of 2.5 M TMAPS to the buffer increases the maximum pressure generated per volt by 2.5 reaching pressures as high as 156 kPa/V. Although this approach looks attractive, one must be careful because of the high concentration of additive (1 M or higher) that is necessary to affect the efficiency of the pump in an appreciable manner. The additive will become part, for example, of the elution buffer of a chromatography column. They argue that the performance of a reverse-phase column is not affected by TMAPS but this is questionable at the concentrations used (1 M). Zwitterions will increase the ionic strength of the solution and affect electrostatic and hydrophobic interactions. Moreover, they will have to be separated from the protein downstream.

Valves

Microvalves are classified into active (with actuator) and passive (without actuator). They are found in micropumps as check valves or in different parts of the microdevice to switch flow direction or to control flow. Shape-memory alloys (Koch et al. 1999) and pH or charge-sensitive hydrogels (Baughman et al. 1991; Beebe et al. 2000) have been used to make microactuators for microvalves (Shoji and Esashi 1994; Shoji, Nakagawa, and Esashi 1990; Shoji 1998; Shoji, Esashi, and Matsuo 1988). Examples include a bulk silicon cantilever (Tiren, Tenerz, and Hok 1989), a suspended polysilicon round disk (Esashi, Shoji, and Nkano 1989), or a molded silicone rubber float (Shoji, van der Schoot, and de Rooij 1992).

Figure 10.4 shows two simple valve geometries: push-down and push-up. The channel is made with a viscoelastic polymer. In push-down valves, the control line flattens down the flow structure, creating a seal. In push-up valves, the control line passes under the flow channel. Pneumatic actuation of the control line causes the membrane to deflect up or down into the flow structure stopping the flow. Push-up valves can be used in deeper channels and require lower actuation pressures than push-down ones.

An interesting concept, because of its simplicity, is to fabricate a valve with a simple screw. Weibel et al. (2005) developed such a device. They embedded small machine screws ($\geq$ 500 μm) into a layer of polyurethane cast above a microchannel fabricated with poly(dimethylsiloxane (PDMS). They start with a microchannel (15 μm in height) made by a layer of PDMS (3 to 7 mm thick). First, a 4-mm diameter hole is drilled. Then, a smaller hole (1 mm) is drilled (1 to 3 mm long). The screw is inserted, the monomer is added, polymerization is started, and the polymer is cured to secure the screw with poly(urethane). As the screw is screwed down its tip presses down a thin layer of PDMS and closes the channel. The advantage of this device is that it allows complete obstruction of a channel of any shape and the channel can stay closed for an unlimited time. The obvious disadvantage is that the valves need to be operated manually. Another advantage is that the system is totally portable, but it is not clear how many closing and opening cycles can be performed without jeopardizing the integrity of the device.

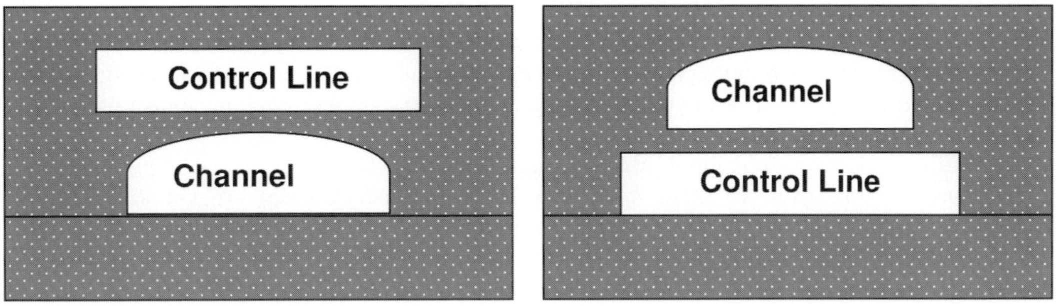

Figure 10.4. Schematic showing two simple valves (Cal Tech microfabrication Center). The flow is perpendicular to the plane of the paper.

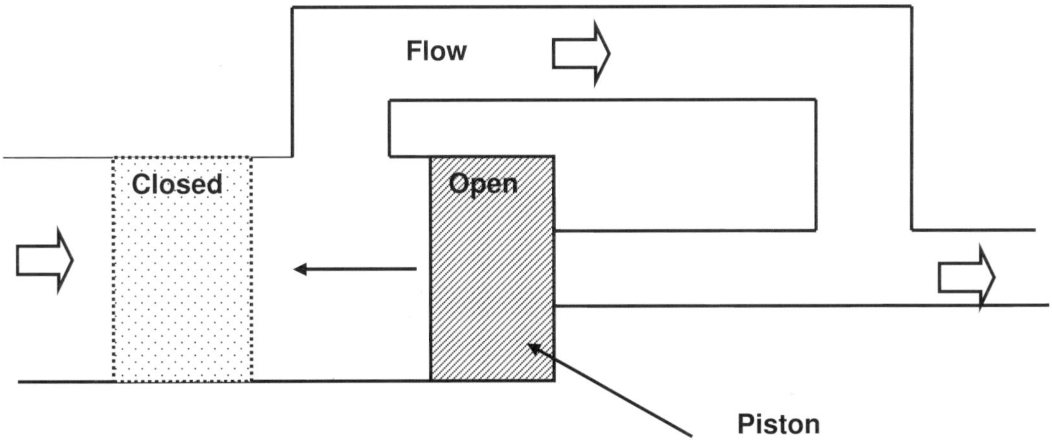

Figure 10.5. Schematic of pressure actuated pump. The piston (for example, a monolith polymer) moves from left to right (open) and from right to left (closed).

An interesting approach to flow control was developed in Sandia National Laboratory (Hasselbrink, Shepodd, and Rehm 2002). Using the same technique that they developed to make membranes (laser induced in situ polymerization) they made monolithic plugs that can control pressures exceeding 34.45 MPa and can be actuated in milliseconds. Different materials may be used to make the plug as long as they do not permanently deform and they are non-conductive to eliminate the movement of fluid because of electro-osmotic flow. Figure 10.5 illustrates the concept for a check valve. A similar architecture was used to design an on/off valve placing the monolith in a control line that forms a T with the main line. The choice of monomer, either acrylates or methacrylates, affects pore size, surface energy, and mechanical strength of the valve (Kirby, Shepodd, and Hasselbrink 2002). A material with the desirable properties was made by a 65:35 mixture of monomer (2,2,2-trifluorethyl acrylate -1,3-butanediol diacrylate, 50:50) and solvent (2-methoxyethanol). This monolithic plug is actuated with a electrokinetic pump. The plug moves forward and closes the channel using positive pressures and moves backward and opens the channel using negative pressures. The plug is able to hold pressures as high as 35.36 MPa, can be actuated by modest pressures (34.47-68.95 kPa), and it is resistant to the acetonitrile water mixtures typically used in reverse-phase chromatography. The equivalent Darcy constant of the plug is equivalent to a 20-cm-long chromatography column. Problems of monolithic plugs are swelling of the polymers upon changes in solvent composition (e.g., methacrylate will swell in contact with water) and potential adsorption of proteins.

Yu et al. (2003) also used monolithic polymers as valves that were actuated by changes in temperature. The plugs were made with poly(N-isopropylacrylamide) cross-linked with 5% methylenebisacrylamide by direct photoinitiated polymerization within the channel. The glass of the microchannel was vinylized to allow the covalent attachment of the plug to its walls. The polymer changes its volume at 32°C; this swelling and subsequent de-swelling happens very rapidly and it is controlled with a thermoelectric element able to change the temperature between 17 and 57°C. The element lasted at least 120 close-open cycles and resisted pressures up to 1.38 MPa.

A completely different approach was followed by Hua et al. (2002) who used electrochemically generated bubbles to regulate flow. The prototype consists of two platinum microelectrodes inserted in the microchip where the flow needs to be controlled. A potential difference between 1 and 8 V is imposed. In the presence of an electrolyte (such as NaCl in picomoles amounts) gases that are formed at the catode and anode (H_2 and Cl_2) produce a bubble whose size is a function of the potential difference. The bubble is able to withstand pressures up to 110 kPa. The bubble can be created in milliseconds and the partial deflation of it (also in the milliseconds range) is enough to restore flux.

Dialysis

One of the simplest microdevices to manufacture is a microdializer. Figure 10.6 shows a schematic of such a device. The substrate can be made by glass or silicon. Two substrates, prepared as described in Figure 10.1, are glued together with a regenerated cellulose membrane in between. One channel receives the feed, whereas the dialyzing buffer is circulated on the other channel. The major obstacle to fabricate an efficient device is to make a very thin membrane. Otherwise, the needed contact time can be so long that equilibrium between the two streams is not realized. The time needed to reach steady state can be estimated from,

$$t \approx \frac{d^2}{D_m} \qquad 10.12$$

in which d is the thickness of the membrane and D_m is the effective diffusivity in the membrane. Assuming a value of 10 x 10^{-13} m^2/sec and a membrane of approximately 50 μm thick, the time needed to reach steady state is $50^2\ 10^{-12}$ /10x 10^{-13} = 2500 seconds or 41 minutes, which will be the minimum time before a sample can be measured in the dialyzate. Therefore, the key in improving this technique will be in developing thinner membranes.

An interesting approach to improve the quality of the membranes is to produce the membrane in place by in situ polymerization. Membranes as thin as 7 μm have been reported; with this thickness the equilibration time would be reduced to only 50 seconds. The membrane is prepared in a short channel (~5 mm) in between two glass slides. A series of silica posts are fabricated at regular intervals along the channel to improve the mechanical strength of the membrane. These membranes resist up to 101.3 kPa, are stable in the pH range from 3 to 8.5, and show negligible protein adsorption (Song et al., 2004). Two types of membranes were fabricated, one of low and the other of high molecular weight cut-off. The former can be used for dialysis and the latter for protein fractiona-

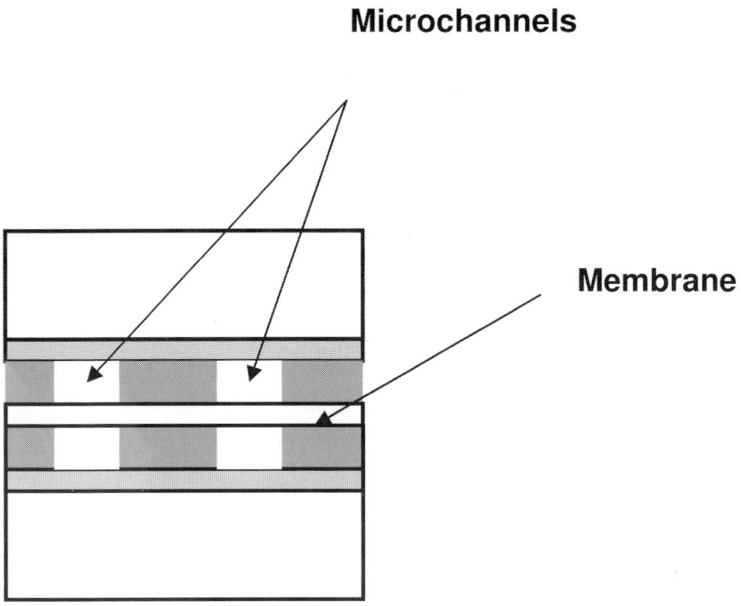

Figure 10.6. Schematic showing a simple micro dialysis device. The feed circulates through the upper channels and the dialyzate is collected in the lower channels. The flow is perpendicular to the plane of the paper.

tion. The feed and the dialysis buffer are brought into contact in countercurrent. A very simple model yields the following for the concentration profile in the sample channel,

$$C_s = \exp\left[\alpha\left(\frac{x}{L} - 1\right)\right] \qquad 10.13$$

in which x is the distance from the perfusion channel inlet and α is equal to,

$$\alpha = \frac{k\ell}{v_s w} \qquad 10.14$$

where v_s is the velocity of flow, k is the mass transfer coefficient for the membrane, ℓ is the length of the channel, and w is the channel width. X is the distance from the perfusion channel. This equation is valid when the flow velocity in the permeate is much higher than the flow velocity in the retenate. The highest extraction will take place at the permeate inlet (x=0). For example to obtain 98% removal we will need a value of α of 4. This can be achieved by increasing the length of the channel by a factor of 4, decreasing the flow by a factor of 4, or increasing the mass transfer coefficient by a factor of two, which is equivalent to decreasing the thickness of the membrane by a factor of 4. These calculations are valid when we start with a value of α of 1.

Another option to prepare in situ membranes has been developed at Sandia National Laboratory (Kirby and Singh 2002). An acrylate membrane 4 to 14 μm in thickness was made by shaping and focusing the 355-nm line of a 12-kHz, 800 ps-pulse, 160 nJ-pulse frequency tripled Nd:YAG laser into the monomer mixture already placed inside the channel. The ability to control the laser sheet thickness is key for the success of this fabrication method.

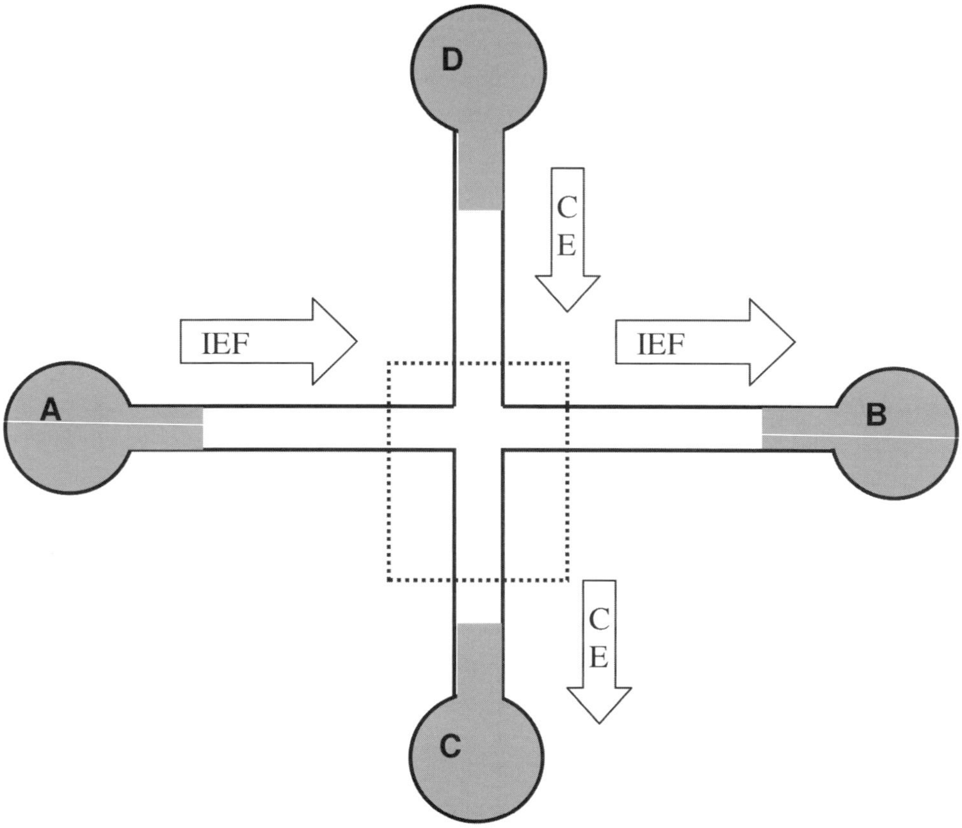

Figure 10.7. Two dimensional separations in a chip. The "bubbles" are buffer reservoirs. The sample is first focused in the horizontal direction. Then the sample is mobilized and picked up in the vertical direction by free solution electrophoresis.

Electrophoresis in a Chip

Capillary electrophoresis is a well established technique and highly automated instruments are available from several suppliers. Separations of very similar compounds are possible in a matter of minutes and the amount of sample needed is on the order of nanoliters. Therefore, the transition from capillary electrophoresis to an on-chip architecture is relatively straightforward. Because the analysis time in bench scale capillary electrophoresis is quite short, it is questionable if there is a need to develop that technique in a chip. One possible advantage is parallelization by developing multiple channels so that several separations can be done at once. Considering the cost of a commercial capillary electrophoresis, the use of multiple machines is not economically feasible. I believe the main niche for the use of electrophoresis in a chip is in two-dimensional separations.

Two-dimensional separations are key in the development of proteomics, in which a large number of components must be separated. The traditional way of doing this is in gel electrophoresis in two dimensions; proteins are separated according to size in one direction and according to charge (isoelectric focusing) in the other. This two dimensional separation can be performed in a chip. Figure 10.7 shows a sketch of a prototype. The two channels cross each other at a right angle. A, B, C, and D are buffer reservoirs. Isolectric focusing is effected from west to east and solution electrophoresis is affected from north to south. Herr et al. (2003) built such a device. It was constructed in acrylic to eliminate the problem of electro-osmotic flow generated by silica (glass) microchannels. The channels were 200 μm wide and 20 μm deep. The IEF channel was 2.8 cm long and the CE channel was 2.5 cm long. Electrodes were placed in the wells and ports were prepared to fill the channel with ampholytes. The sample is injected into the isoelectric focusing (IEF) well by electrokinetic flow. After a voltage is applied and the analytes focused, they are mobilized to the junction by electrokinetic flow where they are "picked up by voltage switching" by regular CE (although the CE channel is filled with ampholytes, they do not have extreme buffer conditions in the CE reservoirs, and therefore, they are unfocused). Although further optimization of the system is needed to achieve the resolution obtained by two-dimensional gel electrophoresis, this approach seems to be a step in the right direction.

Chromatography in a Chip

Chromatography is less straightforward to incorporate into a chip than CE, because it is difficult to mimic the standard components of a chromatography system: a series of pumps, valves, the column, filters, and a detection system. It is apparent that the incorporation of all of these devices into a couple of square centimeters is, at the least, challenging. Still, substantial progress has been achieved.

The first chromatography system in a chip was developed in 1990 (Manz et al. 1990). Since then, reverse-phase, anion-exchange, and cation-exchange chromatography systems in a chip have been developed and tested. de Mello reviewed this topic recently (de Mello 2002). One of the main problems with chromatography is to fill the channel with the gel. For many years, open-channel architectures were preferred because of this reason. Lately, in situ polymerization (of gels or monoliths) have been explored to eliminate the "packing" problem. The future of the technique seems to be in this mode of packing. Therefore, the main uses of monolithic gels seem to be in microdevices, because the dimensions of the channel are so small that the fabrication of the monolith is relatively straightforward. Figure 10.8 shows a column 5x5x10 μm in size. The architecture has been fabricated by reactive ion etching. Observe that the structures mimic the gel beads found in regular chromatography; the channel is about 1.5 μm in size (He, Tait, and Regnier 1998; Slentz et al. 2001). They found a column efficiency of 770,000 plates per meter. The first devices were fabricated in glass, which had blockage problems. More recently, this group has been using PDMS and the new columns seem to be much better than the primitive ones.

There are several ways in which microcolumns can be incorporated into a separation scheme. Most of the work done consists of a mix of microchannel-based columns as a part of a standard laboratory high-pressure liquid chromatography (HPLC) system. The best that has been achieved so far is a complete in-chip chromatography set up able to inject a sample directly into a mass spectrophotometer.

Figure 10.9 shows a completely integrated chromatography device. The upper part consists of two electrokinetics pumps that are connected to two buffer reservoirs to be able to perform gradient elution of the column. The two electrokinetic pumps converge into a mixer that feeds the column (middle of the schematic). The lower part of the figure shows a second set of electrokinetic pumps that are used to seed the sample previously placed in the sample reservoir. In this figure, the black arrows show the active part of the chip during sample injection (the black rectangle at the beginning of the column is the sample). The white arrows show the flux during elution of the column. The eluent exiting from the column may be injected directly into a mass spectrophotometer.

A device following the principles of this schematic was built and tested by Lazar, Trisiripisal, and Sarvaiya (2006). The device was tested by separating a protein digest obtained from MDF7 breast cancer cell line. The cytosolic protein extract was digested by tripsin and prefractionated by using a strong cation-exchange chromatography column. The on-chip column was 2 cm long. Using ammo-

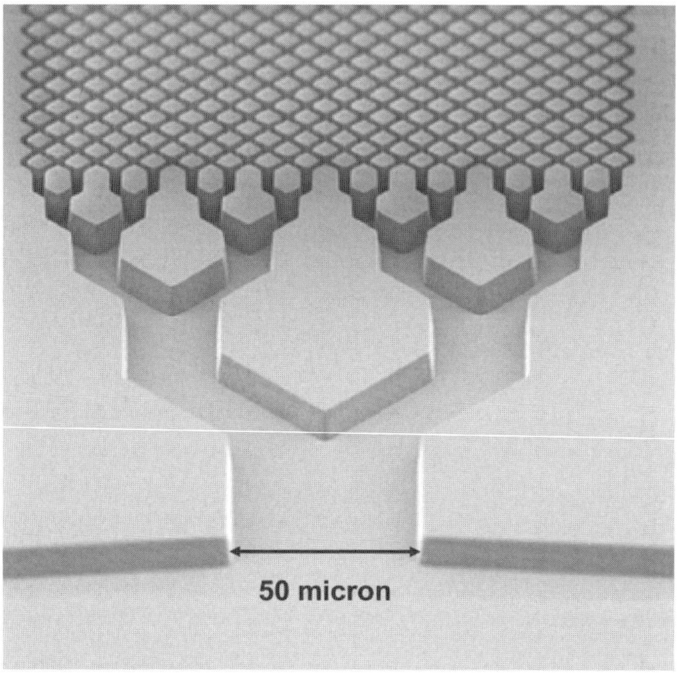

Figure 10.8. Scanning electron micrograph of a micromachined chromatography column (adapted from Slentz, B. E., N. A. Penner, E. Lugowska, and F. Regnier. 2001. Nanoliter capillary electrochromatography columns based on collocated monolithic support structures molded in poly(dimethyl siloxane). *Electrophoresis* 22:3736-44). Copyright by Wiley-VCH Verlag GmbH & Co., with permission.

nium carbonate (15 mM, pH 8) as the elution buffer they were able to find a total of 77 protein matches. This compares reasonably well with a 2-cm column run in a standard HPLC system using the same gel (C18), 91 proteins. However, the on-chip device was not a match for a 12-cm-long column eluted with trifluoracetic acid (TFA) (935 protein matches). The authors argue that the main problem is injection volume (12 μL vs. 1 μL). However, there are other problems associated with this. In order to have enough pressure in the pumps, a relatively concentrated strong electrolyte is needed as elution buffer, so TFA cannot be used. Also, if the column is made longer, the electrostatic pumps may not have enough power to move the fluid. If the column is coiled to decrease the size of the device, the coiling may cause higher degrees of mixing, more dispersion, and therefore, less resolution. Another problem with this device is the generation of gases, which will always be a problem with electro-osmotic flow-driven pumps unless an approach similar to the one described before is used to eliminate them. Considering that this publication appeared in 2006 it is obvious that the fabrication of a reliable and completely integrated in-chip chromatography system is not there yet.

Proteomics

It has been recognized by Freire and Wheeler (2006) that microfluidics can make a contribution to the field of proteomics in the chemical processing, sample preconcentration and cleanup, separations of peptides and proteins, and interfacing with mass spectrometry areas.

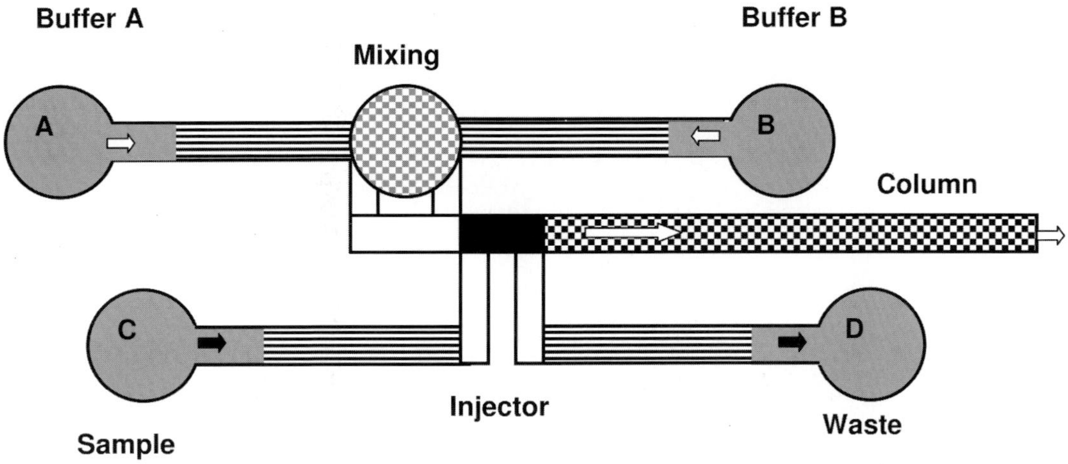

Figure 10.9. Schematic of an integrated chromatography system including buffer reservoirs, injection valve, electro-osmotic pump and column. The eluent can be injected directly into a mass spectrophotometer (adapted from Lazar, L. M., P. Trisiripisal, and H. A. Sarvaiya. 2006. Microfluidic liquid chromatography system for proteomic applications and biomarker screening. *Analytical Chemistry* 78:5513-24). Copyright American Chemical Society; with permission.

After the sample is removed from the gel in two-dimensional electrophoresis, it needs to be digested by proteolytic enzymes or other lytic compounds. This process takes up to 18 hours to complete. Shortening of this time would considerably decrease the total analysis time; from hours to minutes or even seconds. Progress has been made in this direction. For example, Gottschlich et al. (2000) used open-channel devices for digestion and reduction of proteins. Another option is to use trypsin adsorbed (immobilized) in channels; for example, by adsorbing the enzyme in beads and packing them in a microfluidic device. Monolith methacrylate is another option. Up to today, there has not been any effort to integrate the digestion reaction with reduction, alkylation (to prevent disulfide bonds reforming) and digestion. This is a bottleneck, but it will probably be addressed in the years to come.

Sample preconcentration and removal of contaminants is key for the detection of low expression proteins. One technique that is often used in microdevices is isotachophoresis (commonly used in capillary electrophoresis to narrow the bands). Another technique that is relatively easily incorporated into a microchip is dialysis. Finally, the most common type of sample preconcentration and clean-up is solid-phase extraction; nothing but a fancy name for some type of chromatography, mostly reverse-phase. The "column" may be packed with derivatized beads or may be filled by in situ polymerization (monolithic beds). Immobilized metal affinity chromatography also has been used.

Separation of the protein mixture is the bottleneck for the successful implementation of proteomics. Capillary electrophoresis has been quickly adapted to microfluidics, because it is naturally a technique that is done in microdevices (a capillary is not a chip, but its dimensions are comparable). Less common is the use of chromatographic separations at this stage of proteomics. Both CE and chromatography have the disadvantage that they are able to separate compounds in only one dimension (just remember that the traditional "proteomic way" of doing this is by two-dimensional electrophoresis). Two-dimensional chromatography, micellar, and free-solution electrophoresis has been miniaturized. Chromatography and electrophoresis have been implemented in co-axial archi-

Figure 10.10. Photograph of a commercial micro chromatography column. Courtesy of Nanostream (Pasadena, CA). Copyright by Nanostream.

tecture, which have several intrinsic problems like needing phase separations. Therefore, it has been argued that in-chip separations with proteomics in mind is the less mature field and the most likely to grow. It needs to be clarified that chromatography in a chip has advanced beyond prototypes. For example, Nanostreams (Pasadena, CA) sells 12 and 24 parallel microchips equipped with chromatography columns operating in different modes, although the most common type are reversed-phase ones. Figure 10.10 is a photograph of different multi-column chromatography chips and Figure 10.11 shows a picture of a high pressure liquid chromatography instrument able to work with these cartridges.

Ideally, the sample separated in the microdevice should be applied to the mass spectroscope directly. The designing of appropriate injection devices has flourished. Although substantial advances have been achieved, the development of an efficient method to interface a microchip with electrospray ionization (ESI) has remained elusive. Nanoelectrospray seems to be the most promising technique, but there are still some technical challenges that need to be resolved. Matrix assisted laser desorption/ionization (MALDI) may be used instead of ESI, although it is less clear how a microdevice may be connected to it, because the sample needs to be immobilized in a solid matrix before analysis. Closed-channel devices have been proposed, but it is not clear if they represent a true advantage.

Summary

Ingenuity and sophisticated modeling techniques have contributed to the development of the fabrication of bioseparation units in miocrochips. Although great advances have been done in the fabri-

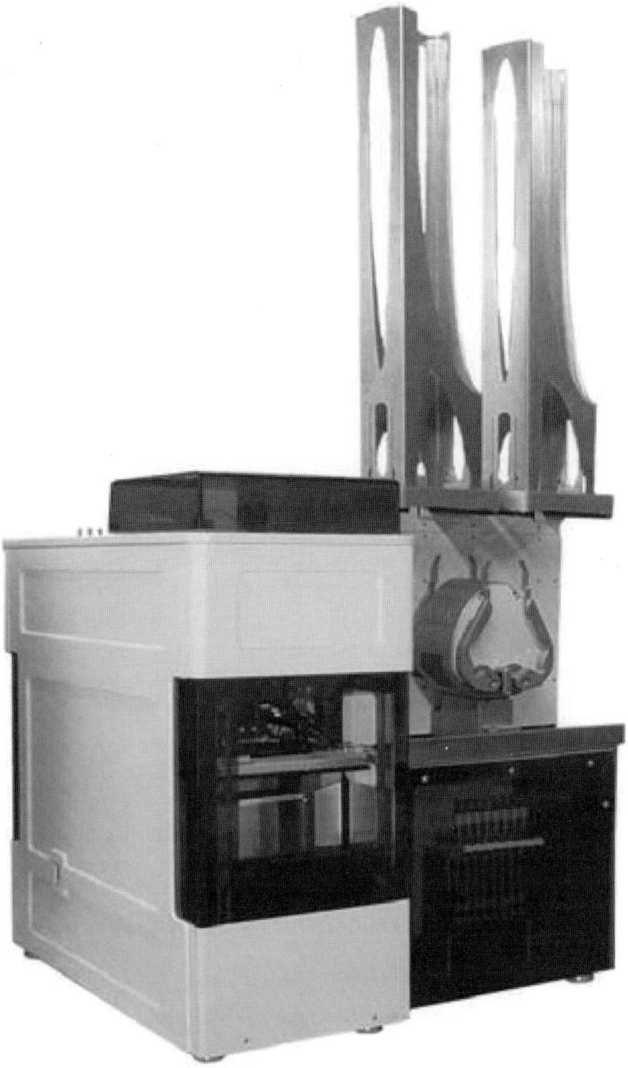

Figure 10.11. Photograph of the hardware used to operate the microchannel chromatography devices. Courtesy of Nanostream (Pasadena, CA).

cation of individual units such as pumps, valves, and chromatography columns, the systems are not yet fully integrated. Still, I believe that it is just a matter of time for microdevices to become common sights in laboratories and industry.

Nomenclature

a	Capillary radious (m)
a_p	pore's diameter (m)

A cross sectional area (m^2)
C_z concentration of zwitterions
d membrane thickness (m)
D_h hydraulic diameter (m)
D_m effective diffusivity in the membrane (m^2/sec)
f friction factor
I electric current (A)
k mass transfer coefficient for the membrane (sec/m)
ℓ length of the channel (m)
L length of the capillary (m)
L_M membrane thickness (m)
P_o poiseulle number
ΔP pressure capacity (Pa)
Q flow rate (m^3/sec)
Re Reynolds number
v_s velocity of flow (m/sec)
$\bar{v}$ characteristic velocity (m/sec)
V_{app} applied voltage (V)
V_{eff} effective pump voltage (V)
w channel width (m)
X distance (m)

Greek Letters

ε permittivity (C^2/J m)
ε_0 permittivity of the solution in the absence of the additive (C^2/J m)
μ viscosity of the fluid (kg/m sec^2)
μ^* normalized viscosity
μ_0 viscosity of the solution at zero zwitterionic solute concentration
σ^* normalized conductivity
σ conductivity (S)
σ_0 conductivity of the solution at zero zwitterionic solute concentration (S)
ρ density of the fluid (kg/m^3)
ξ zeta potential (V)
ϕ electrical potential (V)
ψ porosity
τ tortuosity

References

Baughman, R. H., L. W. Shacklette, R. L. Elsenbaumer, E. J. Plichta, and C. Becht. 1991. Micro electromechanical actuators based on conducting polymers. *Topics in Molecular Organization and Engineering* 7:267-89.

Beebe, D. J., J. S. Moore, J. M. Bauer, Q. Yu, R. H. Liu, C. Devadoss, and B. H. Jo. 2000. Functional hydrogel structures for autonomous flow control inside microfluidic channels. *Nature* 404:588-90.

de Mello, A. 2002. On-chip chromatography: the last twenty years. *Lab Chip* 2:18N-54N.

Ehrfeld, W., K. Golbig, V. Hessel, H. Loewe, and T. Richter. 1999. Characterization of mixing in micromixers by a test reaction single mixing unit and mixer arrays. *Industrial and Engineering Chemistry Research* 38:1075-82.

Esashi M., S. Shoji, and A. Nkano. 1989. Normally closed microvalve and micropump fabricated on a silicon wafer. *Sensors and Actuators* 20:163-9.

Freire, S. L. S., and R. Wheeler. 2006. Proteome-on-a-chip: Mirage or on the horizon. *Lab on a Chip* 6:1415-23.

Gong, Q., Z. Zhou, Y. Yang, X. Wang. 2000. Design, optimization and simulation on microelectromagnetic pump. *Sensors and Actuators A* 83:200-7.

Gottschlich, N., C. T. Culbertson, T. E. Mcknight, S. C. Jacobson, and J. M. Ramsey. 2000. Integrated microchip-device for the digestion, separation and postcolumn labeling of proteins and peptides. *Journal of Chromatography B* 745:243-9.

Hasselbrink, Jr. E. F., T. J. Shepodd, and J. E. Rehm. 2002. High-pressure microfluidic control in lab-on-a-chip devices using mobile polymer monoliths. *Analytical Chemistry* 74:4913-8.

He, B., N. Tait, and F. Regnier. 1998. Fabrication of nanocolumns for liquid chromatography. *Analytical Chemistry* 70:3790-7.

Hua, S. Z, F. Sachs, D. X. Yang, and H. D. Chopra. 2002. Microfluidic actuation using electrochemically generated bubbles. *Analytical Chemistry* 74:6392-6.

Herr, A. E., J. I. Molho, A. Drouvalakis, J. C. Mikkelsen, P. J. Utz, J. G. Santiago, and T. W. Kenny. 2003. On-chip coupling of isoelectric focusing and free solution electrophoresis for multidimensional separations. *Analytical Chemistry* 75:1180-7.

Kirby, B. J., T. J. Shepodd, and E. F. Hasselbrink, Jr. 2002. Voltage-addressable on/off microvalves for high-pressure microchip separations. *Journal of Chromatography A* 979:147-154.

Kirby, B. J., and A. K. Singh. 2002. In-situ fabrication of dialysis membranes in glass microchannels using laser-induced phase-separation polymerization. Proceedings of the TAS 2992 Symposium 6th, Nara, Japan. *Micro Total Analysis Systems* 2:742-4.

Koch, M., N. Harris, A. G. R. Evans, N. M. White, and A. Brunnschweiller. 1998. A novel micromachined pump based on thick-film piezoelectric actuation. *Sensors and Actuators A* 68:98.

Koch, M., K.D. Skrobanek, and S. Miyasaki. 1999. Development of stress-optimized shape memory microvalves. *Sensors and Actuators A* 72:243-50.

Koch, M., A. Evans, and A. Brunnschweiler. 2000. *Microfluidic Technology and Applications.* Taunton, UK: Research Studies Press Ltd.

Lazar, I. M., and B. L. Karger. 2002. Multiple open-channel electroosmotic pumping system for microfluidic sample handling. *Analytical Chemistry* 74:6259-68.

Lazar, L. M., P. Trisiripisal, and H. A. Sarvaiya. 2006. Microfluidic liquid chromatography system for proteomic applications and biomarker screening. *Analytical Chemistry* 78:5513-24.

Liu, R. H., M. A. Stremler, K. V. Sharp, M. G. Olsen, J. G. Santiago, R. J. Adrian, H. Sref, and D. J. Beebe. 2000. Passive mixing in a three-dimensional serpentine microchannel. *Journal of Microelectromechanical Systems* 9(2):190-8.

Lin, C-H., L-M. Fu, and Y-S. Chien. 2004. Microfluidic T-form mixer utilizing switching electroosmotic flow. *Analytical Chemistry* 76:5265-72.

Manz, Y., J. Miyahara, Y. Miura, H. Watanabe, H. Miyage, and K. Sato. 1990. Design of an open-tubular column liquid chromatograph using silicon chip technology. *Sensors and Actuators* B1:249-55.

Mohammadi, B., and J. G. Santiago. 2001. Simulation and design of extraction and separation fluidic devices. ESAIM: *Mathematical Modeling and Numerical Analysis* 35(3):513-23.

Nguyen, N. T., S. Schubert, S. Richter, and W. Dotrel. 1998. Hybrid-assembled micro dosing system using silicon-based micropump's valve and mass flow sensor. *Sensors and Actuators A* 69:85.

Paul, P. H., D. W. Arnold, D. W. Neyer, and K. B. Smith. 2000. Electrokinetic pump application in micro-total analysis systems: mechanical actuation to HPLC. In *Micro Total Analysis Systems,* edited by A. van den Berg, W. Olthuis, and P. Bergveld, pp. 583-90. Dordrecht, The Netherlands: Kluwer Academic Publishers.

Paul, P. H., D. W. Arnold, and D. J. Rakestraw. 1998. Electrokinetic generation of high pressures using porous microstructures. In *Micro Total Analysis Systems*, edited by D. J. Harrison and A. van den Berg, pp. 49-52. Dordrecht, The Netherlands: Kluwer Academic Publishers.

Reichmuth, D. S., G. S. Chirica, and B. J. Kirby. 2003a. Increasing the performance of high-pressure, high-efficiency electrokinetic micropumps using zwitterionic solute additives. *Sensors and Actuators B* 92:37-43.

Reichmuth, D.S., and B. J. Kirby. 2003b. Effects of ammonioalkyl sulfonate internal salts on electrokinetic micropump performance and reversed-phase high-performance liquid chromatographic separations. *Journal of Chromatography A* 1013:93-101.

Richter, M., M. Linnemann, and P. Woisas. 1998. Robust design of gas and liquid micropumps. *Sensors and Actuators A* 68:480-6.

Slentz, B.E., N. A. Penner, E. Lugowska, and F. Regnier. 2001. Nanoliter capillary electrochromatography columns based on collocated monolithic support structures molded in poly(dimethyl siloxane). *Electrophoresis* 22:3736-44.

Shoji S., and M. Esashi. 1994. Microflow devices and systems. *Journal of Micromechanics and Microengineering* 4:157-71.

Shoji, S., S. Nakagawa, and M. Esashi. 1990. Micropump and sample-injector for integrated chemical analyzing systems. *Sensors and Actuators A* 21:189-92.

Shoji, S. 1998. Fluids for sensors systems. *Topics in Current Chemistry* 194:163-88.

Shoji, S., M. Esashi, and T. Matsuo. 1988. Prototype miniature blood gas analyzer fabricagted on a silicon wafer. *Sensors and Actuators* 14:101-7.

Shoji S., M. E. van der Schoot, and N. de Rooij. 1992. A study of a high-pressure micropump for integrated chemical analyzing systems. *Sensors and Actuators A* 32:335-9.

Song, S., A. K. Singh, T. J. Shepodd, and B. J. Kirby. 2004. Microchip dialysis of proteins using in situ photopatterned nanoporous polymer membranes. *Analytical Chemistry* 76:2367-73.

Tiren, J., L. Tenerz, and B. Hok. 1989. A batch-fabricated non reverse valve with cantilever beam manufactured by micromachining of silicon. *Sensors and Actuators* 18:389-396.

Todd, P., D. C. Szlag, L. D. Plank, S. G. Delcourt, M. E. Kunze, F. H. Kirkpatrick, R. G. Pike. 1989. Investigations on gel forming media for use in low gravity bioseparations research. *Advances in Space Research* 9:97-103.

Weibel, D. B., M. Kruithof, S. Potenta, S. K. Sia, A. Lee, and G. M. Whitesides. 2005. Torque-actuated valves for microfluidics. *Analytical Chemistry* 77:4726-33.

Yao, S., D. E. Hertzog, S. Zeng, J. C. Mikkelsen Jr., and J. G. Santiago. 2003. Porous glass electroosmotic pumps: design and experiments. *Journal of Colloid and Interface Science* 268:143-53.

Yu, C., S. Mutlu, P. Selvaganapathy, C. H. Mastrangelo, F. Svec, and J. M. M. Frechet. 2003. Flow control valves for analytical microfluidic chips without mechanical parts based on thermally responsive monolithic polymers. *Analytical Chemistry* 75:1958-61.

Further Reading

Berthier, J., and P. Silberzan. 2005. *Microfluidics for Biotechnology (Microelectromechanical Systems)*. Norwood, MA: Artech House, Inc, p. 373.

Minteer, S. D. 2004. *Microfluidic Techniques: Reviews and Protocols.* New York: Humana Press, p. 247.

Chapter 11
Separation Cascades

Introduction

Sometimes one finds the separation protocol for a desired product already published in the open literature. The process can be adapted to the particular needs and a viable protocol becomes available. Unfortunately, on many occasions, separation strategies for a particular product are not available and they need to be developed from scratch. Decisions are made at coarse and detailed levels. The broad level decisions consist of choosing the units needed and their order to produce a coarse flow-sheet. Detail level decisions include the sizing of the different units and the determination of the operating conditions. After the flow-sheet is completed, it is optimized to maximize product yield and purity while minimizing the cost. It is always intriguing to place ourselves in the minds of those developing new separation processes. Why were some decisions made? Some of them appear obvious, but that is because we are seeing the finished product and not the thinking process leading to it. Unfortunately, there is not much information in the open literature about the strategies followed. It always appears that decisions were made in a sort of arbitrary fashion rather than following a rationale. I believe that the origin of this problem is found in the background of most authors, they work in academia where the exhaustive analysis of probabilities is limited by cost. Therefore, they develop a separations protocol that works, which is not necessarily the optimum one.

There is not a well-established sequence of steps to follow to develop a separations protocol. The following sequence is one of many options: 1) characterization of target, 2) characterization of main impurities, 3) determination of product purity, productivity and cost, 4) developing a preliminary flow-sheet, 5) optimization of the separation steps, 6) simulation of the process for feasibility and economy, 7) scaling up, 8) validation and recovery of product, 9) determination of product cost and productivity, and 10) maintenance and implementation.

If the target is a protein or a nucleic acid, the material presented in Chapter 4 can be used as an introduction to the characterization of the target molecule and main impurities. The impurities present depend on the raw material used and the addition of additives. The most important are nucleic acids, proteins, salts, and introduced contaminants like antifoams and viruses. The desired product purity, productivity, and cost is dictated by the market (see Chapter 1) and by the process cost. The developing of a preliminary flow-sheet is done by a combination of knowledge and ingenuity. Most purification cascades consist of the same type of units in a particular sequence. Most of the time, available flow-sheets can be adapted to the product of interest. Each unit (for example ultrafiltration systems, chromatography systems, or precipitation vessels) need to be designed and optimized. The material presented in the preceding chapters should help in making an informed decision. Simulation of the process can be done with one of the various simulation packages available in the market (for example, Intelligen Superpro or Aspen Batch). It is important to recognize, however, that available software has not reached the maturity that simulation software has reached for the chemical industry. Recently Shanklin et al. (2001) compared those two software packages using as a benchmark the production of a vaccine by Merck & Company. The results were not surprising. Both packages are able to do simple mass and energy balances and they do a good job in predicting economic indicators and scheduling tasks. However, neither package is able to capture the complexity of biological systems because the models used (with a few exceptions) are too simplistic. Validation is done at every single step in the process and some general guidelines were presented in Chapter 9. Finally, maintenance of the different units follows standard procedures recommended by the manufacturers. If the final product is for pharmaceutical use, then a new set of standards needs to be considered. If the scale is large (such as in the case of antibodies) then environmental impacts must be considered.

Process development can also be divided into three "temporal" stages: research, development, and production. During the research stage, it is important to use techniques whose scale-up capabilities are proved and whose vendors may provide the necessary validation information. Most of the time, the product is novel, and therefore, new analytical techniques may have to be developed and validated. The process is scaled up several times from development until full production scale. The first scale up consists of increasing the production rate from a few mg in the research stage to grams needed in the development stage, for example, to produce enough material for clinical trials. During development, the separation process needs to be refined. It is highly recommended to use good manufacturing practices (GMP) during this stage as it is required to use GMP during production anyway. Therefore, validating the entire process as well as all analytical assays during the development stage saves time. Moreover, sterilization must be proven. It is obvious that having process needs in mind during the research phase is very important, because otherwise the entire process may need to be re-designed later.

The number of steps in a bioseparation cascade has such an economic impact on the entire process that it may determine its feasibility. Figure 11.1 shows the overall yield versus the number of steps for 95%, 90%, and 80% yield in each individual step. Ten steps are common, but to illustrate the point we may consider just five. It is challenging to keep 95% yield in each of the five steps; still the overall yield is only 77%. If the yield in each step decreases to 90%, the overall yield is only 59% after five steps. Finally, an 80% yield in each unit has as a consequence an unacceptable 33% overall yield. More than half of the product is lost! These simple calculations highlight the need for optimization of each separation step and the need to use the minimum number of separation units.

Decisions in the developing of the separation process are made early. The separation protocol depends strongly on the host organism used to produce the desired product. Unfortunately, the production section and the separation section of a company do not necessarily interact closely. Therefore, what seems to be the most convenient host from the production of the desired protein may be the most inconvenient one from a separations viewpoint. A classical example is the use of transgenic plants for the production of human proteins. Plants are easy to grow and the time needed for harvest is typically shorter than in the case of transgenic animals. Moreover, the human protein may accumulate in a particular portion of the plant, which facilitates purification. However, in part as a result of environmental concerns and in part because traditionally *Escherichia coli* or yeast have been used, plants (or animals) are not common hosts.

As stated in the introduction, the downstream process includes: 1) harvest of the cells, 2) capture, 3) purification, and 4) polishing and formulation. The selection of the host organism is beyond the scope of this book. It is sufficient to recognize that one may use bacteria, yeast, transgenic animals, transgenic plants, insect cells, and mammalian cells. Table 11.1 summarizes the pros and cons of

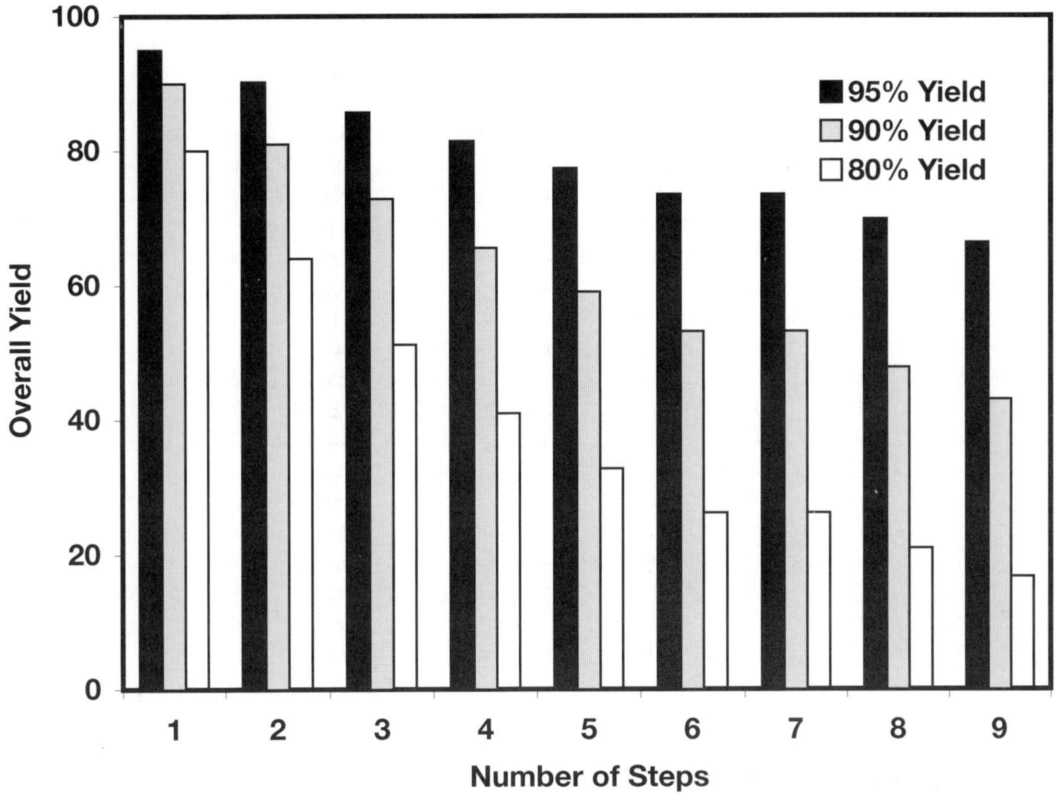

Figure 11.1. Effect of yield in the individual separation steps on the overall yield.

Table 11.1 Effect of Host Organism on Expression Levels, Main Impurities, and Regulatory Issues (adapted with permission from Bioprocess Guide, 2005 GE-Healthcare. Copyright GE-Healthcare).

	E. coli	Yeast	Insect Cells	Mammalian Cells	Transgenic Animals	Transgenic Plants
Expression Level	High	Medium	Medium	Medium	Very high	Low
Time to produce	Fast	Fast	Medium	Slow	Very slow	Medium
Operating costs	Low	Low	High	High	Medium	Very low
Extra-cellular expression	No	Yes	Yes	Yes	Yes	No
Post-translation modifications	No	Yes	Yes	Yes	Yes	Yes
Main impurities	Endotoxins	Glycosylated products	Viruses	Viruses	Viruses and prions	Tannins phenols
Unintended glycosylation	No	Possible	Possible	Possible	Possible	Possible
Host cell protein coexpression	No	No	No	No	Yes	No
Regulatory track record	Good	Good	?	Good	?	?

each. Obviously, the purification train is host specific. A few issues need to be stated. Gram-negative bacteria (such as *E. coli*) express the proteins intracellularly, whereas Gram-positive bacteria and yeast secrete the recombinant protein. That simplifies the separation scheme, but the drawback is low expression levels. Insect cells are attractive because they can do posttranslational modification without the added concerns of viruses such as in mammalian cells. Insect cells release nucleic acids into the media; the consequent increase in the viscosity complicates downstream processing. Transgenic animals may be used and the target protein recovered from the milk. One problem is the presence of viruses and prions.

The type of expression system chosen determines not only the purification train, but also the main validation concerns. The three main hosts are *E. coli*, yeast, and CHO cells; however, insect cells, plants and animals have been used too. If *E. coli* is engineered to secrete the product to the periplasm, the product may contain a higher level of endotoxins than the intracellular product. On the other hand, intracellular proteins in *E. coli* normally contain terminal methionine that may be immunogenic. Yeast is safer than *E. coli,* because it lacks endotoxins. The main problem associated with the use of yeast is to be able to validate the glycosylation pattern, whereas one of the main concerns in using CHO cells is the presence of viruses and other oncogenic contaminants. Insect cells are very expensive to grow but they secrete their product; proteins are correctly folded and glycosylated. Still, the protein may be incorrectly glycosylated and the cell culture may harbor viruses. Finally, transgenic animals or plants produce proteins that are glycosylated and are able to produce large amounts of, although dilute in the case of plants, product. Animals have the inherent problem of virus control but they produce the correctly glycosylated product. Plants, on the other hand, are normally virus free but the product may be incorrectly glycosylated.

Chromatography

It is not possible to describe in detail the fine points of the design and operation of each unit separation used in bioprocessing in this monograph. Specialized books must be consulted. Because of the focal point of chromatographic separations in bioseparation, the following paragraphs describe details about scalability and optimization of chromatography columns.

The selection of chromatography media is done considering three criteria: performance, scalability, and regulatory issues. The performance of a chromatography media is measured by its selectivity and dynamic binding capacity. Some gels may be excellent in laboratory scale, but their performance in industrial scale may be questionable. The scalability of a given media depends on its availability of "industrial grade beads"; that is, beads available in the right size, easy to pack, and with good flow characteristics. To comply with regulatory agencies, the chromatographic material must be commercially available at the desired scale and the material must be usable under current good manufacturing practice (cGMP) conditions. The vendor should be able to supply a comprehensive regulatory file and sufficient in house data to guarantee reproducibility between batches, clean in-place characteristics, and low leachability. The cost of the material is not too critical, partly because of the high return of biotech-based products. However, cost may be an issue in the developing stages done by small ventures.

Two key variables in the operation of a column are the flow rate and the type of gradient to be used. Both need to be optimized in laboratory scale. High flow rates increase productivity but they may compromise resolution and compact the column to unacceptable levels. Therefore, a compromise between resolution and production rate is needed. After the type of gradient has been established, the proper slope must be chosen. Slow slopes normally result in better resolution. However, if the slope is too shallow the resolution may worsen because of axial diffusion effects.

The flow rate and the loading of the column determine the productivity of a column. Usually, columns are operated in the linear region of the adsorption isotherm. Again, the proper loading must be determined experimentally. The normal procedure is to keep the sample concentration constant and change the injection volume; this, in general, does not produce any deleterious effect such as band broadening.

Often, the quality of the separation is strongly affected by the pH and the type of buffer that is used. There is not an established rule to follow. A good starting point is to follow the recommendation of the manufacturers who normally base their suggestion on an extensive database. If two buffers yield the same resolution, use the one with additional benefits. For example, if the choice is between phosphate and NaOH buffers, use the latter because it is bacteriostatic. Another variable to take into account is the molarity of the buffer. If the molarity is too low, the buffer capacity may be exceeded particularly in the region of the column where the protein concentration is at its highest. If the molarity of the buffer is too high, the physical integrity of the gel may be compromised and/or the solubility limit of the protein may be exceeded. Again, a trade-off is necessary.

If you leave a polysaccharide gel in its buffer and at 5°C for a long time, you will notice a slimy deposit on top of the gel. This indicates that either a bacteria or a fungus is growing in the gel. Therefore, a bactericide is normally added to the buffer if the process allows it. Sodium azide and chloroacetamide are commonly used in the laboratory, but they are dangerous at process scale (sodium azide is explosive).

The length of a column has a distinct effect on size exclusion chromatography, the longer the length the better the separation. However, for other chromatography techniques such as ion exchange chromatography or affinity chromatography the column length does not affect the separation; that is, the number of theoretical plates is not affected by column length.

Temperature is a variable that if manipulated correctly may have a profound effect on the resolution of the column. However, use of high or low temperatures is by far more common in laboratory settings than at industrial scale. Temperature may affect the strength of the interaction of the protein with the column, it may change the conformation of the protein, and it does change the viscosity of the mobile phase. If the integrity of the protein is not jeopardized by an increase in temperature, high temperatures are particularly recommended for very large proteins whose viscosity in solution may be so high as to deteriorate the performance of the gel.

Examples

Chapter 1 presented two flow-sheets. One for the purification of monoclonal antibodies from plants and one for the purification of coagulation factor XII from plasma. One more is presented here because it includes a digestion and a precipitation steps that were not present in the other two examples.

Figure 11.2 shows the flow-sheet for the production of CroFab, an antivenom against rattlesnake poison (Protherics K. K. Limited) (Chhatre et al., 2006). CroFab is a polyclonal FAB antibody purified from ovine serum. The feed is ovine serum, which is first filtered to remove cryoprecipitates. Then the antibodies are precipitated by sodium sulfate. A disk stack centrifuge separates the precipitate. The Fc component is cleaved by papain digestion. The product is concentrated by ultrafiltration, and then diafiltrated to reduce conductivity and remove digestion excipients. The total FAB mixture is purified by ion exchange chromatography (the Fc fragment is bound out). After conditioning, the FAB fragments are further purified by an affinity step. The product consists of four different antivenom FAB that are blended together in the final preparation.

Experiments

The experiments described in the following pages cover a wide range of raw materials (plant tissue, animal tissue, and bacteria) and a variety of experimental techniques. Some, but not all, individual

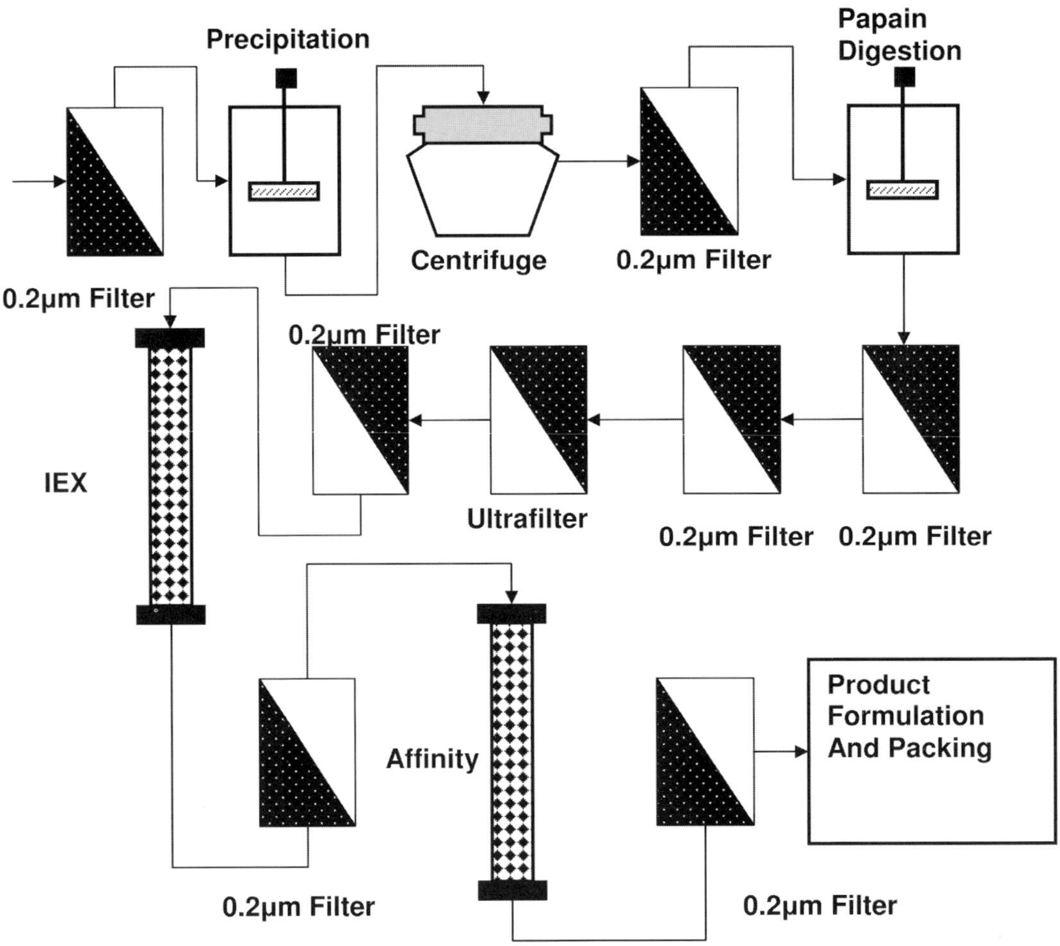

Figure 11.2. Flow-sheet for the production of a polyclonal FAB antibody purified from ovine serum (adapted from Chhatre, S., C. Jones, R. Francis, K. O'Donovan, N. Titchener-Hooker, A. Newcombe, and E. Keshavarz-Moore. 2006. The integrated simulation and assessment of the impacts of process change in biotherapeutic antibody production. *Biotechnology Progress* 22(6):1612-20, with permission). Copyright American Chemical Society.

unit operation experiments have been described in the appropriate chapter. Any experiment that has not been described before is described in detail in this section.

Purification of γ-Crystallins from Calf Lenses

Whole lenses undergo a liquid/liquid phase transition as the temperature is lowered below some critical value (cold cataracts). This critical temperature (-5°C) depends on the particular species and on the age of the specimen. One of the proteins present in the lens (γ-crystallins) undergoes a similar phase transition. There is an interest in the study of solutions of crystallins, because research in that direction is expected to suggest mechanisms for pathogenic or traumatic cataracts.

The α, β, and γ-crystallin subfamilies compose most of the lens proteins in mammals. The α-crystallin subfamily consists of highly polydispersed proteins that constitute -40% of the protein content of the lens. The molecular weight of the α-subclass is approximately 1 million Da. β-crystallins are also polydispersed proteins and have a molecular weight ranging from 50,000 to 300,000 Da. γ-crystallins are a heterogeneous group of low molecular weight proteins, 30,000 Da, with subtle differences in their amino acid compositions. γ-crystallins can be divided into five types (γ-IV, γ-III, γ-II, γ-l, and γ-S).

A number of different separation techniques are employed to obtain γ-crystallins from whole calf lenses. A sequence of steps to purify the γ-fraction that is used routinely in our laboratory is the following (Petitt, Edwards, and Forciniti 1997):

1. Cell disruption
2. Centrifugation
3. Size exclusion chromatography
4. Dialysis
5. Ion exchange chromatography
6. Dialysis
7. Freeze drying

Details of steps 1, 2, 3, 5, and 7 have been given in the appropriate chapters. In short, after disruption is completed and the cell debris is centrifuged out, the supernatant is seeded into the gel permeation column. The γ-crystallin fraction needs to be dialized to change buffers from phosphate to 275 mM acetate buffer pH 5 containing 1 mM 2-mercaptoethanol before seeding it into the ion exchange column. The individual γ-cristallin fractions are eluted at high salt concentrations. Thus, they need to be dialized before freeze drying. Dialysis is described in the following paragraphs.

Preparation of Dialysis Bag

Cut a piece of dialysis bag (a molecular weight cut of 1000 will be enough) large enough for the samples. Run warm water over the bag for 3 hours to wash off coating. Estimate the volume of γ-crystallins. If either sample size is larger than -15 mL then perform ultrafiltration through a 10,000 MWACO membrane to reduce the volume and concentrate the sample.

Protocol

Tie a double knot at one end of the bag. Pour sample into bag using a funnel. Remove all air from bag. Tie a double knot at other end of the bag. Make sure knots are tight, otherwise you might lose your sample in the dialysis buffer. Fill a beaker with l L of 0.040M phosphate buffer at pH 7. Begin dialysis by placing the beakers in a stirring plate. Change the buffer every 8 hours for a 24-hour period.

Isolation of Alcohol Dehydrogenase from Candida boinidii

The objective of this experiment is to isolate alcohol dehydrogenase from *Candida boinidii* by using affinity aqueous two-phase systems.

The isolation and purification of intracellular proteins is challenging because as the cell is disrupted, thousands of contaminants (proteins, nucleic acids, lipids, carbohydrates, and so on) are liberated. A usual procedure to isolate intracellular proteins consists of:

1. Cell disruption
2. Centrifugation to eliminate cell debris
3. Gel permeation chromatography

4. Ion exchange chromatography
5. Affinity chromatography
6. Freeze drying/crystallization

The process is lengthy and the yield is normally quite low (<40%). An alternative process consists of using affinity chromatography immediately after centrifugation. This has problems of column fouling and the need to process large volumes. In this project, alcohol dehydrogenase is purified in a single step by affinity partitioning, following removal of the cell debris by centrifugation.

Chapter 5 described the growth and disruption of *C. boinidii*. Chapter 6 described the centrifugation of the cell debris.

Following centrifugation, the supernatant is used directly in an affinity partitioning experiment. The parameters that can be varied in affinity partitioning are:

1. Dx molecular weight
2. PEG molecular weight
3. Tie-line length
4. Volume ratio
5. pH
6. Concentration of the enzyme
7. Concentration of salts
8. Concentration of affinity ligand

A factorial experiment design (see Appendix) indicates that we should perform 2^8= 64 experiments. The first step is to select the lows and highs for each parameter. Polymer molecular weights are chosen according to commercial availability (Dx 10,000 to 500,000 and PEG from 600 to 40,000). Tie-line lengths can be chosen from available phase diagrams (Forciniti, Hall, and Kula 1991). They should not be too close to the critical point (difficulties in phase separation) or too far from it (very high viscosity). Volume ratio extremes are determined by scaling up considerations. If one phase is too small, then the phases are very difficult to separate by a stack disk centrifuge. The range of pH is determined solely by the stability of the enzyme. Concentration of salts is determined by the stability of the enzyme-dye complex. Finally, the concentration of affinity ligand is chosen according to a minimum necessary for having a one to one by mol correspondence and a maximum determined by prohibitive costs.

Materials

In addition to those materials needed to grow the yeast, the following materials are needed: sodium phosphate monobasic and dibasic, PEG of various molecular weights, dextran of various molecular weights, alcohol dehydrogenase, PEG-red, and nicotine amide dinucleotide.

Preparation of Stock Solutions

Dextran stock solutions are better prepared by adding some water to the solid polymer to make a paste, add the rest of the water and heat at 95°C. Prepare a 50% PEG (8,000) stock solution in deionized water. The buffer concentrations should be 0.040 M. To prepare the stock solution of PEG-red (use as a guideline the results published by Walsdorf, Forciniti, and Kula 1990).

Preparing the Phase Systems

Add to a 15 conical centrifuge tube

-2.5 g of Dx stock solution
-1.0 g of PEG stock solution
-3.0 g of sodium chloride stock solution
-2.5 g of buffer
-1.0 g of cell debris supernatant

or

-2.5 g of Dx stock solution
-1.0 g of PEG stock solution
-3.0 g of sodium chloride stock solution
-2.5 g of buffer
-1.0 g of cell debris supernatant
- X μL of PEG-red

The final phase system composition is 7.5% (w/w) Dx, 5.0% (w/w) PEG, and 0.04 M sodium phosphate buffer. Prepare blanks of the phases; do this by not adding the supernatant.

Mix the contents of the tubes in a rotary mixer for 20 minutes and then centrifuge for 30 minutes at 1,500 g. Carefully pipette out 2 g of each phase. Leave the solution resting and stir again. Inspect the solution to detect density differences along the axial direction of the test tube. Determine total protein content by the Bradford test and dehydrogenase activity with the kit from Sigma (Sigma, St. Louis, MO). The enzyme may be separated from the ligand using dialysis.

Isolation of a Human Antibody from Corn Meal

The transgenic production of antibodies in plants needs to be followed by a robust separation process. Otherwise, the cost advantages associated with the production of the antibody in plants are lost. Because the use of plants for the production of human proteins is relatively new, little is known about the downstream processing units that would present an economic advantage.

We have developed a new process by which human antibodies expressed in corn are isolated to high purity and yield using aqueous two-phase extraction (Lee 2006). We have found that one or two extraction steps, in which the target antibody concentrates in the bottom phase or top phase, followed by a second extraction step, in which the target antibody precipitates at the interface, yields the best results. The optimum purification protocol consists of the following steps:

1. Extraction of the antibody (and contaminating proteins) from the corn meal using a NaCl solution.
2. Addition of PEG and a salt to a concentration high enough to induce the formation of two liquid phases at equilibrium.
3. Removal and disposal of the upper phase (PEG-rich).
4. Addition of PEG to the bottom phase to create a second ATPS.
5. Recovering of the antibody from the new liquid/liquid interface.
6. Removal of the excess salt by diafiltration.
7. Polishing of the product by protein A chromatography.

Materials

Corn meal, human immunoglobulin G (IgG), PEG of various molecular weights, phosphate and citrate buffers, BCA test (Pierce), IgG standards, and protein A columns (analytical and preparative).

Equipment

Standard laboratory equipment, including a rotary mixer, spectrophotometer, HPLC/FPLC systems, and centrifuge.

Extraction of the Antibody from Corn Meal

Weigh 50 g of corn meal and add 500 mL of 150 mM NaCl aqueous solution containing 2 mg/mL human IgG. Of course, the addition of the antibody is not necessary if you are using transgenic corn. Incubate with stirring for 8 hours at 4°C.

Centrifuge the suspension for 30 minutes at 10,000 rpm or filter through a filter paper, and then filter through a 0.45-μm membrane.

Partitioning Experiments

Add a known amount of the phase-forming species in solid form (PEG and potassium phosphate or PEG and sodium citrate and so on) a salt in solid form or from a stock solution, and enough corn meal extract to complete 10 g into centrifuge polypropylene tubes. The pH of the buffer determines, in this case, the partitioning of the antibody into the bottom phase or at the interface. The optimum pH is antibody-dependent. Thus, it needs to be determined experimentally.

Mix the content of the tubes in a rotary mixer for 20 minutes after initial manual mixing and then centrifuge at 2,000 rpm for 20 minutes. The volumes of the top and bottom phases should be noted and samples carefully removed from each phase with a pipette. The top phase and the interface is removed and discarded. More PEG is added to induce the formation of two phases separated by an interface, in which the antibody precipitated. The protein needs to be carefully removed from the interface, resuspended in buffer, and determine antibody concentration by running the phases through a Protein A column calibrated with human IgG.

Resuspend the antibody precipitate in 0.010 M sodium phosphate buffer and dialyze against the same buffer to eliminate the remaining PEG. Run the preparation through a protein A column using the gradient protocol described in Chapter 7 to polish the antibody.

References

Chhatre, S., C. Jones, R. Francis, K. O'Donovan, N. Titchener-Hooker, A. Newcombe, and E. Keshavarz-Moore. 2006. The integrated simulation and assessment of the impacts of process change in biotherapeutic antibody production. *Biotechnology Progress* 22(6):1612-20.

Forciniti, D., C. K. Hall, and M.-R. Kula. 1991. Protein partitioning at the isoelectric point: effect of polymer concentration and polymer molecular weight. *Biotechnology and Bioengineering* 38:986-94.

Lee, J. 2006. Purification of a human antibody from transgenic corn using aqueous two-phase systems. Ph.D. Thesis, University of Missouri-Rolla, Rolla, MO.

Petitt, P., M. Edwards, and D. Forciniti. 1997. A small angle neutron scattering study of proteins at their isoelectric point. *European Journal of Biochemistry* 243:415-21.

Shanklin, T., K. Roper, P. K. Yegneswaran, and M. R. Marten. 2001. Selection of bioprocess simulation software for industrial applications. *Biotechnology and Bioengineering* 72(4):483-89.

Walsdorf, A., D. Forciniti, and M.-R. Kula. 1990. Investigation of affinity partition chromatography using rormate dehydrogenase as a model. *Journal of Chromatography* 523:103-17.

Appendix—Elements of Experimental Design

Introduction

Bioseparation units can be analyzed using mathematical models, and entire separation cascades may be simulated in a computer. These simulations provide valuable process information and may allow for a rational selection of units; however, the final process design and optimization is based on laboratory scale experiments. Using the results of laboratory experiments, the process is scaled up to pilot plant scale and then to industrial scale. Because of the uncertainty in the scaling-up procedures, further testing and re-optimization are often necessary in the pilot scale stage.

Many times, reported experimental data and the conclusions reached from them do not survive the most simple of statistical tests. Often, scientists seem to forget that a single data point is not sufficient to make a generalization. This mistaken generalization habit is evident after reading just a few manuscripts in the open literature. Very rarely are errors or statistical tests reported, for example, in manuscripts that report new separation schemes. In an industrial setting, this is completely unacceptable; however, the practice seems to have escaped the academic world's attention. Scientists constantly compare outputs from different data sets and then draw biased conclusions and gross generalities. Again, this "analysis" is often done by simple observation of a few data points for each population. A more rigorous examination of data sets is required and may show that although there are apparent differences between them, they are not different enough to state that the populations are actually different (usually this is stated with an 80% or 90% confidence interval). These kinds of practices lead to headaches for those who try to reproduce data and who elaborate further hypotheses based on these uncertain conclusions.

Experimental research has been historically designed following the personal experience of the researcher directing the work. This is somewhat risky because the experimentalist may overlook conditions that would have improved, for example, the separation of a particular protein. Experimental design may be used in conjunction with personal expertise to select the experiments that will provide the best information. Unfortunately, many see experimental design as a science that belongs to pure mathematics rather than as a valuable tool for the experimentalist. This is a perception originated, in part, on the dryness of many monographs covering the topic. This chapter seeks to demystify this potentially powerful approach (Kwang-Min and Gilmore 2006; Bensch et al. 2005).

The use of statistical tools in the laboratory is valuable because they allow the intelligent design of the experiments (experimental design), they allow the experimenter to trust his results, and they allow the laboratory to prove its claims to regulatory agencies. Experimental design is based on the use of statistical tools to determine the kind and number of experiments that are needed to maximize information and to prioritize the effect of various experimental variables. Moreover, statistical models can be constructed and their use may be allowed to determine, for example, the coupling of several variables on a process. Statistics are used to ascertain the reliability of theoretical models, to determine parameters that are then fed into theoretical models, and to establish the precision of the analytical techniques used to acquire data. Finally, regulatory agencies require solid experimental evidence about the process used to manufacture a new product. Therefore, the data reported to the federal agencies must have statistical significance. It is impossible to introduce elements of experimental design and statistical treatment of data without a brief introduction to statistics. Thus, in this Appendix, I briefly review some concepts that should always be present in the mind of those working in bioseparation. The review is not meant to be rigorous and the reader is referred to some excellent books for an in-depth treatment of this subject.

Some Definitions

Statistics deals with populations. A population is an infinite set of data characterized by its mean, standard deviation, and skew. The population mean, for example, is the average of a series of measurements to determine the protein concentration at the exit of a chromatographic column. The standard deviation reflects our uncertainty in the reported value. The mean (or average) of a population is equal to the sum of the individual values of a population divided by the number of members of the population. In statistics, it is common to distinguish between populations (an infinite set) and sample (a finite set that should represent the population). This distinction is not just semantic because some statistical tools are exact for populations but not for samples. The experimentalist should be aware of this difference and try to determine if the sample actually represents the population. Of course, true populations are never available in experimental work. The mean of a population is defined by,

$$\mu = \lim_{n\to\infty} \frac{1}{n} \sum_{i=1}^{n} x_i \qquad \text{A.1}$$

The mean of a finite sample, $\bar{x}$, is just the sum over all data points, x_i, divided by the number of data points, n. The mean must not be confused with the median, which is the mid-point of a population. If the population is not normally distributed, the media may be very different from the mean. The standard deviation of a population, σ_p, is,

$$\sigma_p = \lim_{n\to\infty} \sqrt{\frac{\sum_{i=1}^{n} (x_i - \mu)^2}{n}} \qquad \text{A.2}$$

This expression is valid only for a very large number of data points, something very rarely available in the laboratory. The equivalent expression for a sample is, σ,

$$\sigma = \frac{1}{n-1} \sqrt{\sum_{i=1}^{n} (x_i - \bar{x})^2} \qquad \text{A.3}$$

Notice that both quantities are the same if n is a large number. An associated quantity to the standard deviation is the relative standard deviation, which is equal to the standard deviation divided by the average. The relative standard deviation is known as the coefficient of variance, CV,

$$CV = \frac{\sigma_p}{\mu} \approx \frac{\sigma}{x} \qquad \text{A.4}$$

The relative standard deviation is often used to determine the "goodness" of an analytical test. If the relative standard deviation is ± 2% the assay is "good." If the relative standard deviation is larger than 5%, something is just not right.

A population is always distributed in some fashion with respect to the value of the independent variable. Different functions may be used to mimic the actual distribution; these functions are called probability distribution functions. They can be classified into discrete and continuous. An example of a discrete distribution is the binomial distribution. The continuous version of the binomial distribution is the normal or Gaussian distribution. Most of the discussion that follows applies to normal distributions and exceptions to this are noted. It is important to remember that there is a large menu of distributions (discrete and continuous) and the reader should refer to specialized textbooks if the actual distribution of experimental data is not Gaussian. The normal or Gaussian probability distribution function is,

$$pdf_G = \frac{2}{\sigma\sqrt{2\pi}} \exp\left(-\frac{\left(x - \overline{x}\right)^2}{2\sigma^2} \right) \qquad \text{A.5}$$

in which σ^2 is the variance. This probability distribution function has some unique properties. It is symmetric about its mean, the mean is also its mode and median, the inflection points of the curve are at one standard deviation from the mean, and the skewness and the excess kurtosis are both equal to zero.

The Chi-square probability distribution is widely used in statistical significance tests. The probability density function is,

$$pdf_\chi = \frac{\left(\frac{1}{2}\right)^{k/2}}{\Gamma(k/2)} x^{\frac{k}{2}-1} e^{-\frac{x}{2}} \qquad \text{A.6}$$

in which k is the degrees of freedom and Γ is the gamma function. If x_i are normally distributed random variables with means $\overline{x}$ and variances σ_i^2, then the random variable is distributed according to the Chi-square distribution according to,

$$\chi_k^2 = \sum_{i=1}^{k} \left(\frac{X_i - \overline{x}_i}{\sigma_i} \right)^2 \qquad \text{A.7}$$

in which k is the number of degrees of freedom (the number of independent random variables with means $\overline{x}$ and variances σ_i^2). The parameter in Chi-square statistics is the number of degrees of freedom. Values of critical Chi-square distributions for various degrees of freedom can be found in numerous textbooks and Web sites. The calculated value of Chi-square is compared with the tabulated value (null hypothesis). If the value is higher than the tabulated value the null hypothesis is true.

The Chi-square distribution has a mean k, a median k-2/3, a mode k-2 for k>2, a variance 2k, a skewness equal to $\sqrt{8/k}$, and an excess kurtosis 12/k.

Distributions are often not symmetric around the mean; the distribution may be biased at lower or higher values from the mean. The skewness of a distribution measures its asymmetry with respect to the mean. It is defined by,

$$\Gamma = \frac{\mu_3}{\sigma^3} \tag{A.8}$$

in which μ_3 is the third moment of the mean. For a sample of n values, the sample skewness is given by,

$$\tau = \frac{\sqrt{n}\sum_{i=1}^{n}\left(x_i - \bar{x}\right)^3}{\left(\sum_{i=1}^{n}\left(x_i - \bar{x}\right)^2\right)^{3/2}} \tag{A.9}$$

Errors

Any experimental measurement has an error associated with it. Because errors cannot be eliminated, the goal should be to minimize them or, at least, quantify them. The errors in the measurements determine the accuracy of an analytical test. In other words, how close is the measurement to the true value. Errors may be absolute or relative. The absolute error of a measurement is equal to the mean (or average) of a series of measurements minus the "accepted" value of the quantity. The intrinsic problem of this definition is that, in general, we do not know the accepted value. Therefore, the mean of a large number of measurements often is taken as the accepted value. The relative error is equal to the difference between accepted and measured values divided by the accepted value and multiplied by 100.

Experimental errors can be systematic or random. Systematic errors are called a bias. They have a fixed value and they are caused by instrumental errors, personal errors, and method errors. It is not trivial to assign a cause to a bias. Inexperienced experimentalists have a strong tendency to assign the biases in their measurements to instrumental errors. However, most of the time systematic errors are associated with personal errors (e.g., incorrect use of precision pipettes, lack of focus during measurements, poor understanding of the instrument) or method errors (wrong experimental design). Random errors are a consequence of undetermined variations in the measurement and they result in imprecise data. Random errors can be treated statistically, whereas systematic errors need a more drastic solution; that is, an instrument must be fixed or a person needs to be retrained. If the distribution of random errors is Gaussian, the random error is equal to the difference between the mean of a small number of measurements (let us say two or three) and the true value (or the mean of an infinite, 20 or more, number of samples).

Reporting Experimental Data

A good analytical test must be precise, accurate, specific, and robust. In addition, it is desirable that its input/response be linear. Specificity is the ability of an assay to distinguish between the analyte and similar compounds. An analytical test is characterized by its limit of detection, LOD,

$$LOD = 3 * noise \tag{A.10}$$

For example, a standard UV spectrophotometer has a noise of approximately 0.003 units of absorbance. Thus, any reading below 0.010 is below the noise. The limit of quantification, LOQ, is a more important quantity. It is defined by,

$$LOQ = 10 * noise \qquad \text{A.11}$$

Again, for a typical absorbance reading, any value less than 0.030 is not useful for quantification. Finally, the range of the analytical assay is the interval between its upper and lower limits of detection.

In the absence of systematic errors, we can calculate the probability that the true mean of a population is bounded by some specific limits. For example, we may make the claim that our confidence for a mean is 95%. The confidence limit is formally calculated from,

$$X = X_{measured} \pm z\sigma_p \qquad \text{A.12}$$

for one sample. The extension to N samples is,

$$\mu = \bar{\mu} \pm \frac{z\sigma_p}{\sqrt{N}} \qquad \text{A.13}$$

in which z is the number of standard deviations that separate the true mean and the sample mean. The value of z is determined by our choice of "confidence." For example, 95% confidence has a z-value of 1.96. Therefore, if we have a single sample, we normally express our result at 95% confidence by,

$$X = x_{measured} \pm 2\sigma_p \qquad \text{A.14}$$

Sometimes, we cannot approximate the population standard deviation by the sample standard deviation because the number of samples is very small. To avoid this difficulty, we define a new statistical parameter call t,

$$t = \frac{X - \overline{X}}{\sigma} \qquad \text{A.15}$$

t is a function of the confidence level and the number of degrees of freedom. We present our measurement as,

$$X = \bar{x} \pm \frac{t\sigma}{\sqrt{N}} \qquad \text{A.16}$$

For example, for three degrees of freedom (a triplicate) and 95% confidence, t is equal to 3.18. Compare this to z=1.96 for the same confidence interval.

Most often, any analytical technique will consist of a series of steps or measurements, which may have created a random error. The overall error must reflect the accumulated uncertainties of the successive measurements. To calculate the overall standard deviation we use,

$$\sigma = \sqrt{\left(\frac{\partial f}{\partial x}\right)^2_{a,b,c....z} \sigma_x^2 + \left(\frac{\partial f}{\partial y}\right)^2_{a,b,c.....z} \sigma_y^2 .... \left(\frac{\partial f}{\partial a}\right)^2_{b,c,......z} \sigma_a^2} \qquad \text{A.17}$$

This approach to calculate standard deviations is the most commonly ignored statistical tool, and the consequences of this omission may be catastrophic.

Statistical Tests

Sometimes it is necessary to compare two or more populations. For example, suppose we want to desalt a protein solution by using diafiltration. We have at our disposal three different membranes and want to assess the dialysis efficacy of each by measuring the conductivity of the retenate after five volumes of buffer have been exchanged. Each experiment is repeated ten times. Therefore, we can calculate the mean conductivity for each test and the standard deviation of the sample. The random error may be so large that the distributions, although Gaussian, overlap with each other because they are so broad (Figure A.1). If we look at the means, we may conclude that membrane number 1 is better than the other two because it yields the larger reduction in conductivity. Will our conclusion be the same if we take into account the standard deviation of the populations? Different tests may be conducted to assess these results and we discuss them below.

One of the most common types of variance analysis is Anova. The specific type of Anova analysis depends on the number of samples in the populations. The simplest Anova analysis is Single Factor. It tests the hypothesis that means from two or more samples are equal. Anova analysis is useful in analyzing, for example, the robustness of an analytical chromatography test. An extension of the single-factor Anova analysis is the two-factor Anova with replication that includes more than one sample for each group of data. Finally, we can perform Anova two-factor without replication, which does not include more than one sampling per group, and it will test the hypothesis that means from two or more samples are equal.

Correlation analysis measures the relationship between two data sets that are scaled to be independent of the unit of measurement. The population correlation calculation returns the covariance of two data sets divided by the product of their standard deviations. The correlation analysis determines whether two ranges of data move together—that is, whether large values of one set are associated with large values of the other (positive correlation), whether small values of one set are associated with large values of the other (negative correlation), or whether values in both sets are unrelated (correlation near zero).

Covariance is a measure of the relationship between two ranges of data. Covariance analysis calculates the average of the product of deviations of data points from their respective means. The information obtained from covariance analysis is the same as the one obtained from correlation analysis.

The F-test is a statistical test in which the test statistics have an F-distribution if the null hypothesis is true. It is typically used to test the hypothesis that the means of multiple normally distributed populations are equal (if all have the same standard deviations) It may be used to test standard deviations but it is not relative to distributions that are not normal. The F-test statistic is calculated from,

$$= \frac{\dfrac{RSS_A - RSS_B}{A}}{\dfrac{RSS_B}{n - B}} \qquad \text{A.18}$$

in which n is the number of observations, RSS is the residual sum of squares, A is the number of unrestricted coefficients in model A, and B is the number of coefficients restricted by model B.

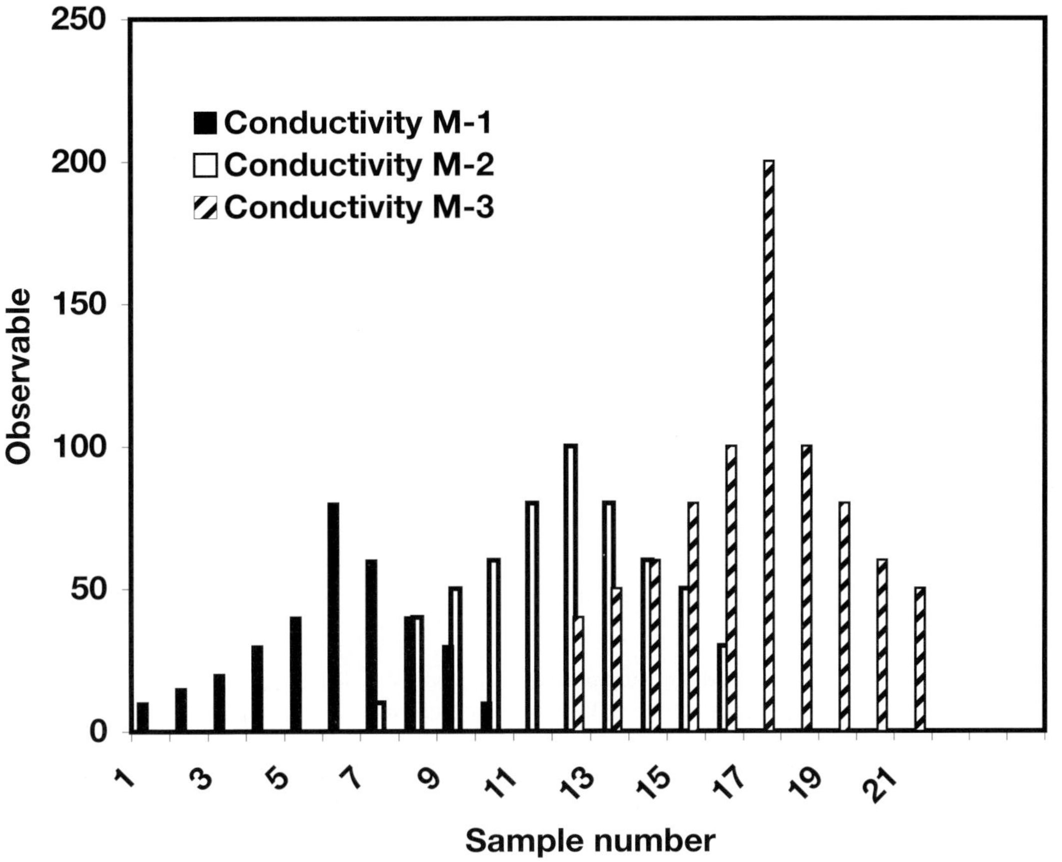

Figure A.1. Plot of an observable (conductivity) from three different dialysis membranes vs. sample number.

The student t-test is one of the most common statistical tests used to compare the means of two normally distributed populations. Because it was originated in the monitoring of beer brews, it is a test very close to the hearts of bioprocessing engineers. Different forms of the test are used depending on the nature of the two samples: independent or paired. For example, if we were comparing the dispersion of two different chromatography columns or the three membranes included in Figure A.1, the samples would be independent. On the contrary, if we are comparing two duplicates of the same sample (two runs of the same sample in the same column with the same elution protocol) the populations will be paired. If the calculated t-value is higher than the chosen threshold for statistical significance (normally 0.05 level, 95% confidence) then the two means are different. The t-value is a measure of how large a coefficient in a regression is with respect to its standard error.

The t value for independent samples of unequal size is calculated from,

$$t = \frac{\overline{x_1} - \overline{x_2}}{S_{\overline{x1}-\overline{x2}}} \qquad \text{A.19}$$

Table A.1. Statistical Comparison of Three Dialysis Membrane.

	Membrane 1 vs. Membrane 2	Membrane 2 vs. Membrane 3	Membrane 1 vs. Membrane 3
Mean	33	56	82
t-value	−2.04	1.54	2.97
t-critical	2.1	2.144	2.16
p-value	0.05	0.14	0.01

in which,

$$S_{\overline{X1}-\overline{X2}} = \sqrt{\frac{(N_1-1)S_1^2+(N_2-1)S_2^2}{N_1+N_2-2}\left(\frac{1}{N_1}+\frac{1}{N_2}\right)} \qquad \text{A.20}$$

in which N-1 is the degrees of freedom for sample 1 or 2. For dependent samples, the t value is calculated from,

$$t = \frac{\overline{x_D}\sqrt{N}}{S_D} \qquad \text{A.21}$$

in which $\overline{x_D}$ and S_d are the average and standard deviations of the differences. The key in the application of the t-test is to determine if the data points are normally distributed; this can be assessed by normality tests. Application of the t-test to the three membranes shown in Figure A.1 yields the results summarized in Table A.1. These results show that the difference in performance of membranes 2 and 3 is statistically insignificant, whereas the difference in performance of membranes 1 and 3 and in between 1 and 2 are statistically significant and borderline significant respectively.

The t-test also may be used to determine if biases (non random errors) are introduced in the measurement. We first calculate the difference between the sample mean and the population mean using,

$$\bar{x} - \mu = \pm\frac{t\sigma}{\sqrt{N}} \qquad \text{A.22}$$

The values of σ and t may be replaced by σ_p and z if the number of samples is large enough. If the experimental value of the left-hand side is larger than the one calculated by the right-hand side, then biases may be present. Of course, to do this we need to know the true value of the mean, which is often impossible!

The p-value is the chance of getting a larger t-value by chance alone. A small p-value indicates that a coefficient in a regression is large with respect to the noise. For example, a p-value of $p < 0.05$ suggests a significance at the 0.05 level (or 95% confidence). Table A.1 includes p-values for the three membrane pairs.

Experimental Design

The objective of experimental design is to minimize the amount of experiments while maximizing the information obtained from them. Often, experimentalists do experimental design without knowing that they are doing it. For example, experience may suggest that there are only three important independent variables and that it is important to find out how the process responds to high and lower values of those variables. Other times, experimentalists do an unnecessarily large number of experiments because of poor planning; that is, they perform as many experiments as they can hoping for the best. It is common practice to fix all variables but one with the expectation that the effect of that variable would be isolated. This is a faulty line of thought, because the experimentalist is learning about the effect of one variable at a particular condition. What will happen as the other variables are allowed to change is anybody's guess.

One of the reasons experimental design is not widespread is that experimentalists feel overwhelmed by a statistical jargon that seems to lead nowhere. Research is still a single individual activity and researchers often feel tempted to follow their own instincts, which can lead to good, but not optimal results. It is also not a very efficient approach to problem solving.

Experimental design is a collection of predetermined settings of experimental factors. Factors are really variables, such as pH, buffer composition and concentration, temperature, and so on. Each combination of settings is called a run and a response variable or observation is a measure of performance. In other words, we do the experiments and obtain results. Of course, any experimental design will fail if the experimentalist does not ask good questions and if he is unable to separate noise from signal. Nothing can replace the good quality of the questions we ask. They are based on the knowledge of the process extracted from literature and from our own experience. Lack of separation between signal and noise is a common mistake made by the inexperienced experimentalist. It is also common to blame a particular analytical technique for the noise. In fact, the largest source of noise is either the experimentalist who has not selected the optimum conditions and who is not careful enough in his experiments. The instrument itself imposes a limit as to how good a measurement can be.

Experimental design is a cyclic activity. The number of times we go through the cycle depends on how much we know about the process. The first step is to set up a series of screening experiments to identify important variables. The second step is to optimize the experiments by building a predictive model. We need to be careful here because the so-called model is nothing but a nonlinear regression of the responses. Finally, the results need to be confirmed.

Because of the large number of variables that may affect experimentation with biological systems, experimental design is particularly useful. Two types of experimental designs will be discussed here, full factorial and fractional factorial. Each can be two-level or mixed-level. A two-level factorial design uses an upper and lower bound (two levels) for each of the variables. Mixed-level factorial designs use an upper, middle, and lower level for each of the variables; it is typically used to determine curvature (non-linear dependence of the dependent variable with the independent ones). In fractional factorial design one selects a subset of possible experiments. For example, if six variables are identified there are 2^6=64 possible combinations (64 experiments) using a two-level full factorial design. A fractional experiment may consist of sixteen experiments. The idea is to explore high and low values of each value and observe the change in the effect or results. This type of design is commonly used for screening experiments. In summary, there are 2k combinations of variable settings in a k-factor, two level, full factorial experiment.

In the absence of interactions, variables should have an additive effect to the response. Most times, synergistic (positive) or antagonistic (negative) interactions between variables exist. One way of determining the presence of interaction between two variables is to plot the response versus one variable using the other as a parameter. If the lines are parallel, there is not interaction between the variables. If the slopes of the lines are not the same, then there is interaction between the variables. For example, both ionic strength and pH affect the swelling of ionic exchange resins but the effect of ionic strength depends on pH. Therefore, if we plot swelling versus ionic strength using the pH as a parameter the lines will not be parallel. It is important to determine the magnitude of that interaction, which is calculated as follows: First, the average of the runs in which both factors are at extrema are calculated (high-high and low-low). To that quantity we subtract the average of the runs with mixed factors (high-low and low-high). The higher the number we get the stronger the coupling between variables.

Fractional factorial designs are very popular because they are more economical. The risk, of course, is to lose information about the interaction between pairs of variables. If there are k variables and a p-fractional design is desired, then the number of runs is,

$$= 2^{k-p} \tag{A.23}$$

The resolution of a factorial experiment design refers to our ability to determine whether the interaction between variables exists. Experimental designs are classified according to their resolution. A resolution V design allows the determination of all interactions (full factorial). They are experimentally expensive, but they yield the most information. A resolution IV design does not allow confounding main effects with interactions between variables. Finally, a resolution III design allows confounding between main effect and factor interactions, but it does not allow confounding between main effects. Experimental design is never done with a resolution below III because the information is not valuable. Similarly, experimental design is never done with a resolution higher than V because the information obtained, that is, three factors interactions, is often impossible to interpret and most three factors interactions are weak. Resolution V designs are called clear signal designs because they yield unambiguous observations.

Three stages are followed to choose a particular design. First, we need to select the experimental variables. Experience is difficult to replace but one alterative is to use Ishikawa diagrams (Ishikawa, 1986). For bioprocessing, the number of variables for the different unit operations is relatively well defined. The recommendation is to stick to resolution V designs. Resolution III results can be used for screening experiments (and still have good signal). We can look for the appropriate design for example in the appendix of the book by Haaland (1989).

An important piece of information obtained from experimental design is the estimation of the magnitude of the effects. For example, we know that both ionic strength and pH affect swelling of a chromatography gel; which one is more important? The simplest way of obtaining that information is to construct Pareto charts (Ishikawa 1986), which are bar plots of the difference between the high and low effects for all of the variables. It allows for the identification of high and low impact effects, but it yields ambiguous results for a factor of intermediate impact. A better choice is to use normal plots in which the normal probability score is plotted versus the size of the effect; points that do not fall in a straight line are statistically significant. Interaction plots were described before. Cube plots are useful to analyze sets of three variables. The response at high and low levels for each factor is plot in each of the vertices of the cube. Therefore, phases of the cube correspond to two-factor interactions.

Most experimental designs include several variables and up to two-variables interactions. As stated before, it is possible to have three and higher order interactions but they are less useful, because their interpretation is normally ambivalent. The results of an experimental design must be analyzed for the statistical determination of important variables and for the existence of two-variables interactions. The best way of obtaining statistically meaningful information is by replicating the experiments. A common technique to achieve (nearly) the same result without replication is to prepare a Pareto chart (a bar plot of factors and factor-factor interactions versus the size of the effect; that is, the difference between higher and lower average over the other factors. The observed trends can be fitted with a statistical model (a linear combination of one-factor and two-factor interactions). For example suppose that we have four factors (pH, I, T, and A [additive]) and that the Pareto plot suggests that only PH-I and T-I are important two-factors interactions. Then, the statistical model will be,

$$\left[Observable\right] = \alpha + \beta * pH + \gamma * T + \delta * A + \varepsilon * I + \phi * pH * I + \varphi * T * I \tag{A.24}$$

The parameters of the model are calculated directly from the Pareto chart. This model is then used to predict the experimental outputs and the prediction is analyzed using analysis of variance. The important quantities of the ANOVA analysis are the t-value and the p-value for each coefficient. The t-value is a measure of signal to noise (the higher the better). The p-value is the chance of getting a higher t-value by chance alone. The smaller the p-value the more significant the effect of the factor is.

For any number of variables it is up to the experimentalist to decide the sample size or number of independent experiments (at the end that choice will determine the resolution of the experiment). The power of an experimental design (the ability to identify correctly important factors) is a function of the sample size. We must note that the experimental techniques used determine the optimal signal-to-noise ratio. If the signal to noise is small, we can repeat the measurement several times to reduce the error. The power of an experimental design is equal to one minus the probability of missing an important factor. The specificity of an experimental design for finding only important factors is the significance of the design; that is, a significance of 10% means that we have 10% chance of calling an unimportant factor a signal. The sample size will determine the power and significance for a given fractional factorial design. The minimum that we should require is a power of 80% with a significance of 10%. For these conditions the minimum detectable effect (effect/experimental error) for two level experiments and sample sizes of 8, 12, 16, 24, and 32 will be 2.0, 1.6, 1.4, 1.1, and 0.9. For eight samples, we will be able to detect an effect no better than twice the experimental error. If we want to detect an effect smaller than the experimental error we will need thirty-two samples or change the analytical technique. One alternative is to perform one particular measurement a large number of times (let us say ten or more). This data set can be statistically analyzed to determine the error in our measurements. Then, the remaining samples can be done only in duplicate and the results reported as the mean of the two samples plus minus two times the standard deviation of the statistically valuable sample.

Summary

First, never trust experimental reports that do not have an error associated with them. Second, never accept for granted a conclusion reached in a manuscript that may not be supported by the statistical significance of the data. Finally, experimental design may save time and money in the laboratory and at the same time provide valuable information about the effect of the process variables. That information can be applied to true models for the process.

Nomenclature

k	degrees of freedom
n	number of data points (observations)
N	number of samples
RRS	residual sum of squares
$\bar{x}$	mean of a finite sample
x_i	data point
$\bar{x}$	mean of the sample
z	number of standard deviations that separate the true mean and the sample mean

Greek Letters

σ	standard deviation of a sample
σ_p	standard deviation of a population

σ^2 variance
τ skewness of a sample
μ mean of the population
μ_3 third moment of a distribution
Γ skewness of a distribution
$\Gamma(k)$ gamma function

References

Bensch, M., P. S. Wierling, E. von Lieres, and J. Hubbuch. 2005. High throughput screening of chromatographic phases for rapid process development. *Chemical Engineering and Technology* 28:1274-84.

Haaland, P. D. 1989. *Experimental Design in Biotechnology.* New York: Marcel Dekker, Inc.

Ishikawa, K. 1986. *Guide to Quality Control*, 2nd edition. Tokyo, Japan: Asian Productivity Organization.

Kwang-Min, L., and D. F. Gilmore. 2006. Statistical experimental design for bioprocess modeling and optimization analysis. Repeated-measures method for dynamic biotechnology process. *Applied Biochemistry and Biotechnology* 135:101-15.

Index

Page numbers followed by an f denote a figure. Numbers followed by a t denote a table.